# A Taste for Writing

## COMPOSITION FOR CULINARIANS

Vivian C. Cadbury

# A Taste for Writing

## COMPOSITION FOR CULINARIANS

Vivian C. Cadbury
Associate Professor
The Culinary Institute of America

Photographs by the author

THOMSON
DELMAR LEARNING

Australia Canada Mexico Singapore Spain United Kingdom United States

A Taste for Writing: Composition for Culinarians

by Vivian C. Cadbury

**Vice President, Career Education Strategic Business Unit:**
Dawn Gerrain

**Acquisitions Editor:**
Matthew Hart

**Managing Editor:**
Robert Serenka, Jr.

**Assistant to the Director of Learning Solutions:**
Anne Orgren

**Director of Production:**
Wendy Troeger

**Senior Content Project Manager:**
Glenn Castle

**Director of Marketing:**
Wendy E. Mapstone

**Channel Manager:**
Kristin McNary

**Art Director:**
Joy Kocsis

**Cover Design:** Joseph Villanova



Printed in the United States of America
1 2 3 4 5 6 7 XXX 10 09 08 07

For more information, contact
Thomson Delmar Learning,
5 Maxwell Drive, Clifton Park,
NY 12065
Or find us on the World Wide Web at http://www.thomsonlearning.com, or www.hospitality-tourism.delmar.com



Library of Congress Cataloging-in-Publication Data
Cadbury, Vivian C.
A taste for writing: composition for culinarians / Vivian C. Cadbury.
p. cm.
Includes bibliographical references and index.
ISBN-10: 1-4180-1554-7
ISBN-13: 978-1-4180-1554-1
1. Cookery–Authorship. I. Title.
TX644.C33 2008
808'.066641–dc22

2006103312

## NOTICE TO THE READER

# Table of Contents

# Preface

A *Taste for Writing* offers comprehensive coverage of topics in composition and grammar with a unique focus on food and cooking. The process of brainstorming ideas, for example, is compared to that of peeling a potato, while the commas that surround interrupters in a sentence become a server's hands carrying a tray. The many student writing samples, which exemplify points of composition in each chapter, are often food-related, addressing topics from mother's home cooking to working the line. In addition, the text is illustrated with fifty photographs of culinary students at work in the kitchens, bakeshops, classrooms, and gardens of the Culinary Institute of America, as well as with numerous tables and diagrams.

The book encourages its audience to develop a taste for reading as well. In addition to the student writing samples, a small but delectable group of food-centered essays, stories, and poems by professional writers of various backgrounds appears in an appendix. Examples from the work of three major food writers—M. F. K. Fisher, Anthony Bourdain, and Jeffrey Steingarten—are intermingled with poems by Lucille Clifton and Galway Kinnell. "Tortillas," by artist, poet, and community leader José Antonio Burciaga, is a special gift, while excerpts from David Mas Masumoto's *Epitaph for a Peach*, Shoba Narayan's *Monsoon Diary*, and Eric Schlosser's *Fast Food Nation* add a delicious variety to the collection.

As a classroom text, *A Taste for Writing* can provide material for lecture and discussion, as well as activities, readings, and written assignments designed to help students understand concepts and practice skills. Instructors may follow the order of chapters as presented or devise a sequence that better suits their needs. For example, some chapters may be more appropriate as reference material, while others may be omitted altogether, according to the purpose and scope of the course. The instructor's manual reviews a number of these options. Intended primarily for college freshmen, *A Taste for Writing* can also be used in a high school English class or as a reference book.

## THE HISTORY OF THE BOOK

In 2001, I began teaching writing to students in culinary arts and baking and pastry arts at the Culinary Institute of America. Although my previous teaching experience had been with students of differing interests and career goals, I had tried to engage them in the writing process by using music and film in the classroom. My new students, however, shared an interest, indeed a passion, for food. What could I do with that? The text and exercises in this book are peppered with answers to that question. Indeed, it was in trying to make writing attractive and accessible to these passionate culinarians that the idea for this book was conceived.

The first chapters (now Chapters 18 through 22) were written as a course handout to offer my culinary students an explanation of and practice with some grammar basics while at the same time trying to engage their interest with references to food (their passion) and to literature and film (*my* passion). As I began to work with Thomson Learning on the book, the concept grew to include additional topics in grammar and mechanics and to add an entire second half on the composition process. Material on persuasive writing, research, business writing, and English as a second language was added in response to the suggestions of initial reviewers.

The last element to arrive was the collection of food-based essays, stories, and poems by professional writers. The pieces themselves are grouped together at the end of the book, but each chapter contains exercises that ask students to relate the chapter content to the readings.

## ORGANIZATION OF THE BOOK

The first three units of *A Taste for Writing* address the writing process. In Unit 1: Getting Started, the three chapters explain the purposes and ingredients of writing. Chapter 1 presents writing as a form of communication that has a unique relationship with its audience. Chapter 2 focuses on the writing process itself, a messy affair somewhat like washing and peeling a potato. In Chapter 3, we talk about the "shape" of writing and look more closely at the parts of an essay. In Unit 2: Developing the Main Idea, the focus is on developing ideas through various methods, such as telling a story (Chapter 4), working with descriptive details (Chapter 5), using examples (Chapter 6), and comparing two or more items (Chapter 7).

Unit 3: Revising Your Writing goes more deeply into the most important step of the writing process: *re*-writing. Chapter 8 explores ways of evaluating and improving the content and organization of an essay. Chapter 9 addresses the importance of the introduction and conclusion, while Chapter 10 outlines the use of transitional expressions that move the reader from one idea to the next. Finally, Chapter 11 talks about finding the best words for a particular essay.

The next two units explore different applications and special situations in writing. The first two chapters of Unit 4: Understanding the Rhetorical Modes add two more writing methods or rhetorical modes to the repertoire. Process Analysis (Chapter 12) explains *how* events happened or how to do something. A recipe is an example of process writing. Cause and Effect (Chapter 13) addresses *why* an event occurred or why a certain condition exists. While most of the writing up to this point has been informative, Chapter 14 expands the purpose of communication to include persuasion. Last, Chapter 15 deals with some of the special approaches and techniques we use to write about literature and film.

The two chapters in Unit 5: Supporting Your Ideas with Research explain some of the basic methods and principles of using research in an essay, from finding and evaluating your sources (Chapter 16) to incorporating their information appropriately in your text (Chapter 17).

The next three units in the book cover basic grammar and punctuation with a view toward improving editing and proofreading skills. Unit 6: Editing for Grammar reviews the parts of speech (Chapter 18) and addresses basic sentence structure (Chapter 19), sentence fragments (Chapter 20), run-on sentences and comma splices (Chapter 21), and subject-verb agreement (Chapter 22). Unit 7: Editing for Style and Usage explores additional aspects of verbs (Chapter 23), as well as pronouns (Chapter 24) and modifiers (Chapter 25). Chapter 26 lays out the principles of parallel structure. Unit 8: Proofreading the Final Draft describes the process of correcting the punctuation and spelling. Chapter 27 covers such sentence basics as end marks and capitalization. Chapter 28 explains the use of commas and semicolons, while Chapter 29 reviews additional punctuation, for example, parentheses and italics. Chapter 30 focuses on the process of proofreading itself and offers information on spelling and commonly misused words.

Finally, Unit 9: English as a Second Language covers topics of special interest to students for whom English is not the first language, such as the use of articles and the progressive tenses, though the material may be useful to any writer.

The book also includes seven appendices. The first three have to do with proofreading: Commonly Misspelled Words, Spelling of Selected Culinary Terms, and Commonly Misused Words. The next three appendices cover different applications of writing: Basics of Business Writing, Creating a Persuasive Menu, and A Sample Research Paper. Appendix VII, A Taste for Reading, collects nine food-centered essays, stories, and poems by professional writers from various backgrounds.

## SPECIAL FEATURES

### Student Writing

The book contains samples of writing from over sixty students who attended the Culinary Institute of America between 2001 and 2006. It is important to note that these writers were simply completing a class assignment and did not intend their essays for publication. Some essays were written over a period of days, while others were composed in a testing situation within a time limit of ninety minutes. While I have edited some of the essays, I do not present them as "perfect" but rather as "tasty": they are included to exemplify a particular point of composition, such as using sensory details or tying the conclusion to the introduction, and especially to model open and engaging communication. While some readers may not always agree with the stylistic choices in the samples, I hope they will appreciate the effort and intent of the student authors.

I have also used student writing as the basis for some of the grammar quizzes. Here I am trying to simulate the editing phase of the writing process in which students will read paragraphs of text written by their peers (rather than teacher-designed exercises) and correct the grammar and word usage. In order to provide this more genuine, inviting environment for practice, I received permission from the writers to introduce specific types of errors—sentence fragments, for example—into their texts. I owe a special thank you to these students for allowing me to do so!

## Chapter Exercises and Activities

The exercises in the text are designed to help students practice the concepts and skills presented in the chapters. Instructors may use the exercises for individual or group work, in or out of class, collected or not, graded or not. Students may read and/or grade each other's work. Each chapter on composition concludes with a summary of its main points, exercises that relate the chapter content to the professional writing samples in Appendix VII: A Taste for Reading, and ideas for writing. The chapters on grammar include a quiz as well as a summary of main points.

## SUPPLEMENTARY MATERIAL

The instructor's manual explains the philosophy behind this textbook and offers both general suggestions for teaching and supplemental activities for each chapter. "Finding the Beating Heart" addresses students' feelings about writing, while "Approaching Grammar" describes hands-on activities for explaining such concepts as sentence fragments and commas with interrupters. The manual reviews information about writing assignments, exercises in grammar and mechanics, and the readings in Appendix VII. In addition, the manual provides various models of a course outline and grading rubrics, as well as reproducible handouts related to revision and editing. Sections on journals and peer review describe approaches to supplementary activities.

The manual then provides an overview of each chapter, noting especially any difficulties this author has encountered in teaching the material. Ideas for classroom activities include questions for discussion, often based on specific film clips; topics for supplementary reading and research; and suggestions for group activities, including the use of manipulatives. Finally, the manual offers notes on selected figures in the text, as well as answer keys for and comments on the chapter exercises.

# Acknowledgments

In thinking back over all who helped make this book possible, I am flooded with vivid memories of my own early teachers—Florence Stein, Dick Mullen, David Bourns, Mary Carruthers, Moreen Jordan, Dan Lindley, and Gloria G. Fromm. I see each one of them before me now, and my heart is overflowing with affection and gratitude.

I am also profoundly grateful to my friends and colleagues at the Culinary Institute of America, particularly those in the writing department. Richard Horvath formally reviewed this book, in addition to his valuable guidance from the next cubicle, and Sharon Zraly has been from the beginning of my teaching here a guide and a support. In addition, thanks are due to Stephen Wilson, Robert Biebrich, Adam Williams, Christine Crawford-Oppenheimer, and Erin Decker for their assistance with individual chapters, and I am grateful to Kathy Merget and Denise Bauer for their encouragement.

A special acknowledgment is owed to Marjorie Livingston, who suggested the title *A Taste for Writing*, and to Tim Ryan, who proposed the term *Culinarians*, as well as assisting me with the photographs taken on campus. Shortly before he became the President of the Culinary Institute in 2001, Dr. Ryan was an early reader of the student writing about peaches and apricots that appears in Chapters 5 and 7, and I am grateful for all his help with this project.

I am especially appreciative of those around campus who have helped me in both my teaching and my writing, including Professors Bonnie Bogush, Elizabeth E. Briggs, Richard J. Coppedge, Jr., John Fischer, Thomas Griffiths, Vincenzo Lauria, Anthony J. Ligouri, Claire Mathey, Michael Pardus, Nicholas Rama, Johann Sebald, John Storm, and

Jonathan Zearfoss. For their help with permissions, I am grateful to the mailroom staff and to the offices of the President, the Registrar, and Alumni Affairs. Sue Cussen, Nathalie Fischer, and Bruce Hillenbrand provided important guidance early in the process.

Thanks to Matthew Hart, the acquisitions editor at Thomson Learning who first believed in the book, and to Anne Orgren, my developmental editor, who walked me through the process with intelligence and flexibility.

I am particularly indebted to the students who allowed me to reprint their essays and snap their pictures. In addition, I am grateful to Wonderland Farm Market of Rhinebeck, New York, for permission to take photographs in Chapter 7 and Appendix VII.

Throughout this labor, I have been sustained by the counsel and friendship of Karen Kasius, Anne Henry, Lisa Henderling, and Linda Levine.

Finally, I have been particularly blessed in my family—my parents Chris and Mary Cadbury and my sons Anatole and Maksim Malukoff—without whose patience and support this book could not have been written and to whom it is lovingly dedicated.

Vivian C. Cadbury<br>August 2006

The author and Thomson Delmar Learning would like to thank the following reviewers:

Michael F. Courteau
The Art Institutes International Minnesota
Minneapolis, MN

Jo Nell Farrar
San Jacinto College Central
Pasadena, TX

Stephen C. Fernald
Lake Tahoe Community College
South Lake Tahoe, CA

Richard Horvath
The Culinary Institute of America
Hyde Park, NY

Nancy McGee
Macomb Community College
Warren, MI

John Miller
The Art Institute of New York City
New York, NY

James W. Paul II, MS, CCE, FMP
University of Nebraska at Kearney
Kearney, NE

Laima Rastenis
Cooking and Hospitality Institute of Chicago
Chicago, IL

Wayne Smith
Mesa State College
Grand Junction, CO

Lance Sparks
Lane Community College
Eugene, OR

Jorge de la Torre
Johnson & Wales University
Denver, CO

Sharon Zraly
The Culinary Institute of America
Hyde Park, NY

# About the Author

Vivian C. Cadbury has taught English at both the high school and college levels and has published educational materials on classic novels and the films based on them. At Ulster County Community College she taught study skills and information literacy in addition to English 101. She came to the Culinary Institute of America in May 2001, the third hire in the newly established writing department that now comprises eight full-time and over a dozen adjunct instructors. She is a member of Conference on College Composition and Communication (CCCC) and the National Association for Developmental Education (NADE), as well as a Certified Hospitality Educator (CHE).

Professor Cadbury earned a bachelor's degree in English from the University of Illinois at Chicago (UIC) and completed the requirements for certification in grades 7–12. During her student teaching at Lane Technical High School in Chicago, she was interviewed with her class on Studs Terkel's radio program, which can be heard over the Internet at *Studs Terkel: Conversations with America.*[1] As a graduate student at UIC, she specialized in the teaching of English and composition before completing her master's degree in English Literature, following which she studied Renaissance literature for two years at Oxford University in England.

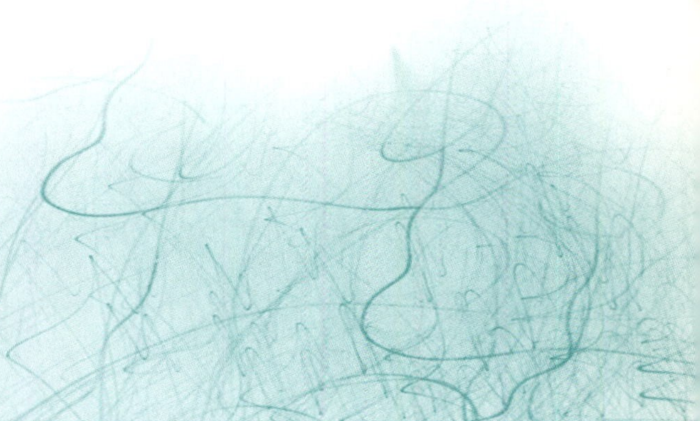

# Introduction: Recipe for Writing

Food has never been more popular in the United States than now in the twenty-first century, and neither has food writing. The number of students in culinary programs continues to grow, along with the country's appetite for cookbooks and restaurant reviews, articles on finding an effective diet or selecting the right wine, even murder mysteries whose detectives solve crimes while simultaneously cooking (or at least eating) a gourmet meal.

Written communication is addressed to a particular audience for a particular purpose and, to be effective, is written in a style appropriate to both. In this way it can be compared to preparing and serving food. Particular eaters have particular purposes for eating, for example, to fill up, to get a healthy diet, to celebrate a special occasion. Various types of restaurants are designed to serve these purposes: the fast food restaurant, the family-style diner, the fine dining establishment. Writing, too, takes various forms, such as stories, letters, and essays. Further, just as we sometimes cook just for ourselves, we sometimes write for ourselves alone—perhaps in a journal, perhaps in a letter we don't intend to mail.

**STUDENT WRITING** | Gerardo Vela Meza

Writing is a therapy. It's a way to take things out of your system. Watching those thoughts come out of your brain and onto a piece of paper is like taking a shower and feeling all the dirt wash away. Once I was really mad at someone who had hurt me badly, so I grabbed a

Students develop a taste for writing in the cloistered garden at the Culinary Institute of America.

piece of paper and a pen and started writing, nothing grammatically correct, just swearing on a blank sheet. When I finished, I went to the kitchen, turned the stove on, and set the paper on fire to let it burn. I felt so good afterwards without having to talk to anyone. Writing is great therapy.

Writing is also a way of thinking. And writing doesn't just *record* our thoughts—it actually stimulates *new* ideas and insights.

## STUDENT WRITING | Idan Bitton

"Writing is thinking" means to me that we go through a whole thinking process as we write about something. If we pick any topic we want to extend our thinking about, we should just start writing about it. We can find ourselves with whole new ideas about it when we are done writing. Writing can take us places we don't expect to go. If we are having a hard time thinking about something, we should start writing about it. As we write, we can encounter our most deep and hidden thoughts about the topic. We can realize things we never thought our brains held.

---

Writing can reflect and develop our thoughts, and it can communicate those thoughts to others. Writing can also be a form of therapy or entertainment. But it isn't easy. Although we may have plenty of ideas, we can't always get them down on paper. Sometimes we have no ideas at all and stare helplessly at a blank sheet, unable to write a single word. In this book we will look at techniques designed to help with these very problems—how to find out what we think and how to write it clearly. Our skills will improve with experience and coaching, and ideally our self-assurance will grow as well. In the following journal excerpt, a student notes a change in her attitude.

**STUDENT WRITING** | Astrid Sierra

Looking back to my past classes, I never enjoyed writing because I thought I could never hear my own voice. But now with all this practice, I am able to write a paper on whatever subject with confidence and pride. I know how to proofread it and what errors I am looking for, and I can correct them without doubting myself.

---

"I could never hear my own voice," she writes. No wonder she had no taste for writing! With practice, however, she begins to feel "confidence and pride" in her papers, just as she has felt confidence and pride in her cooking.

This textbook was written for culinarians, for those who take pride in the kitchen and its processes. And, surprisingly, the process of cooking has some intriguing parallels with the process of writing. While this is not the first book to note that cooking and writing have some common features,[2] *A Taste for Writing* develops the comparison more fully. Both cooking and writing involve processes that take certain ingredients and turn them into something new. Both can be messy at times, with pots and potato peels piling up in the sink, or reams of rough drafts littering the desk. One culinary student had this to say about the similarities between writing and cooking:

**STUDENT WRITING** Robert A. Hannon

These similarities include the need to pay acute attention to detail, the certainty that only hard work will result in assured success, the need to train, study, practice, and do research, and the importance of being able to work under pressure. Other somewhat more subtle similarities exist: both pursuits involve simplicity of tools and equipment, both rely on a foundation of tradition passed along through literature, and both are vitally dependent on criticism. Paradoxically, they both have rich, almost rigid traditions; however, to be successful, the professional cook and the professional writer must acknowledge these traditions while striving to be creative, thereby standing out in similarly competitive fields of endeavor.

This book asks its readers to "train, study, practice, and do research" and to be open, thoughtful, and creative in putting their ideas on paper. I am very proud of the students who have done just that, some of whose writing appears in these pages. Whether or not these culinarians enjoyed the composition process, they learned to appreciate its difficulties and—usually—its rewards. They developed a taste for writing.

# A Taste for Writing

## COMPOSITION FOR CULINARIANS

Vivian C. Cadbury

# Mise en Place

## Getting Started

UNIT 1

*chapter*

1

# Cooking Up Communication

Writing is a type of **communication**, a way of recording and sharing information, feelings, and ideas. Of course, writing is not the only kind of communication, nor is it necessarily the most effective kind for certain situations. We might identify two basic kinds of communication: **verbal communication**, that is, using words, and **nonverbal communication**, without words. In addition to writing, speaking is a significant form of verbal communication, and we'll talk more about that in a moment. Let's think now, though, about nonverbal communication, about how information can be communicated without words.

Remember Dirty Harry, the outspoken, bend-the-law San Francisco detective played by Clint Eastwood? One of my favorite nonverbal sequences occurs in a Dirty Harry film from 1983 called *Sudden Impact*. A judge has just reprimanded Harry for searching a car without the proper warrant, and the case has been dismissed. Filled with indignation, Harry heads for his favorite diner to get a cup of coffee. He's looking down at his newspaper as he enters the diner and doesn't notice the uneasy expressions on the faces of the waitress, the cook, and some of the customers—though we in the audience see them. As he remains absorbed in the paper, the waitress, Loretta, puts the coffee cup on the counter and begins to pour in sugar from the glass jar. She pours and she pours, all the time glancing nervously up at Harry and then around the diner. On and on she pours, while the audience

Student and teacher use a combination of verbal and nonverbal communication as they evaluate a gluten-free blueberry cornbread muffin.

begins to chuckle and Harry remains oblivious. Finally she stops, and Harry takes the cup and walks out the door. Still reading the paper, he takes a few steps, sips the coffee—and spits it out. He looks back at the door of the café, and someone inside turns the Open sign to Closed.

As it turns out, there is a robbery in progress in the diner, and Harry has gone in and out again without noticing a thing. But the coffee gets his attention. Without saying a word, and without attracting the attention of the three men holding the customers at gunpoint, Loretta has communicated very effectively. Dirty Harry understands the message and returns to the diner to stop the robbery. And when Harry stares down the barrel of his gun, the suspect understands that Harry will shoot. He doesn't even need to hear the famous lines. The expression on Clint Eastwood's face speaks more clearly than his verbal command. The coffee wasn't telepathic, but people are clever and effective communicators without words.

## SPEAKING VERSUS WRITING

Speaking—which, like writing, is a form of verbal communication—is interesting because, in addition to the words that are exchanged, there is a great deal of nonverbal communication going on at the same time. As we're talking, we can watch the facial expressions, hand gestures, and body language of the audience and make any necessary adjustments to improve communication. Similarly, the audience can watch us as we speak and can interpret the accompanying nonverbal cues as well as the words themselves. Some people find talking much easier than writing because of this nonverbal support.

When it comes to writing, however, it's as if a brick wall has been erected between the reader and the writer (see Figure 1.1). The writer has an idea in her head that she wishes to share. She writes a text that puts this idea into words, then passes it on to the reader. In most cases there is no direct contact between the two, no opportunity to ask questions or check comprehension. Therefore the writer must make a pretty good guess about what the reader already knows about the subject, what questions the reader might wish to ask, what

FIGURE 1-1 In written communication, it's as if a brick wall separates the writer from the reader.

vocabulary will be most clearly understood, what style will be most appropriate.

This difference between writing and speaking can be very clearly seen in directions or recipes. It's one thing to be in the kitchen and learn a new dish under the direct instruction of the chef, who can see what you do understand and what you do not understand, what you're doing correctly and what you're doing incorrectly. It's quite another to have only the written recipe with no opportunities for questions or corrections. Have you learned more about cooking from reading recipes or from working under the supervision of an experienced cook, whether it was your mother, a high school teacher, or the boss at your first job?

Since it doesn't allow for questions or corrections and must work without the assistance of body language and facial expressions, writing needs to be very clear, specific, and complete. Good essay writing means clear communication. Our purpose in this course is to improve the effectiveness of our written communication, to make our texts do a better job of reflecting what is in our minds.

**EXERCISE 1.1** **Speaking versus Writing**

Write a set of instructions for blowing a bubble or tying a shoe. Read the instructions to the person next to you. Are they effective? What would you change? How would these instructions have been different if you had been speaking instead of writing?

## UNDERSTANDING YOUR AUDIENCE

Communication involves three components: the writer, the message (essay), and the audience (reader). To make the message more effective, we must know something about its intended audience. Different audiences will have different backgrounds, needs, and interests, and writers must take these things into account as they design their papers. Would your succulent filet mignon be a success with a vegetarian? Would your scrumptious white chocolate macadamia nut cookies be a treat for someone with a nut allergy?

We already know a lot about differences between audiences. When you were growing up, weren't there some friends or family members who were more likely to give you money for ice cream, while others could be counted on to listen sympathetically to your complaint about a teacher at school? Discovering how to "work" an audience to achieve our purpose—the ice cream, the sympathetic ear—is an important step in the development of communication skills.

Difficulties can arise, however, when we meet a *new* audience. In the film *Miss Congeniality*, Sandra Bullock plays FBI agent Gracie Hart, whose focus on her job has left her neither time for nor interest in clothes or hairstyles—or even table manners. Among her peers in law enforcement, these idiosyncrasies are taken for granted. At the FBI, an agent's mental and physical skills are more important than her appearance or "charm." But when Gracie Hart is asked to go undercover as a beauty pageant contestant, her new audience is not so forgiving. Meeting her for the first time at lunch in an elegant restaurant, pageant consultant Victor Mellings (a hilarious Michael Caine) is dismayed by her tousled hair and shapeless clothes. His face glazes over with shock as he watches her plunge a forkful of steak in a pool of catsup and then jab it into her mouth. His jaw drops when he sees her walk, a movement that in his eyes is as primitive and unattractive as a dinosaur's. In other words, Agent Hart has not communicated effectively with her new audience.

Fortunately, Mr. Mellings and his team of highly trained beauty professionals can help her. They put mousse in her hair and gloss on her lips. They tweeze and wax and polish. They clothe her in a tight dress and high heels. They coach her on how to move and speak, even on how to cry with joy if she wins! Forty-eight hours later she arrives at the pageant, transformed into a believable contestant, though still occasionally stepping on someone's foot in a dance routine or making an insensitive remark. Yet the real change has taken place beneath the surface through her attempts to meet the expectations of this new audience. As she starts to understand and appreciate the other contestants, she begins to share her more of herself, which in turn makes her communicate more effectively. When, at the end of the film, she wins the title "Miss Congeniality," it is a recognition of her open and effective communication.

Let's move the discussion to a restaurant scene. Imagine that you've been asked to evaluate the menus of two very different properties. One is a successful yet humble family diner in a suburban setting, the other an ambitious, high-end establishment located in a major city. Now suppose each restaurant offered a pasta and cheese entrée. Would the menu descriptions be similar? How would a chocolate cake be described in each establishment? Look at the examples in Figure 1.2.

The customers at the diner expect basic food items, solid and familiar, with a homemade feel. We often see family names in their menu

**FIGURE 1-2** **Menu descriptions at a high-end restaurant and a family diner.**

*Bistro Urbano*

*Auattro Formaggi Macaroni. Fontina, mild cheddar, mozzarella, and smoked gouda sauce with freshly made pasta. Tossed with asparagus, tomatoes, and olive oil.* $18.95

*Flourless dark chocolate truffle cake served with raspberry compote and a cognac glaze.* $6.50

Downtown Diner

Classic Mac & Cheese ........................$8.95

Auntie Laura's Homemade Chocolate Cake ........$3.50

descriptions: Grandma's chocolate chip cookies, Uncle Bob's BBQ ribs. Customers at the trendy urban restaurant, however, expect more details in their menu descriptions, including specific ingredients that are out of the ordinary, such as fontina and flourless cake. There's no right or wrong here—just effective communication with two different audiences.

### EXERCISE 1.2 Understanding Your Audience

Briefly describe three separate restaurants of different types. What do the customers at each restaurant expect from their food? Now, write three different menu descriptions for a ground meat item, each one designed for a particular audience.

Just as Agent Gracie Hart with her clothes and hair tried to meet the expectations of the beauty pageant audience, and just as restaurants with their menu descriptions try to meet the expectations of their customers, writers are wise to anticipate what their readers will expect from them. In terms of content, we want to give readers enough information to make our ideas clear without going into unnecessary detail. In terms of style, we want to use a level of formality that is appropriate for the audience. If we're writing an e-mail to a good friend, for example, we unconsciously use a vocabulary she can understand, and we probably don't have to be as careful about grammar and punctuation as we would be in writing an essay for English class. College professors, on the other hand, do expect that we will use the conventions of standard written English in terms of word choice, grammar, and punctuation.

## WHY WRITE?

We've looked a bit at the differences between writing and other kinds of communication and at the audience and its expectations, but what makes us try to write in the first place? What is our purpose? In writing that e-mail to a friend, our purpose may be to make dinner plans, or to ask a favor, or to share a funny story. The writing is part of the workings of the friendship. Or we may write letters with the purpose of carrying out business, such as applying for a job or a scholarship. Sometimes our writing is entirely personal, for example, a journal entry or a poem. In college, we often write as part of an assignment. In other words, there are several different reasons or purposes for writing, and these will influence our choice of content, organization, vocabulary, grammar, and punctuation.

A piece of writing is often said to have one or more of the following **purposes**: to inform, to persuade, to entertain. Are you trying to pass on information, for example, how to lower the fat content of a three-course meal? Are you trying to persuade the reader to do something, such as invest in your new restaurant? Or, are you simply trying to entertain, to tell a funny or exciting story? Essays may have one of these purposes, or they may contain some blend of two or three.

As you plan a new piece of writing, you will begin to shape it depending on your purpose and how you might best achieve that purpose with a particular audience. If your story is about a narrow escape with a hot saucepan, for example, your essay might have the purpose of informing the reader on kitchen safety as well as entertaining her with your antics in the kitchen. If the reader has restaurant experience, you might spice up your prose with a little culinary slang: "I was weeded that night!" If not, your tone might be more serious: "Second-degree burns require immediate medical attention."

In this first student example, the writer's purpose is to share information about her visit to a storeroom of fresh produce, a visit she calls "Walking in Heaven."

**STUDENT WRITING** Seoyoung Jung

Among the vegetables what welcomed me first was a giant eggplant. Usually I have a very small and slim eggplant in my country, so I was shocked with the size of it. I thought to myself I might lose my appetite when it was on the plate because it was literally HUGE indeed.

This next example has a persuasive tone.

**STUDENT WRITING** Justine A. Frantz

Every occupation entails some sort of civil duties. Some of these duties are quite obvious, such as a soldier's duty to defend his or her own country; however, other jobs include dealing with less obvious societal issues. Ordinary people may not think that chefs have much to do with society other than feeding the general public, but there are more responsibilities in being a chef than meet the eye. Food-related problems are rapidly increasing world-wide, and who better to help conquer these issues, such as obesity, eating disorders, and world hunger, than people who have a strong passion for and understanding of food.

This chef-instructor has an attentive audience in his restaurant class.

Finally, in the student essay that follows, note how the content and style achieve the purpose of entertaining the audience.

## STUDENT WRITING Vincent Amato

My mother is the best cook in the world. People are always saying that about their moms, but this time it's true. When I was a child, coming home to a nice, hot homemade meal was an everyday event. My mother was even in cookbooks. The recipe was called Addie's Sukiaki. Delicious! On Sundays my sisters would help my mother cook all day long. I smelled the fresh sauce, gravy as Italians would call it, cooking on one burner, meatballs frying next to parmesan-crusted chicken cutlets on another. Italian bread was baking in the oven while linguine was boiled until *al dente*. The smell was so good it made the stomach ache with hunger.

It's actually hard to choose one dish that can be described as my favorite. Usually everything my mother made, even meat loaf, was delicious. When my father cooked was when we had to worry. He was into cooking different things like tripe or pig's feet. When he did cook, which wasn't much because my mother didn't let him, he would make enough to feed an army. My mother would usually bribe him with his favorite meal just so she didn't have to clean up his mess. All she had to say was "Ned, I'll make your favorite," and he agreed. This made us kids happy for two reasons, one being we hated pig's feet, and the second that we loved what my mother would make, Chicken Supreme.

Chicken Supreme starts with frying the chicken in a special batter. Sorry, I'm not allowed to give specific ingredients! After the pieces are fried, they are put into a deep baking dish and smothered with heavy cream, garlic, and mushrooms. To loosen the sauce, chicken broth is added, and the dish is baked for an hour. My mouth is watering just writing about it! After that hour, the sauce should be boiling. Serve it hot so the chicken stays moist and delicious.

My mom used to let me help her cook Chicken Supreme every time she made it. This might be the reason it is one of my favorites. Each time she tells me the story of her and her mother doing the same thing when she was a child. My mother would help my grandmother cook this same dish. Whenever my mom cooks Chicken Supreme, she cries. She never really talks after she has told the story. She just cooks, and thinks about her mom.

When a meal is passed down through generations in this manner, it has more meaning. Someday I hope to have kids, and we will cook Chicken Supreme. The same stories will be told, and hopefully the meal will be just as delicious. Maybe my children will love this dish just as much as I did as a kid.

---

## EXERCISE 1.3 Achieving Your Purpose

What is one of your favorite childhood dishes? Write a paragraph in which you simply inform the reader about the dish. Next, write a paragraph in which your primary purpose is to entertain rather than inform. Finally, write a paragraph in which you attempt to persuade the reader to try this dish. What differences do you notice between the three paragraphs in terms of content and word choice? Explain.

## WHAT MAKES GOOD WRITING?

Sometimes student writers may think their job is to write "what the teacher wants." And perhaps there is some truth in that. Students usually do have the goal of getting a decent grade, and teachers do have somewhat different ideas of what constitutes a "good" essay. However, if writers don't put their real thoughts and feelings into the essay, readers just won't be that interested. In the film *Finding Forrester* (see Chapter 2), Sean Connery muses: "Why is it the words we write for ourselves are always so much better than the words we write for others?"[3] Teachers are somewhat of a captive audience; they have to read every writing assignment. But in the outside world—the world of cover letters and résumés, restaurant reviews and cookbooks—readers may lose interest if you don't make a connection with them.

We all know people who are great storytellers, people with an eye for detail, a sense of humor, the ability to keep us enthralled from beginning to end. We seek these people out and buy them a cup of coffee just to hear their latest tale. On the other hand, we also know people who are so vague or boring or negative that we walk quickly in the opposite direction when we see them coming. The same is true of writing. There is some writing that we enjoy thoroughly, that is vivid and detailed, funny or suspenseful. There is other writing that is dull, pretentious, wordy, insincere—and we just put the book down.

Think for a minute about the different people you've met recently. Some approach you with a warm handshake and an open smile, volunteer a few bits of information about themselves—not bragging, just get-to-know-you stuff—and seem equally interested in hearing something about you. There's a *genuine* quality to these people and to their interaction with you. In contrast, we occasionally meet people who *lack* that quality of genuineness, whose expressions seem somehow false, or who flood us with amazing "facts" about themselves. We may sense they're manipulating us, or are just mentally elsewhere. And that's not appealing.

So, what makes good writing? The answer is, some of the same things that make good cooking: fresh ingredients, specialized skills, and a passion to reach the audience. Our purpose in this course is to find the freshest ingredients for our writing, the freshest ideas, examples, descriptions; acquire particular skills, such as brainstorming, revising, editing; and connect our own interests and passions to the content of any given writing assignment.

Every piece of writing begins with a need to communicate, that is, with a purpose. Further, every piece of writing must take into account the background and experience of the audience. Between the purpose and the audience lies the writing itself, the message, and this message must be "cooked" in such a way that it can be digested by the customer.

## RECIPE FOR REVIEW Cooking Up Communication

**Communication** has three components: the *writer,* the *reader,* and the *text* (see Figure 1.1). **Writing** and **speaking** are types of **verbal communication**; that is, they consist of words. However, while speakers can also use **nonverbal communication**, writers must rely on their words alone. It is especially important, therefore, that writers try to understand the **audience** of their communication so as to choose the most effective details, vocabulary, and style. Writers don't have a chance to interpret the reader's response or to answer questions.

Communication has one or more of the following purposes:

1. To inform
2. To entertain
3. To persuade

Good writing is specific, sincere, and meaningful.

## A Taste for Reading

1. Choose three of the readings in Appendix VII and answer the following questions: Who is the intended audience of each piece? How do you know?
2. In terms of purpose, compare the excerpts from *Kitchen Confidential* and *Fast Food Nation.* Explain how differences in content and style between the two pieces reflect their different purposes.

## Ideas for Writing

1. Write an essay or paragraph in which you discuss the differences between writing and speaking. Is one easier for you than the other? Explain.

2. Like writing, movies also have one or more purposes, including to inform, persuade, or entertain. Write a paragraph describing a movie that primarily informs, another movie that primarily persuades, and a third that primarily entertains.

3. Who is a person you think of as an effective communicator? What makes him or her effective? Write an essay or paragraph in which you explain your answer.

4. What makes a good dish of food? List as many characteristics as you can. Then list as many characteristics of a good piece of writing as you can think of. Compare the two lists. Are there any similarities? Explain.

chapter 2

# The Writing Process

Although it is the end product of writing that is consumed (through reading), writing is all about the *process*. Just like cooks, writers begin with raw ingredients, shape and season them, and plate them appealingly. Just like cooking, writing can take a shorter or a longer time, and can even be quite messy. Understanding the process helps relieve fears about writing and improves the quality of the end product. Experienced cooks walk confidently into the kitchen and know where to begin. Similarly, experienced writers know how to get started, and they realize that writing—like cooking—can get messy.

The messiness of the process often comes as a surprise to students. Perhaps they imagined themselves dropping rounded tablespoons of words onto a pristine sheet of paper as if they were dropping spoonfuls of chocolate chip cookie dough onto a baking sheet. However, the writing process is not exactly like a recipe for cookies. You don't always have to follow the same steps in the same order. In fact, there's no guarantee that following the steps will produce an essay the first time. It's not like knowing the cookies will be done after ten minutes in a 350° oven. What the writing process *does* is to give us a place to start and specific techniques to follow, especially if things are not going smoothly.

## THE STEPS OF THE WRITING PROCESS

The writing process consists of five steps:

1. **Brainstorming** ideas and details
2. **Outlining** the main points
3. Writing a **rough draft**
4. **Revising** the focus, content, organization, and details
5. **Editing** the grammar, usage, spelling, and punctuation

These steps fall roughly into two parts. When you start on a piece of writing, you need to get the ideas flowing; brainstorming, outlining, and writing the first complete draft should proceed without anxiety about the final outcome and especially without worrying whether each sentence sounds good or each word is spelled right. These concerns will be taken up later in the process.

The second part begins whenever you've finished the rough draft. It's the process of "revision," literally *re-seeing* the paper. At this point you use the more critical half of your brain and evaluate whether the rough draft in fact says what you want it to say, whether you need to add more supporting details or delete repetitive or irrelevant material, reorganize the paragraphs, rewrite the introduction, and so on. You ask yourself whether the main point is clear and whether it is effectively summarized in the conclusion. Finally, you look hard at each sentence and check the grammar, usage, spelling, and punctuation.

### STUDENT WRITING | Jonathan A. Campbell

Writing and cooking have many similarities. Even though cooking is done in a pot and writing is done on paper, the process is the same.

When beginning the cooking or writing process, preparation is important. Before you start writing, you must be organized. Having a topic, supporting ideas and strong examples is essential. Before you start cooking, you must have your mise en place: "Everything in place." This means you should have all your ingredients, equipment, a recipe and anything else that will help you complete your product. When developing a paper, you need time to transform it into a good paper, a strong paper. Revising is needed to reach a strong paper. During the cooking process, time is also essential. A fully developed, full-bodied, full-flavored product is reached through simmering.

The last step in the cooking and writing processes is fine tuning. In the cooking process, fine tuning is done through adjusting seasoning and consistency. Salt, pepper and a thickening agent could help perfect your product. The final step to producing a good paper is checking for spelling and clarity. If not done, your paper will be weak and lack purpose. Even though writing and cooking have different mediums, the similarities are there; you just have to look closely.

Many students find that focusing on the steps of the process rather than trying to jump to the end product helps them work more effectively and write a better paper. You don't want to serve a Thanksgiving turkey that's underdone or an essay that's underdeveloped.

Like peeling a potato in the early stage of cooking, brainstorming can be a messy part of the writing process, and you may not use all of the information or "peels."

## FINDING FORRESTER

The film *Finding Forrester*[4] offers a charming illustration of the two major phases of the writing process. The film is about Forrester, a successful, older writer, and Jamal, a high school student whose teacher is highly critical of his writing. As the two men become acquainted, they begin to help each other. In a funny early scene, Forrester places two typewriters on a table, sits down, and quickly types a page of script. Jamal himself is trying to produce something really impressive and is therefore unable to type a single letter. "Is there a problem?" Forrester asks. "I'm just thinking," Jamal defends himself—certain that thinking is a valid part of the writing process. But Forrester shakes his head. "No thinking," he declares, "that comes later."

Forrester encourages Jamal to brainstorm, to free his mind from the constraints of editing, from the constraints of the sequence-oriented and analytical part of his brain, and to get *something* down. Once the words are released into a rough draft, the story can be developed and polished during revision. "You write your first draft with your heart," he continues, "and you rewrite with your head."

Time passes, and Jamal sits helplessly staring at the blank page. The sun is setting and the room darkens, but Jamal doesn't move. "Writer's block" has paralyzed his fingers. Then Forrester has an idea. He brings out a copy of one of his old stories called "A Season of Faith's Perfection."

"What's this?" asks Jamal.

Forrester hands him the story. "Start typing that," he instructs, "and when you begin to feel your own words, start typing them. Sometimes the simple rhythm of typing gets us from page one to page two." Slowly, tentatively, Jamal begins to type. "*Punch* the keys, for God's sake!" demands the author, and, as Jamal does so, "You're the man now, dawg."

The image of Jamal frozen in front of the typewriter in the fading light is an excellent illustration of the need for brainstorming. Jamal is so worried that the very first word he writes won't be perfect that he is unable to move. The solution Forrester offers is "free writing," that is, writing a certain number of pages, or writing for a certain length of time, without giving a thought to the shape or quality of the final product, or to grammar, spelling, and punctuation. The idea is simply to unlock the words and let them spill onto the page. Later on, some of these words might be useful for the first draft. However, even if no interesting ideas or details are revealed in this exercise, the free writing has the effect of stimulating the brain, of heating up the oven.

Beginning a writing project is often difficult. Perhaps our focus is too much on the end product rather than on the process. Like Jamal, we might be paralyzed by the pressure to perform. Even without anxiety, we still might be "at a loss for words," unable to "think" of anything to say. We need to find a way to get started.

## PEELING THE POTATO

Imagine that you are making a potato dish, say whipped potatoes or potatoes au gratin. Your first step looks nothing like the end product, right? You're scrubbing and peeling the rough brown potato, and it doesn't look anything like the final spoonful of creamy white. Yet you don't experience a moment of anxiety. Your focus is on a series of familiar steps rather than on what those steps are expected to produce. You have learned to have confidence in the *process*.

Let's look at writing in the same way. Although you know that eventually you need to end up with a two- to three-page essay with an introduction, body, and conclusion, you also know that it doesn't happen all at once; there are steps to go through and transformations that must take place. Thus you may begin the process with a brainstormed list of ideas or a page of free writing that bears no resemblance to the finished essay, just as the unwashed potato bears no resemblance to a mound of whipped potatoes on a beautiful china plate.

## BRAINSTORMING IDEAS

Now suppose we *do* have confidence in the process, and we're *not* worrying about the final result, but we can't think of a single idea or even a single word to write down. Peel the potato. You probably have some kind of topic, more or less specific. Pick up your pencil, or open your laptop, and start writing. Let's say the topic is "a good restaurant experience." Here we go:

> I need to write about a good restaurant experience but I can't think of any just at this moment in fact I can't think of much of anything but I need to keep on writing so what shall I pick I think I went to several restaurants last month but I was kind of low on cash so I'm not sure there was a good one oh wait a minute what about Mesa Grill

This is probably the kind of free writing that Forrester pounded out on his typewriter to Jamal's amazement, a kind of stream-of-consciousness that clears away clutter in the brain and often leads to the germ of an idea: Mesa Grill. It's as if we were making stock and periodically skimming the fat off the top. Free writing can be done relative to *time*—that is, "write steadily for five minutes"—or to *space*—"fill up a page of notebook paper."

### EXERCISE 2.1 Free Writing

Like Forrester and Jamal, write or type a single page—without stopping and without "thinking"—on the following topic: Tell the story of a funny or frightening experience in your life.

Another effective type of brainstorming is to make a list. For example, we might jot down all the details we remember about the visit to Mesa Grill. While we won't be as "free" to write filler as we were in free writing, we must be careful not to slow down this part of the process by trying to evaluate the quality of these details. Remember, we don't always use the whole potato in the end. We're still in a messy, creative phase where we're just trying to keep the words flowing. Here's a list of details about the Mesa Grill:

| | |
|---|---|
| New York City | two floors |
| pictures on the walls | Monday night |
| Mexican fusion | lamb taco |
| deep fried oyster | grouper with vegetables |
| timely service | dessert plate |
| Spanish guitar music | Guy and Helene |

As we look back over this list, the specific food items seem quite promising, for example, the lamb taco and the deep fried oyster. Some of the points relating to the décor might also be effective if properly developed with specific details. Other details may turn out to be irrelevant, such as the fact that it was a Monday night and that Guy and Helene were there.

### EXERCISE 2.2 Listing

Brainstorm a list of details about the people, places, and things in the funny or frightening story you began in Exercise 2.1.

A third way of getting some ideas down on paper is often called webbing or clustering. We might put *Mesa Grill* at the center of the web to start us off. Then in a series of branches circling round that detail we might write *New York City, Mexican fusion, dessert plate.* By extending each branch, *New York City* might lead us to *Manhattan* and *Fifth Avenue,* while *dessert plate* might lead to *soft-centered chocolate cake* and *warm peach tart.* For many students, this method of brainstorming can be more attractive and effective than the more traditional outline.

### EXERCISE 2.3 Webbing/Clustering

Choose a topic, such as chicken or noodles, and write it in the center of a circle. Then fill in the branches with ideas and details you associate with that central word (see Figure 2.1). Work quickly, and do not judge the value of your choices; you're just peeling the potato here.

FIGURE 2-1 An example of webbing or clustering.

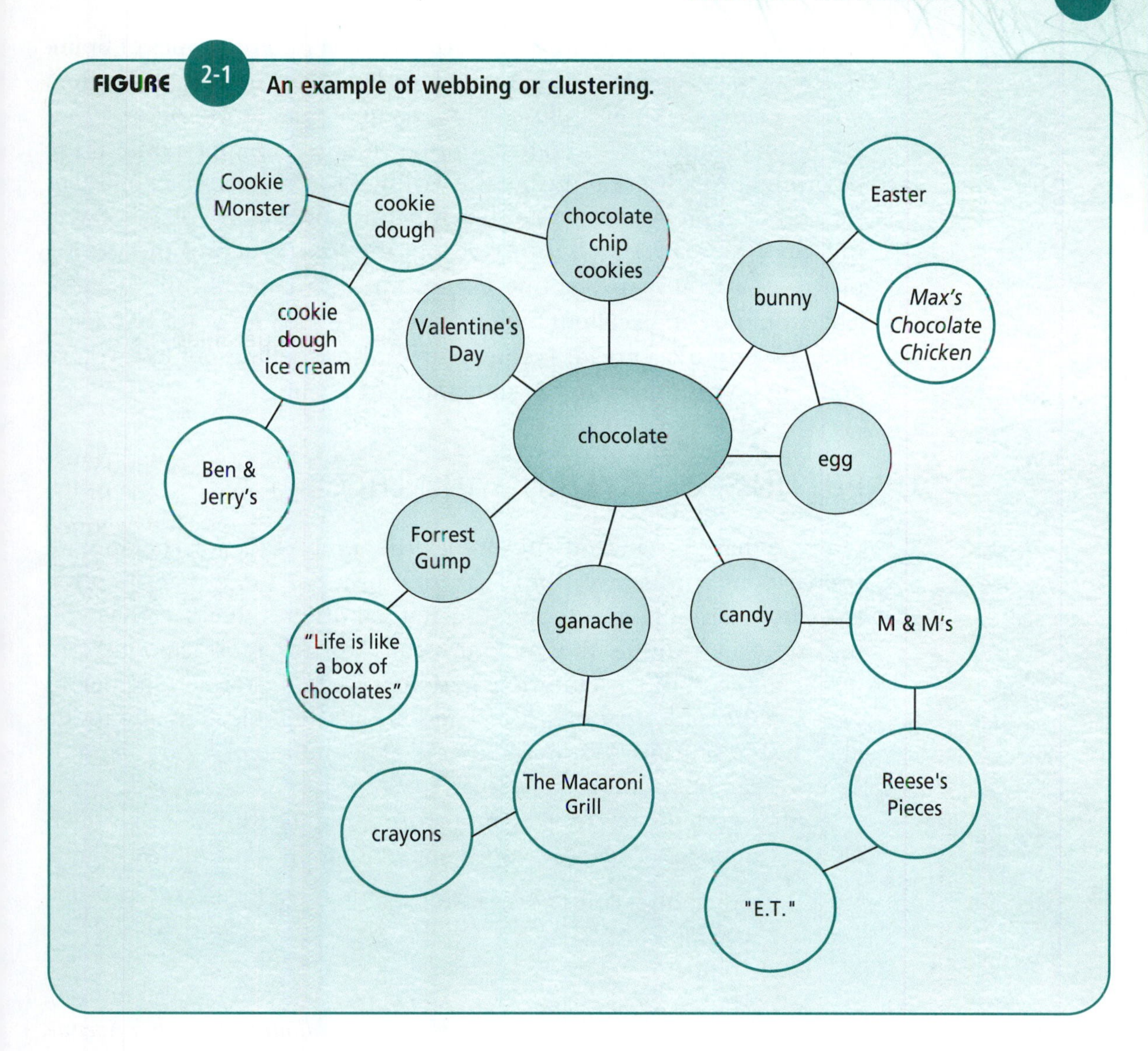


If you often find pictures easier to understand than words, drawing can be an effective method of freeing up your thoughts. Diagrams or charts can also get your thoughts moving. Again, if you choose to brainstorm like this, work freely and do not judge the usefulness or quality of each picture. That comes later.

A last and very effective method of brainstorming—though not one that is appropriate to every classroom situation—is talking with someone about the topic. The stimulus of the interaction and of the sounds themselves can lead us rather quickly to some interesting ideas and vivid details. It is useful to experiment with different types of brainstorming to get an idea of what might work for each individual. Further, if one technique seems to lose steam, try another—and another.

A final comment on brainstorming—it doesn't happen only at the beginning! At any point during the writing process it may be helpful to stop for a moment and reenter this very free and creative phase of writing. Suppose you're well into revision, the fourth step of the writing process, and the paper is fine except for a rather dull introduction. This would be an excellent time for some free writing, or for bouncing ideas off another person. If your potato dish is selling out fast, you're going to go right back to the sink and start peeling.

## PUTTING EVERYTHING IN PLACE

Once we have the ingredients, we lay them out ready for cooking; we prepare our *mise en place.* In the writing process, a step that often follows brainstorming is outlining the general order of ideas so that writing the rough draft—the cooking—can proceed more smoothly. In our example of the restaurant experience, we brainstormed the details *New York City* and *Mexican fusion.* These details might lead us also to the chef and to our own visit to Mesa Grill. The outline of a paragraph or an essay can be as informal as this list: *location, cuisine, chef, visit.*

Other students may prefer a more formal outline with Roman numerals and capital letters. Here is an example using the same information:

I. Introduction—Visit to Mesa Grill
  A. Location
  B. Cuisine
  C. Chef
  D. Visit

The finished paragraph appears below.

Taylor Jones

Mesa Grill in New York City fuses American cooking with Mexican themes. Located on Fifth Avenue in Manhattan, it serves American cuisine with a Mexican flair. Chef and owner Bobby Flay knows all about how to twist American cuisine because of his background in the studying of Mexican food. I visited Mesa Grill two weeks ago while spending the day in Manhattan.

Flow chart of full "Mesa Grill" essay.

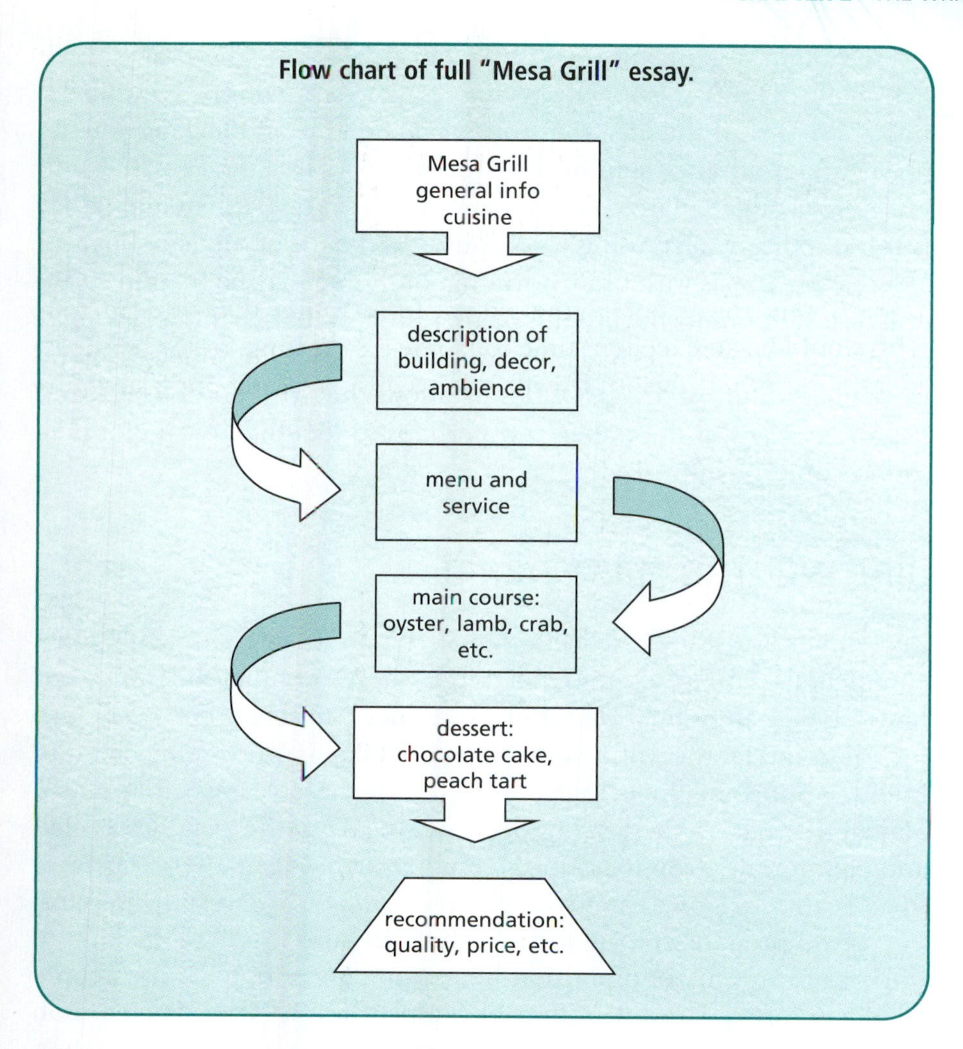


Another way of getting the paper organized is to create a flow chart or some other type of diagram. Whatever the method, the advantage of outlining before we begin to write the rough draft is that we have a clearer sense of our direction.

## EXERCISE 2.4 Outlining

Look at your brainstorming work from Exercise 2.1. Where should that story begin in order to be clear to a reader who doesn't know you? Where should it end? Briefly outline the main events of the story, that is, the main parts of the essay, in an informal list, formal outline, flow chart, or whatever format is helpful to you.

Although organizing your ideas before you start the draft can be extremely helpful, it is not always necessary to write down an outline at this point. And if the idea that you *should* organize is blocking you, let it go and try another step in the process. You can always go back and rearrange your sentences and paragraphs—that's what "cut and paste" is for, whether you're using a computer keyboard or an actual pair of scissors! Finally, as with brainstorming, outlining may be helpful at several different points during the writing process. For example, you may choose to outline the main points of an essay *after* you have gone through a couple of drafts. Outlining allows the writer to step back and get an overview of the paper and perhaps to identify how the organization could be improved.

## WRITING THE FIRST DRAFT

Remember Forrester's advice to Jamal? When we begin to write a first complete draft of a paper, we need to *feel* it. It's not that the topic itself has to be an emotional one. But we do need to feel a *connection* with the topic and let the words spill out freely. Like brainstorming and outlining, writing this first draft is still part of the creative, often messy, playful first half of the writing process. We've got the potatoes peeled and cut; now we drop them in the boiling water. Again, we're far from the elegant swirls of creamy whipped potatoes, but we trust the process.

As you prepare to write the first draft, be sure you have the kind of work space or environment that helps you concentrate. Some people need absolute quiet while others like to play music—from Bach to Bob Marley to Beanie Sigel—and still others like the chaos of a kitchen table in the midst of family activity or the friendly sounds and smells of a coffee shop. If you have not already determined which environments are helpful to you, do some experimenting.

You also need to find the right "tools." Sometimes this may be determined by the assignment. If you have an in-class writing activity or exam, for instance, you probably cannot use your computer and listen to Aretha Franklin on the headphones. Assuming you have options, however, try whether a pencil feels better than a pen, or whether typing helps your thoughts flow more freely. I find that I like to free write on the laptop but outline with a pencil. Typing seems to open up my ideas while the slight resistance of the graphite on the paper helps me concentrate on organization. Each person will have a unique combination of materials and environment that makes writing a little easier.

Very important during the drafting phase is what you're *not* paying attention to. You are definitely not paying attention to grammar, spelling, and punctuation. And again, you are trying not to worry about the end product. The potatoes that you have peeled and cut need to cook before they can be whipped into their final form. Let's look at a draft of the second paragraph of the essay on Mesa Grill.

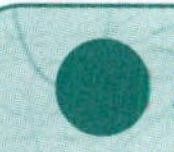

**STUDENT WRITING** Taylor Jones

### Rough Draft

The restaurant is a medium sized building that fits one hundred people, with two floors. As I walked into the restaurant the feel was warm and exciting. The dinning room had beautiful pictures on the walls and viberant color's on the celling of green an orange. When I walked in I felt as thought I was in Mexico. Music of Spanish guitar's were playing throughout the building. Although it was a loud restaurant due to all of the customers. I was still able to make conversation with your fellow patrons.

The student has done a nice job with the flow of detail and has wisely paid no attention to grammar, usage, spelling, or punctuation. In fact, it is not until the final phase—editing—that these categories would be examined and the corrections made.

**STUDENT WRITING** Taylor Jones

### Final Draft (edited for grammar, spelling, and punctuation)

The restaurant is in a medium-sized building with two floors and fits one hundred people. As I walked in, the feel of the restaurant was warm and exciting. The dining room had beautiful pictures on the walls and vibrant colors of green and orange on the ceiling. When I walked in, I felt as though I were in Mexico. Music of Spanish guitars was playing throughout the building. Although it was a loud restaurant because of all of the customers, I was still able to make conversation with the other patrons.

Remember that, like brainstorming, writing the rough draft is still very free. You are still far from the finished product. Keep peeling.

**EXERCISE 2.5** **Drafting**

Take time now to write the first draft of your story about a funny or frightening experience in your life. Find the right environment, assemble your writing materials, and start writing. Don't think about the end product, and don't worry about the grammar, spelling, and punctuation. You're just cooking the potato, remember; it's far from ready for plating.

## ADJUSTING THE SEASONING

The most unpredictable part of the writing process is revision. Here we begin a more "serious," analytical phase, *evaluating* the flavor and *adjusting* the seasoning. The initial revision may be quite complex because you're looking at two large and fundamentally important aspects of the paper: the content and the organization. Let's look at the following example of a rough draft.

 **STUDENT WRITING** Noriko Yokota

### Rough Draft

My cute little dog, bikky, is very greedy for food as like other beagles are always hungry. His place in our living room is on the sofa where overlooks everything so that he won't miss a single change to get a little tiny share.

One day my mom happened to left sandwiches, which were my share for lunch, on the table. My dad went out to do the garden after lunch, mom also went out to golf.

When it got a little dark and everybody home, we found nothing left but a plate on the table.

We looked immediately at Bikky, he seemed to be satisfied. She said to him, "I know you did, bad doggy!"

She scolded him strongly. my dad hadn't said anything, hadn't done anything till Bikky finally ran away under the sofa.

My dad suddenly said, "please don't scold him anymore. It was me who ate the sandwich."

First the student re-reads the essay as if seeing it for the first time. She asks, Is it the right story? Is the main point clear? Are there enough supporting details? Should parts of the essay be eliminated because they are repetitive or irrelevant? It's often helpful to read the essay aloud at this point, or to have someone else read the paper aloud. Errors that we might miss when looking at the page become brutally clear when we hear them.

In this example of Noriko's paper, the story is a good one for the assignment "Tell the story of a funny or frightening experience." However, she is advised to add more details, such as the type of sandwich—"vegetable and ham" —and what exactly happened when her father "confessed."

**STUDENT WRITING** Noriko Yokota

### Revised Draft

But when Bikky finally ran away under the sofa with his tail between his legs, dad suddenly said "please don't scold him any more. . ." Mom and I looked at each other in astonishment. He started to open his mouth and said, "It's not Bikky who ate the sandwich. It's not him . . ." then he seemed to become defiant, embarrassed a little, and confessed, "It was me who ate the sandwich on the table."

The added details make the scene much clearer—and much funnier.

A second major area we want to look at as we begin to revise is the organization, including the order of sentences and paragraphs, the effectiveness of the introduction and conclusion, and the presence of appropriate transitions. In our example, the organization is generally quite clear. However, Noriko is advised to expand the conclusion. Compare the original ending—"My dad suddenly said, 'please don't scold him anymore. It was me who ate the sandwich.'"—with the revised conclusion.

**STUDENT WRITING** Noriko Yokota

### Revised Draft

Mom and I bursted into laugh at first, but after a while we started to feel miserable to Bikky. Mom apologized to him with many kisses and hugs. But It was my dad who most should apologize to Bikky so he gave a countless cookies to him. We have an usage "You have fortune even you're in misfortune." Bikky experienced that personally.

---

This revised ending completes the story and even adds a moral.

Bikky's story was in good shape at the end of the first draft; revision was mostly a matter of adding details and information. However, sometimes the first draft just doesn't work at all. This is why revision is the least predictable step of the writing process. One paper might be done in two drafts; others may take four or more. We should understand that the number of drafts is not a measure of our skill or of the quality of the end product.

Revision is difficult because we must change gears. While we were brainstorming, outlining, and writing the rough draft, we allowed ourselves a great deal of freedom. We wrote from the heart, and we intentionally refrained from evaluating our work. But during revision we need to step back and take a cool, critical attitude. We need to read the paper now as if it belonged to someone else. Most important, we need to be ready to get rid of words, sentences, paragraphs—even entire drafts—that don't work, and this is difficult because we often become very attached to the words we've produced with such effort!

Let's think about cookies for a moment. Suppose we weren't paying attention and added salt instead of sugar to a batch of chocolate chip cookies. Could we fix those cookies? Could we sprinkle sugar on top and serve them to the customer? No. In this case—as with some rough drafts—we have to throw the cookies out and start again. Revision is so important and so complex that we will talk about it again in several of the units that follow.

### EXERCISE 2.6 Revision

Look back over the rough draft of your essay on a funny or frightening experience. Is it the right story? Do you give enough background to make the story clear? Do you need to add more information or more descriptive details? Are there repetitive or irrelevant details that can be omitted? Does the story have a beginning and an end? Use the answers to these questions to make changes in content and organization, and write a second draft of your essay.

This student in a restaurant class wipes the rim of a plate before service; similarly, we check the spelling and punctuation of an essay before submitting it to a reader.

## PLATING THE ESSAY

Once the food is cooked, it can be plated. Similarly, once the essay's content and organization are in good shape, we turn to editing, that is, correcting the grammar, usage, spelling, and punctuation. This is best done by reading the paper one sentence at a time, from the capital letter to the period. In Noriko's essay there were a couple of comma splices, some problems with English usage, and several errors in spelling and punctuation, most of which were corrected in the final draft.

**STUDENT WRITING** Noriko Yokota

### Final Draft

My cute little dog, Bikky, is very greedy for food. Like other beagles, he is always hungry. His place in our living room is on the sofa where he overlooks everything so that he won't miss a single chance to get a little tiny share.

One day my mom happened to leave vegetable and ham sandwiches on the table which were supposed to be my share for lunch. After lunch, Dad went out to do the garden, and Mom went out to golf.

When it got a little dark and everybody was home, we found nothing left but a plate on the table, although I hadn't had lunch yet. We all looked immediately at Bikky. He seemed to be satisfied licking his lips. My mom, who is most strict with manners, said to him, "I know you did it, bad doggy!" While she was chiding Bikky, Dad did nothing. He seemed indifferent to what was going on now.

But when Bikky finally ran away under the sofa with his tail between his legs, Dad suddenly said, "Please don't scold him any more." Mom and I looked at each other in astonishment. He started to open his mouth and said, "It's not Bikky who ate the sandwich. It's not him. . . ." Then he seemed to become defiant, embarrassed a little, and confessed, "It was me who ate the sandwich on the table."

Mom and I burst into laughter at first, but after a while we started to feel miserable about Bikky. Mom apologized to him with many kisses and hugs. But it was my dad who most should apologize to Bikky so he gave countless cookies to him. We have an adage "You have fortune even when you're in misfortune." Bikky experienced that personally.

---

The revised essay provides more information, more attractive specific details, and generally correct grammar, spelling, and punctuation.

**EXERCISE 2.7** **Editing**

Look over the final draft of your essay on a funny or frightening experience and correct any errors in grammar, spelling, and punctuation.

## CLEANING UP

There is one last part of preparing your essay for the reader, and that is **proofreading**. It's as if we've cooked a beautiful meal and plated it up on beautiful china—but need to wipe the edge of the dish before serving it to the customer. Once again we read the paper sentence by sentence, and at this point we sometimes start with the last sentence and read backwards. In this way we can catch tiny typographical errors without getting caught up in the flow and *meaning* of the essay. If we typed the essay on a computer, we would also want to use spell check one last time (see Chapter 30).

One of the last aspects of an essay to be revised may be the title (see also Chapter 10). While not necessary in every case, a good title catches the reader's attention and often highlights some important aspect of the essay. In the same way as a menu description identifies a product of cooking, the title of an essay names the product of another complex process: writing.

### EXERCISE 2.8 Adding a Title

With a partner or small group, brainstorm a list of ten possible titles for Noriko's story. Share the list with the rest of the class. What's the best one? Why? Now, brainstorm a list of possible titles for your own story. Choose one and write it on your final draft.

## RECIPE FOR REVIEW The Writing Process

### Steps of the Writing Process

1. **Brainstorming**: Free write, make lists, create webs or charts, draw pictures, talk to classmates, do research, do experiments
2. **Organizing**: Put main points in order, use a list, outline, or diagram
3. **Drafting**: Write a rough draft, focus on the main idea, get the whole thing on paper, ignore spelling and grammar
4. **Revision**: Read the essay with new eyes: Is the main idea clear? Is it developed with sufficient details and examples? Is it logically organized? Does the introduction focus on the subject and get the reader's attention? Does the conclusion restate the main idea?
5. **Editing**: Check word usage and grammar, proofread for spelling and punctuation

## A Taste for Reading

1. Professional writers often have a particular schedule or process, for example, writing for four hours in the morning and revising in the afternoon. Read an essay by a professional writer that describes the writing process, such as Donald H. Murray's "The Maker's Eye."[5] What did you learn about the writing process?
2. Choose one of the writers whose work appears in Appendix VII and find out all you can about his or her writing process.

## Ideas for Writing

1. Write a paragraph in which you explain your own writing process and assess its effectiveness.
2. Write a short essay about a person, place, or restaurant that you know well. Use all five steps of the writing process.
3. Discuss the similarities (if any) that *you* see between writing and cooking (or baking). Use specific examples to illustrate your points.
4. Think of another "process" that you know well—perhaps training for a sports event, sewing a costume, or making a set of shelves. Write a paragraph or short essay in which you explain the steps of this process in the proper order.

chapter 3

# The Shape of Writing

Dining at a restaurant is an experience that has a certain predictable and, therefore, satisfying shape. Imagine yourself at a nice restaurant, anticipating a healthful and delicious three-course meal. First, the server sets down an aromatic tidbit, perhaps a bit of bruschetta with tomato and garlic. Do you remember those cartoons in which the goofy dog or greedy rabbit, enraptured by the visible waves of a delicious odor, literally floats above the ground, drawn irresistibly by the nose?

Once the appetite has been whetted by the starter, the main course provides the largest share of nutrition and calories. Essentially it satisfies the customer's hunger, as well as pleasing her senses. However, we don't usually like to jump up from the table right after finishing our *coq au vin*. It's nice when the plates are cleared and we linger over a cup of coffee or a bite of chocolate. We take a moment to look back over the meal just finished and slowly prepare to leave the restaurant.

Just like that three-course meal, an essay has three parts (see Figure 3.1). The **introduction** is like an appetizer. It should get the reader's attention, whet her appetite for the essay, and give some idea of the subject. The introduction may also state the main point of the piece, although writers sometimes save that for the end. The entrée or main course of the essay is often called the **body**. It contains all the information about the subject, plus supporting details and examples. Just as the main course is the largest part of the meal, the body is the

**Like this delicious aroma that draws the dog after it in a rapturous dream, the introductory paragraph of an essay should draw the reader into the story.**

largest part of the essay. Finally, the **conclusion** is like that after-dinner drink or dessert. It provides time for the reader to think back over the substance of the essay, to savor its style. Like dessert, the conclusion helps the reader to frame the experience and to move on.

## THE SHAPE OF AN ESSAY

In writing a story or an essay, you are creating an experience for the reader in the same way that you would create a dining experience for a restaurant customer. It is your job to prepare customers for the meal, satisfy their appetites, and ultimately send them on their way. Like the meal, the essay has a "shape" that is expected and, when

**FIGURE 3-1** The Shape of Eating.

**MENU**

**Appetízer**

The Introduction

**Entrée**

Body Paragraphs

**Dessert & Coffee**

The Conclusion

delivered, satisfying to the customer. Not only is this a shape that we can *see*, a shape in space like a menu or an outline, but it is also a shape in *time*, one that unfolds as we eat or read.[6]

Think about the shape of the popular television series *Law & Order.* Each episode begins with the discovery of the crime, often with someone stumbling over a body. Then the detectives investigate and arrest a suspect. That's the first half. In the second half, the prosecuting attorneys take the suspect to trial, weather a twist or two, and hear the verdict, and the show often closes with a pithy, cynical one-liner from McCoy. The shape of this show is predictable, almost stylized, but highly effective. Similarly, an essay has a certain rhythm and feel—introduction, body, conclusion—that is familiar and comfortable to the reader at the same time as it effectively delivers the information.

As you read the following student essay, notice that it has a "shape." It has a beginning, a middle, and an end. You can almost "see" it—an arc, perhaps. You can somehow "feel" the build-up to the story and then the wrap-up at the end.

**STUDENT WRITING** Nelson Tsai

I am obsessed with food. However, I grew up in a traditional Taiwanese family where chefs are not well respected. My parents wanted me to become a doctor or lawyer, even though food and cooking have been my passion ever since I was a kid. I continued to think about a career in the food industry throughout my college life. I love food so much that I can't stop thinking about it, and this leads into a funny experience I encountered a couple of months ago.

On a beautiful summer afternoon, I got back home from the market with my parents. While thinking about the tender braised beef short ribs I was going to prepare for dinner, I got out of the car, locked the car doors, and walked to the back door of the house. Then I tried to open the house door with the car remote. I struggled for a long time without realizing what I was doing. After twenty seconds of delay, I finally came to the awareness of how foolish I was.

From this incident and my daily attitude towards food, my parents saw and felt my determination and my passion for the food industry. Their hearts were finally shaken after I worked two years full-time as a line cook at Le Chien Noir Bistro. Furthermore, now they understand that cooking is hard work, but the reward is even greater. I was electrified when my mom said she would support me one hundred per cent spiritually. I thank my parents for their support, and I am sure they will be proud of me in the future.

The introductory paragraph draws the reader in with a strong opening statement: "I am obsessed with food." It then lays out the essential dilemma in the writer's life: he wants to turn his obsession into a career in the food industry, while his parents want him to choose a different path. The middle paragraph tells the story of a particular moment that highlights the writer's exclusive focus on food: the "tender braised beef short ribs" are so important that he cannot pay attention to something as mundane as opening a door! It's a funny scene, and as we read it, we sense that the next step will be some reflection on the *meaning* of that scene. And this is what happens. In the last paragraph, the writer explains that this moment was pivotal in helping to resolve the dilemma outlined in the introduction. When his parents recognized the power of his obsession, "their hearts were finally shaken" and they decided to support his pursuit of a culinary arts degree.

In addition to its classic and satisfying three-part shape, this paper contains four essential *ingredients* of a good essay. It develops a single **main idea**: The writer's parents finally decide to support his pursuit of a culinary career after witnessing his single-minded passion for food and his determination to work with it. Second, this main idea is developed or explained by one or more **supporting points**. For example, the writer's single-minded passion is illustrated by the funny story about the car keys. Each supporting point is then fleshed out with **specific details**, such as examples, descriptions, or factual information. While the writer is thinking about "the tender braised beef short ribs" that he is planning to prepare for dinner, there's a "twenty-second delay" before he realizes that the car's remote won't open the door of the house. Finally, a number of **transitional expressions** move us smoothly through the essay. The "beautiful summer afternoon" introduces the car keys story while "from this incident and my daily attitude towards food" leads us to the resolution of the writer's dilemma and the end of the essay.

**EXERCISE 3.1** **The Shape of an Essay**

Choose one of the student essays in this book or one of the pieces in Appendix VII: A Taste for Reading. Describe how it does or does not have the three-part shape we've talked about.

## PARAGRAPHS: THE MINI-ESSAY

Paragraphs share many of the ingredients of an essay—the statement of the main idea, the supporting details, and the transitions that move the reader from one idea to the next. Paragraphs may also have sentences that act as introductions and conclusions, although often not as formally as those in an essay. The last line of a paragraph may be more of a *transition* than a conclusion. Consider the following paragraph from a student's essay about his visit to a meat fabrication class:

**STUDENT WRITING** Karl Hacker-Spagnola

As the tour continued, the chef's voice got more excited with every step, as if he was anticipating something. By the time we entered the first of the two walk-ins, he was like a six-year-old on Christmas Day. He grabbed and handled meats, anxiously holding them up and asking the students questions about each piece. The students crowded around him, feeling his intensity as he moved from shelf to shelf until he came to a Smithfield ham. This one item seemed to excite the chef the most, as he fumbled with a student's knife to cut open the tanned colored sack, revealing a second layer of packaging. He ripped open the parchment paper to reveal a mold-spotted hunk of meat. The sight of this stimulated the senses. The taste of the mold swarmed the mouth. The visually disturbing sight made our skin feel as if a layer of grime had just developed. It disturbed many of the students that were flocked around him. Even after the chef explained to the class that the mold was not bad due to the curing process, many would never forget what a Smithfield ham was and, worse, many would never try it.

Like an essay, this paragraph has three parts: introduction, body, conclusion. The first sentence tells us that the chef leading the tour is looking forward to something, and we too feel a sense of anticipation. As an introductory sentence, it is highly effective. What *is* the chef looking forward to, we want to know. The body of the paragraph then offers a brief narrative (the unveiling of the Smithfield ham) interspersed with vivid details about the meat's appearance and aroma, as well as the reactions of the students to this unfamiliar product. Finally, the concluding sentence summarizes these reactions and predicts an outcome: the students will not want to taste a Smithfield ham in the future. Although this line is not the end of the essay, it does wrap up the main idea of this particular paragraph: the chef's eager anticipation and the students' alarm.

In addition to its shape, the paragraph shares its main ingredients with an essay: a main idea, supporting points, specific details that explain or illustrate those points, and transitions that show the reader how all the parts are connected. The paragraph above is about the chef's anticipation of the delectable items in the walk-in. To support the main idea, the writer documents the chef's increasing excitement, shows his eagerness to uncover the Smithfield ham, and explains the students' rather different reaction. Look at the specific details:

Chef's excitement

- "like a six-year-old on Christmas Day"
- he "grabbed" the meats, "anxiously holding them up and asking the students questions"
- the students feel "his intensity as he moved from shelf to shelf until he came to a Smithfield ham"

Eagerness to uncover the ham

- the ham "seemed to excite the chef the most, as he fumbled with a student's knife to cut open the tanned colored sack"
- he "ripped open the parchment paper"

The students' response

- the "taste of the mold swarmed the mouth"
- the "visually disturbing sight made our skin feel as if a layer of grime had just developed"

Notice the detail: The chef is not just a "kid"; he's a "six-year-old on Christmas Day." Further, the ham has a "second layer of packaging" made of "parchment." The choice of verbs also adds energy and specificity to the paragraph: grabbed, crowded, fumbled, ripped, swarmed, flocked.

Finally, notice the transitional expressions. The paragraph is developed chronologically, that is, according to the sequence of events, and the transitions guide the reader through that series: "as the tour continued," "by the time we entered the first walk-in," "as he moved from shelf to shelf," "even after he explained to the class."

Paragraphs are like mini-essays, and each paragraph within an essay needs special attention. In the same way, although part of a larger "plate," each food item should be individually cooked and seasoned.

Like paragraphs in an essay, each item in this display may have been prepared differently, yet it still contributes to the overall design.

## EXERCISE 3.2 The Ingredients of a Paragraph

Study the paragraph below. Then explain how it uses the four ingredients discussed above: a main idea, supporting points, specific details that explain or illustrate those points, and transitions that show the reader how all the parts are connected.

When I was twelve years old, electricity was just established in my town. We didn't have any money to buy it, but my neighbors did. Torcila and Luis were their names. They lived in this small brick house with small windows that looked like a box with small holes on the sides. Anyway, the small window was big enough to watch through it the big television they had in their living room. Every day after school my brothers and I used to run to see this program that came on at 1:00 p.m.—*Tom and Jerry*. We weren't the only ones who wanted to watch this program, though; twenty others wanted to watch, too, so whoever got to the window first would have a better spot.

René S. León

## "THE BEATING HEART"

Where does the shape of writing come from? We might say that it comes from the outline, from the careful planning of the content of each paragraph, and certainly the organization of an essay plays an important role in our sense of its shape. Perhaps the outline is like a skeleton; it provides an underlying form for the soft tissues that cover it. Within that skeleton, however, lies the central force that brings the organism to life: the beating heart.

Let's use a story to make this point clearer. *The Shipping News,*[7] a Pulitzer prize-winning novel by Annie Proulx—as well as a film starring Kevin Spacey, Judi Dench, and Julianne Moore—is about a man named Quoyle, a rather depressed man, actually, whose life is cheerless and drab. His parents don't seem to care much for him, and he doesn't have a job that interests him. One day he is sitting in his car at a gas station when a wild and beautiful woman jumps in beside him. It is Petal, and before the day is over she has announced that she will marry him. They have a little girl, Bunny, and Quoyle has some interest in his life—but at a price. Petal remains wild as well as beautiful and is soon pursuing other men. Quoyle is dismayed but helpless.

When Bunny is about six years old, however, events take a dramatic turn. In order to finance a getaway with a new man, Petal sells Bunny on the black market and is immediately killed in a car accident. Investigating officers find a receipt for Bunny in Petal's handbag and return the little girl to her father. Meanwhile, Quoyle's irritable parents have locked themselves in the garage with some carbon dioxide fumes, after leaving a message for their son that he's been a continual disappointment. Yet before Quoyle is completely overwhelmed, his father's sister, The Aunt, appears on the scene and moves the grieving man and his daughter to the old family home in Newfoundland, an island off the coast of Canada. And this is all within the first seventeen and a half minutes of the film!

In Newfoundland, Quoyle finds work on the local newspaper, and it is with his first story that we receive some advice on the writing process. Quoyle is dispatched to cover a car accident: two people killed when their pickup hit a moose. Given his recent history, this fledgling reporter is understandably upset, even vomits at the scene, but dutifully produces some five pages of typewritten text. His editor, who is as irritably critical as Quoyle's own parents, tosses the pages into the trash, exclaiming "If I'd wanted *War and Peace,* I would have hired William bloody Shakespeare!" Quoyle's story is too long, the editor means to say, though he needs to check his facts. *War and Peace* is a nineteenth-century Russian novel that could not have been written by sixteenth-century English poet and playwright William Shakespeare.

Quoyle is at a loss until Billy Pretty, a much kinder man and a good teacher, picks the story out of the trash and offers Quoyle some constructive criticism. Billy reads, "The policeman ate breakfast at the Cod Cake Diner before he arrived at the accident scene," his voice rising in disbelief. Quoyle gives him the look of an eager but not too bright puppy who fails to understand why he hasn't pleased his master. We in the audience, however, see that the detail is irrelevant. Two people are dead; it doesn't matter where the policeman had breakfast!

Billy kindly takes Quoyle to walk along the waterfront and gives him the straight dope. "Your spelling is fine, and I've seen plenty worse grammar, but finding the *center* of the story, the beating heart of it—that's what makes a reporter." To find that beating heart, he asks Quoyle to take a look across the bay and make up a headline. Quoyle tentatively advances the following:

**Horizon Fills with Dark Clouds**

Billy counters with a more dramatic clause:

**Imminent Storm Threatens Village**

Quoyle's headline misses the point: the horizon is filled with dark clouds *because* a storm is coming, and that storm poses a threat to life and property in the village. That's the meaning behind the dark clouds, and Billy's headline encompasses that important idea. Yet Quoyle is still confused. "What if no storm comes?" he asks.

**Village Spared from Deadly Storm**

replies Billy. The power of the story, the beating heart, the significance—these still stem from the fact that life and property have been threatened, even though the threat did not materialize.

Our own writing needs a beating heart as well—an idea, an image, a feeling—that will give it shape and meaning. Sometimes this idea is clear even before we begin to write; many times, however, we have to go looking for it.

## FINDING THE MAIN IDEA

The main idea of a formal essay often consists of two parts: the topic or subject of the paper and the point we intend to make about it. In Billy Pretty's first headline, for example, the subject is the imminent storm and the point is that it is threatening life and property in the village.

imminent storm (subject) + threatens village (point)

Billy's second headline might be said to have the same subject, the storm, but a different point.

deadly storm (subject) + does not harm village (point)

We might think of these headlines in another way: the village is the subject, and the two different headlines highlight its imminent danger or its fortunate escape.

How did Billy arrive at these headlines? He probably began to ask himself questions, as a good reporter does, questions such as "What do those clouds mean? If a storm is coming, what will happen? Will the village be at risk?" In looking for the center of our own stories, asking questions is an excellent start.

Suppose we've been given the topic of fast food restaurants and families, for example. We might begin by asking, How do fast food restaurants affect families? Do they make meals easier when parents are in a rush? Do they provide the necessary nutrition for growing children? Do they offer an appropriate role model in terms of healthy food choices? The answers to these questions might lead us to the following statements:

Fast food restaurants + can be a help to busy working parents.

Fast food restaurants + encourage both parents and children to eat poorly.

Among all the questions that you ask yourself, look for the one that is most important to you. The answer to that question is your main idea and—if it's an important one—it will be the "beating heart" of your essay, pumping energy and focus into every sentence. In many formal essays, the sentence that sums up the main idea is called the **thesis statement**; in a paragraph, it is called the **topic sentence**.

Not all essays will have such a formal thesis statement, however. A story, for example, may center on a problem or a conflict. Remember Noriko Yokota's essay about her dog Bikky in Chapter 2? The story revolves around Bikky's reputation as a greedy eater and the mysterious disappearance of the ham sandwiches: "I know you did it, bad doggy!" In contrast, Taylor Jones's description of Mesa Grill seeks to capture a feeling or impression: "I felt as though I were in Mexico." In doing a research paper, the main idea may be not so much a thesis *statement* as a particular *question*. Travis P. Becket (see Appendix VI) writes: "If organics is best for the soil and the body, can it be utilized at a level that will feed everybody?" Finally, the power of an essay may grow from the emotion rather than the logic behind the main idea, as in the opening line of Vincent Amato's essay in Chapter 1: "My mother is the best cook in the world."

Once you have a sense of the main idea, you explore and develop it with stories, details, examples, comparisons. In this way you "support" your ideas or "prove" your points. Let's look again at the thesis statements about fast food restaurants. The first example might be supported with details of a busy family's schedule for the week. Again, asking questions is useful. How much time is needed for planning meals, shopping for groceries, and cooking? What if families replaced one or two weekly meals with takeout from a fast food restaurant? Might that time then be spent either on other household chores, such as laundry, or in family activities, such as a game of baseball or Monopoly? The second example, with the same subject but a different main idea, might include nutritional comparisons—a Happy Meal compared to a meal of whole grains, fresh vegetables, and free-range poultry. Developing the main idea is the topic of the next unit in this book, and it is an important one. Without the "beating heart," however, an essay is just words.

In Nelson Tsai's story earlier in this chapter, the heart begins to beat with the very first sentence: "I am obsessed with food." The next sentence reveals the central dilemma: Will he reject the advice of his parents and turn this passion into a career? Nelson's dilemma is, as

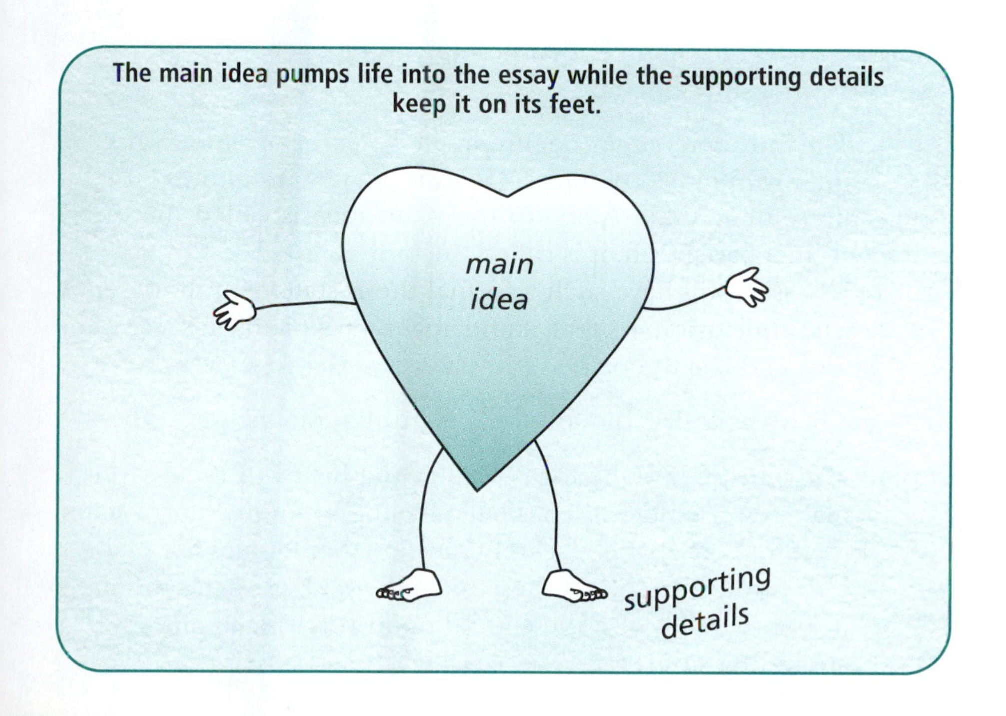

Billy Pretty says, "the *center* of the story, the beating heart," and it controls the shape of the essay much like a thickener controls the consistency of a gravy. It makes the whole paper *cohere* or stick together. It determines which details are important and which are not. And it is the details that make the reader see what you see, know what you know. When you find the beating heart and allow it to infuse your writing with passion, and when you use specific details that are relevant to that central idea, you can shape the reader's experience.

**EXERCISE 3.3** **Finding the Beating Heart**

Pick one of the following topics: cooking shows, the drinking age, movie gore, playing sports, first dates. Ask yourself a series of questions about the topic. Which is the most important? Now write a paragraph in which your answer to that important question becomes the "beating heart" of the story.

## RECIPE FOR REVIEW The Shape of Writing

An **essay** is a series of paragraphs that develop a **main idea**. A **paragraph** is a series of sentences that develop a single idea.

An essay has three parts:

1. The **introduction**, or beginning, lets the reader know what the writing is about and creates an interest in the subject.
2. The **body**, or middle, gives all the necessary information and explanation about the subject.
3. The **conclusion**, or end, summarizes or explains the main idea and lets the reader know the experience is over.

In addition, an essay has the following four ingredients:

1. The **main idea** is the focus or "beating heart" of the essay; it may be a problem or conflict, a feeling or impression, a statement of belief, or a research question. Sometimes the main idea is formulated as a **thesis statement**, which states the subject of the essay and the central point you wish to make about that subject.

2. **Supporting points** develop the main idea or prove the thesis statement.
3. **Specific details** explain or illustrate those points.
4. **Transitional expressions** make clear how one idea is connected to the next.

## Finding the Main Idea

1. Sometimes we begin writing with the main idea already in mind, whether it comes from an assignment or from our own need to communicate. Other times we have to discover the main idea as we go.
2. To find this "beating heart," it can be helpful to ask yourself questions about the topic until you find one that feels especially important.
3. When you've answered that question, you've got the passion that will give shape and energy to your writing.

## A Taste for Reading

1. Read "Food is Good" from Anthony Bourdain's *Kitchen Confidential* in Appendix VII. Explain how the first paragraph contains the four ingredients listed above.
2. Read "The Furin" from David Mas Masumoto's *Epitaph for a Peach* in Appendix VII. What is the "beating heart" of this piece? Describe or draw the "shape" of the essay.

## Ideas for Writing

1. In a paragraph or short essay, tell the story of a moment that had an impact on your life. Perhaps you realized something, or made a decision. After you write a draft, decide whether it has the necessary ingredients. Record your evaluation in a paragraph.
2. Make your own comparison between the shape of a meal and the shape of an essay or story. Then write a paragraph or short essay about your conclusions.

3. Conflict is a powerful shaping force in a story. Describe the conflict in Karl's paragraph on the meat fabrication class.

4. Choose a movie, story, or poem that interests you. Write a paragraph in which you describe its "shape." Or, if you prefer, draw a *diagram* of its shape.

# Cooking

## Developing the Main Idea

UNIT 2

chapter

4

# Telling a Story

In a way, everything we communicate is a story. Telling stories is a basic impulse for human beings. Some of our first sentences were stories ("Mommy, I fell down!"), and much of our conversation with family and friends has to do with the story of "what happened today." Stories are also entertaining to a larger audience, from books read by millions to movies seen around the world.

Telling stories can serve "serious" purposes as well. For example, think of various laws that have been passed with a person's name attached to them, such as Megan's Law.[8] There is a story behind that law, the story of a seven-year-old who was brutally murdered by a convicted sex offender who had been released on parole. When a celebrity suffers from a medical condition, his story can raise both awareness and funding for research. Michael J. Fox, for example, has worked on behalf of those suffering from Parkinson's disease. The story of a particular individual can highlight a general problem, increase understanding, and motivate action.

Many types of writing and many individual pieces of writing contain elements of a story. Even a very short story can add significant interest, emotion, and information. Storytelling can also be an important part of the writing *process*. When you first approach a piece of writing, it's often helpful to begin as if you're telling a story to a friend. The impulse is so natural that it can frequently move us past any initial hesitation or blockage we might have in writing an "essay."

Finally, although an essay might not recount a series of events—as the stories of J. R. R. Tolkien, Stephen King, and Annie Proulx do—it *does* take the reader on a journey through a series of thoughts. And, like a story, an essay has a beginning, a middle, and an end.

## THE INGREDIENTS OF A STORY

A story answers the question *What happened?* It recounts a series of events. Many of our early sentences reflect that impulse to tell what happened: *I fell down* or *Bobby hit me.* When we tell a story to someone in person, we have an idea about who that person is, what that audience is like, and therefore how much background information we need to provide and what "style" or vocabulary will be most appropriate. The audience also can ask for more information. With a written story, however, we don't always know much about the audience, and we must think carefully about what the reader needs to know in order to understand the events we're about to describe. Don't assume that the audience knows anything about you or any other aspect of the story. Thus, while one important ingredient of a story is the events that happen, another is the background information that makes those events clear and meaningful.

In telling a story, you will probably want to explain who the characters are, where the story takes place, and what happens (see Figure 4.1).

**FIGURE 4-1 The Ingredients of a Story.**

**The Characters**

Who are the characters in the story?

How are they related?

How old are they?

What are they like?

Furthermore, who is *telling* the story? Is it one of the characters speaking in the first person? Or is it someone outside the story using the third person?

**The Setting**

**Where** and **when** does the story take place?

What does the area look like? If the setting plays an important role, be sure to describe it more carefully.

**The Plot**

**What** happens?

What's the problem? How is it resolved?

Stories may use the past, present or future tense.

*The Point*

**Why** do the events happen? Why is this story important? Stories may *suggest or imply a* reason for the events, as well as explain why the events are important.

And in some way you will want to let the reader know why the story is important.

Let's look at an example.

**STUDENT WRITING** Brendan Cowley

Back in the late 1970s, my grandparents bought and remodeled a pub in Farmington, Michigan. Our long-time day manager has been a fiery Irishwoman named Carol. She and I get along just fine, but occasionally we like to play pranks on each other. Usually they're nothing serious, but this time I might have gone too far!

One day I step into the walk-in and notice a primal cut of beef that still has the head on it. As soon as I see this, the gears in my mind shake the dust off and start to turn. My plan is complete. This will be the prank to end all pranks. The pure horror of it will send shivers down even the bravest men's spines.

After breaking down the primal, I remove the head and stick it upside down in the freezer. Twelve hours should congeal the blood into a thick pudding. The plan is to put the head in the beer cooler. The beer cooler is Carol's cooler, and she gets mad if anyone messes with it. Every day around 8:00 a.m. she goes down and takes inventory. I put the head upright on a sheet tray on top of a keg. I open the mouth and pull out the tongue so that is flopping around like a wet noodle. The trap is set; it's a waiting game now.

Exactly at eight o'clock the next morning she comes downstairs and walks into the cooler. I'm standing behind the line waiting to hear a blood-curdling scream, but there is only silence. I peek around the line into the cooler, and I see Carol standing there, face to face with this bloody cow head. It couldn't be more perfect. The blood has started to thaw, so it is dripping out of every orifice it can find. I can see this red jelly-like substance coming out its ears, nose, eyes, mouth, and the severed section of its neck.

Finally Carol snaps out of her trance. I see this twisted face come over her, as if someone has poked her with a hot skillet, and she runs out of the kitchen. After she leaves, I go take a closer look. There is blood everywhere. It's all over the keg and all over the floor, and it's seeping out into the main kitchen area.

It takes me about an hour to clean up all the mess. Then I go upstairs to gloat. I find out that Carol was so disturbed by the sight of the cow head that she got sick in the bathroom and went home. Upon hearing that, I realize that maybe I've gone a little too far. Unfortunately that is the end of the pranks.

---

The first paragraph of the story introduces us to the two main characters: the narrator and Carol. We find out that the narrator's family has owned a pub in Michigan for some thirty years and that Carol has been employed there for a long time. We learn that that Carol is "fiery," although she gets along well with the narrator, and that they like to play pranks on each other. This first paragraph also zeroes in on the main idea of the story: "This time I might have gone too far."

The chain of events is clear; we know what happened and when. After explaining the background of the story in the first paragraph, the writer follows the events in simple chronological order—that is, in the order they occurred—and uses a number of transitional expressions to assist in keeping the sequence understandable, such as "*after breaking down the primal,*" "*exactly at 8 o'clock,*" and "*then I go upstairs.*" Not all stories are told in sequence, of course. Such films as *Pulp Fiction* and *Memento* demonstrate that stories can move backward, forward, almost sideways, and still be clear to the audience—at least after we've seen them five times!

In telling about the "primal" prank, the writer includes another important ingredient of a good story: details. We don't just know what happened; we can *see* it, from preparing the bloody cow head to watching its effect on Carol and cleaning up the mess. The details are particularly gruesome and effective: "I put the head upright on a sheet tray on top of a keg. I open the mouth and pull out the tongue so that is flopping around like a wet noodle. The trap is set." The image of the noodle is an especially nice touch! And how about this gruesome image: "The blood has started to thaw, so it is dripping out of every orifice it can find. I can see this red jelly-like substance coming out its ears, nose, eyes, mouth, and the severed section of its neck." The description of Carol's reaction also contains effective details: "I see this twisted face come over her, as if someone has poked her with a hot skillet, and she runs out of the kitchen."

Brendan's story also contains excitement and anticipation. How does he do this? First, he prepares us even in the very first paragraph for something unusual: "This time I might have gone too far." The anticipation is heightened in the second paragraph with these provocative statements: "This will be the prank to end all pranks. The pure horror of it will send shivers down even the bravest men's spines." The suspense is maintained by the rhythm of the story. We don't get to the punch line too quickly; we dwell on the horrific details, down to "waiting to hear a bloodcurdling scream." Finally, like the narrator, we "peek around the line into the cooler" and "see Carol standing there, face to face with this bloody cow head." Good timing is a terrific bonus in storytelling. Without actually stopping—as we might do in telling the story to a live audience—the writer uses sentences that continue to give some information but at the same time create pauses in the *action*. "It couldn't be more perfect," he writes, before we actually get to the climactic moment:

> It couldn't be more perfect. The blood has started to thaw, so it is dripping out of every orifice it can find. I can see this red jelly-like substance coming out its ears, nose, eyes, mouth, and the severed section of its neck.

The use of the first person, *I*, and the present tense also contribute to the immediacy and drama of the story. Look what happens with a change to the *third* person and the *past* tense:

> Finally Carol snapped out of her trance. Brendan saw this twisted face come over her, as if someone had poked her with a hot skillet, and she ran out of the kitchen. After she left, Brendan went to take a closer look. There was blood everywhere. It was all over the keg and all over the floor, and it was seeping out into the main kitchen area.

It's still detailed and exciting, but it's over. The blood isn't all over the floor *right now.*

As the story comes to an end, the writer tells us what happened to Carol: "She got sick in the bathroom and went home." The reader always wants to know what happened in the end. Just think of movies like *Remember the Titans* and *Apollo 13*, in which lines of text just before the closing credits tell us in a nutshell what happened to the characters we've become attached to. In addition to telling us what happened to Carol, Brendan writes "that is the end of the pranks." This particular prank is significant because it's the one that ended the series. The audience likes to know why the story is important. For example, we learn that *Remember the Titans* is based on the true story of a turning point for race relations in Virginia and that Jim Lovell never did make it to the moon.

**EXERCISE 4.1** **Ingredients of a Story**

Think about some of your favorite stories in books or film. What keeps you interested? Can you use any of those elements in your own writing? Explain.

## WRITING AND REVISING A STORY

Sometimes the elements of a story—the setting, characters, and events—are so clear in our minds that jumping directly into writing can be as effective as brainstorming. Perhaps a cross between freewriting and drafting is effective. You may wish to write to yourself—or you may wish to imagine yourself telling the story to a friend over a plate of Pad Thai. At some point, of course, you will need to identify the particular audience of your story and ensure that your readers have the information they need.

As you're writing, try to relive or imagine the story as vividly as possible. Use details to help the reader see and hear what's going on (see also Chapter 5). As with all drafts, don't stop and edit yourself critically. See the story in your mind as you write it on the page. Perhaps you can imagine an expressive audience, telling you with a look when more information or explanation is required. It's easy to assume that the reader knows what you know, so instead you must learn to think outside your skull. Figure out who your audience is, and anticipate what background information will help him understand the story you want to tell.

A story is a sequence of events, and one thing your audience needs to know is what that sequence is. Thus it is important that the verbs in a story reflect the order in which things happen by sticking consistently to one **tense**, whether it be past, present, or future. If an action occurs out of the main sequence, however, the verb must indicate that, too. For example, in the story about the primal cut prank, most of the verbs are in the present tense, and the sequence of events unfolds in that time frame. The very first sentence, though, is in the past tense because it indicates a time some 25 years earlier. Another variation occurs in the last paragraph: "I find out that Carol was so disturbed by the sight of the cow head that she got sick in the bathroom and went home." *Find* is in the present since it is part of the main character's sequence of events. However, *was, got,* and *went* are in the past tense because those things happened *before* the narrator found out about them.

Perhaps you can imagine an expressive audience, telling you with a look when more information or explanation is needed.

## EXERCISE 4.2 Using Consistent Tense

Read the following paragraph (adapted from an essay by Caitlin Crowley) and correct any inconsistencies in verb tense.

> As you walk along, you found the rough brown potato sacks, the cold plastic wrap of lettuce, and the damp cardboard. One box is holding leeks, long rubbery shocks of white and green. Next to them were the mustard greens, with their small, thin, deep green, frilly leaves that are still glistening with water. As you pick off a piece and chew a leaf, you experienced a hot mustard taste in your mouth, but only for a second.

Another aspect of the story that must be consistent is point of view. The story about the primal cut is told from the **point of view** of a narrator who is also a participant in the action. Stories and essays are frequently told in the third person, of course, as in "Adam's first taste of sushi was eye-opening." In fact, some instructors will require that you use the third person for your essays. Stories may also, as in Exercise 4.2 above, be told in the second person: "You experience a hot mustard taste in your mouth." This use of the second person creates an interesting and unusual texture to the story, and, as the story continues in the following example, a delicate sense of humor.

**STUDENT WRITING** Caitlin Crowley

The nut cooler when you open the door has three sides filled with stainless steel shelves containing boxes and clear bins filled with nuts. Pecans, pistachios, cashews, walnuts, even sunflower seeds are stacked all the way to a grown man's head. You can hear the sound of the nuts being poured into bags. If an almond sliver happens to land in your mouth, the thin, crisp sliver with its tan skin greets your taste buds with a rich nutty flavor.

I especially enjoy the author's sly observation "If an almond sliver *happens* to land in your mouth," as if the nuts have a life of their own. The second person works well here, and, more important, it is used consistently throughout the essay. Whichever point of view you choose, stay with it. Do not switch back and forth between "I" and "you," for example (see also Chapter 24).

When you're telling a story, you want to give the reader enough detail—but not too much. As we saw in Chapter 3, Quoyle's first story for the newspaper contained too much irrelevant detail. The story was about the violent death of two people and a moose, yet Quoyle gave us a couple of sentences about the policeman stopping for breakfast at the Cod Cake Diner on his way to the accident scene. Use details to create pictures, smells, and tastes in the reader's mind. Even the smallest details can make a difference. For example, what picture do you get in the following sentence: *I ate some fruit.* Not much of an image here, right? What about the next sentence?

I ate an apple, thinly sliced and arranged in concentric circles on a dark blue plate.

Now we *see* the fruit.

Details often provide essential information on character and setting, and keep the reader focused on what is important. Don't use details simply to fill space or because you want to toss in some big words. Keep details specific and relevant.

Finally, be sure your reader understands the point of the story, its beating heart. You may state it directly—"This will be the prank to end all pranks"—or you may choose to let an image convey your point, as in the story below.

**STUDENT WRITING** | Moises Ortega

## My Father

My dad has always been a serious person. He doesn't laugh or smile too often. He also doesn't show very much emotion to his friends and family. I've never seen my father crying, angry, stressed, or even confused about something. My father would show his affection to my brothers and me by slapping us behind the head. It was his way of saying, How was your day? or Everything will be all right, I am here for you. I always did wonder why he was like that, but he would never tell me.

I remember one time when I asked my father for a hug, he just looked at me with a crooked eye, gave me a slap on the back, and went on his way. I thought that I would never see him show some emotion, until one day I woke up screaming with pain in my stomach. My parents rushed me to the hospital, and the doctor found out that my appendix had to be removed. So the next day, the day of my surgery, the doctor made a mistake and cut my stomach wrong, and they could not close my wound. About three hours later, they finally sealed the wound, and I was asleep in my hospital room. That night I woke up crying because of all the pain. The nurse gave me some pills to calm me down, but the pain was too strong. The doctor had to come down into my room, reopen my wound, and reseal it while I was awake.

A day later I slightly woke up. It was about four in the morning, and I saw my father over me dripping tears. My father was praying to God, telling God not to take his only big boy, that he would rather take his own life to let me live. After a few minutes, my father wiped the tears off his face. He bent over and, not knowing I was barely awake, he whispered something in my ear that will always stay with me. Then he gave me a blessing and a kiss on the forehead, and left the room before my family came.

Months later, after I had fully recovered and everything was back to normal, I thought about the night when I saw my father emotional. I asked him what had happened that night. He did what he always did—looked at me with that crooked eye, gave me a slap behind my head, and went on his way. My father then left for work, but right before he went into his car he turned around and gave me the biggest smile he's ever given anyone.

---

By the time we've reached the end of the story, we understand all the power and delight of that smile. Having laid the groundwork in the earlier paragraphs—"He doesn't laugh or smile too often"—the writer wisely lets this image speak for itself. The writer also withholds some information from us: "He whispered something in my ear that will always stay with me." Like the concluding image of the smile, this undefined "something" adds a delicacy to the story that is both sophisticated in its technique and touching in its effect.

As it happens, this final draft of the story is very close to the first one. It is clearly a moment that the author has thought about. Perhaps it is a story he's told on other occasions. There were nevertheless some interesting revisions. First, the original draft contained a sentence about the author's mother: "My mother, on the other hand, is a very emotional person, but that's another story." Even as he was writing the words, he realized that they belonged somewhere else. Second, where the original reads simply "he just looked at me," the final draft adds "he just looked at me with a crooked eye." That tiny detail hints at something behind the father's look, for example, an impulse to express his feelings that is barely controlled and so appears as a "crooked eye."

Another change is as simple as one word. Where the first draft says "I saw my father over me *shedding* tears," the final reads, "I saw my father over me *dripping* tears." The second choice is less common and somehow more poignant. The tears are not noisy, yet they cannot be stopped. Finally, a small change to the concluding paragraph has the effect of tying the story together with the repetition of a phrase. In the first draft, when asked what happened that night, the father "looked at me weird." In the final version, however, the father "looked at me with that crooked eye," repeating the effective phrase from the second paragraph and preparing us for the final image, as the crooked eye is unraveled by a big smile.

### EXERCISE 4.3 Analyzing the Ingredients of a Story

Who are the characters in the story "My Father"? What do we know about them? We aren't told where or when the story takes place—does it matter? Explain. What is the problem in the story? How is it resolved?

## USING STORIES IN YOUR WRITING

Stories can be a part of all kinds of writing, from books about punctuation[9] to restaurant reviews. Think about how often a review includes the *story* of the meal—characters, setting, plot—as well as a description of the food. Telling this story is an effective way of involving and entertaining the reader. Stories can be a part of other kinds of food writing as well. M. F. K. Fisher, for example, is renowned for her ability to weave her own story through an essay on food (see her essay "Borderland" in Appendix VII, for example). In the following passage, a student tells the story of how sushi led him to major in culinary arts.

**STUDENT WRITING** Adam Miller

The first time I ever had sushi I was six years old. At that time we lived in Cape Cod and were on vacation in Florida. We had gone to Epcot for the day, and my grandmother brought me to a small Japanese restaurant. She ordered a cucumber roll and a tuna roll, and shared them with me. Even though they were simple rolls, they had a big impact on me, and now sushi is my favorite dish. . . .

Sushi is extremely important to me. If I had never had it, I probably wouldn't have opened my eyes to culinary. As I started eating more of it, I became interested in making it and learned how from books. In the eighth grade I did a report on the history of sushi, and I made five different rolls in front of my class. My teacher loved it so much he had me do it again, but in front of the entire eighth grade. Sushi opened my eyes to the art of cooking.

### EXERCISE 4.4 Using Stories in your Writing

Find a restaurant review that contains a story. What is the setting? Who are the characters? What are the main events?

## TAKING A RISK

Every day I encourage students to be open and genuine in their writing, to write about what's important to them, to take risks. It isn't always easy. To reveal our thoughts may make us feel vulnerable, while to relive our pain may seem intolerable. No one has impressed me more with her courage in this respect than Lori Vrazel, a student who, when asked to write about a moment that had changed her life, allowed herself to revisit a moment of terrible pain, to remember the smallest details, to press forward and continue writing and rewriting until the paper said what it needed to say.

## STUDENT WRITING | Lori Vrazel Kallinikos

### The Story of a Moment

I'll never forget that cool autumn evening that changed my life forever. It was a beautiful fall sunset with the sun's crimson glow illuminating the sky and the crisp southern winds whipping past. As I walked into my house, I found it busy with activity. However, as the phone rang, everyone grew quiet. My mother began to speak with a smiling face, but that faded away. An unsettling silence was all that was left from the commotion of before. My family members gathered around the kitchen table with looks of woe across their faces. Suddenly, we noticed our mother's eyes had filled with tears which began to fall like sparkling rain drops. Then she muttered, "Aaron. He's been in a car accident. He's not responding." That was all she could manage to say, and a flood of emotion engulfed my family.

The emotions that I saw from my family members on this day were unlike any that I had ever seen from them before or since. For example, my sister began to despair and tears formed in her eyes. This was the first time that I had ever seen her so vulnerable. My brother began to pace back and forth out of anxiety. However, I could not build up enough strength within me to even move, so I just sat down and watched the others.

Just as the sun finished setting and happiness disappeared, Dad arrived. I found my father's display of emotion to be the most unsettling for the simple fact that he hadn't shown any to begin with. He simply rushed around gathering clothes for his journey to the hospital. I could tell that he was trying to stay strong, but his heart's pain got the better of him, specifically when he was about to get into his car to leave. As my father stood near the narrow doorway to the dark carport, his eyes filled with tears and he told me, "Lori. Don't worry. I'll bring him home to us." Then my mother and he swiftly got into the Lincoln and drove away.

Seeing my father cry made everything seem real to me, and I finally reacted. I got up from the large round kitchen table and walked outside, ignoring all of the sorrow within the house. I needed to be alone, and I was alone outside in the garden. Everything was dark and dreary as I strolled over to the white bench swing that was covered with flowers. It was here that I realized that everything was going to be different. I realized how abruptly life is taken, and with this understanding I found myself wanting to change. I sat down and quickly said a prayer asking God to please let Aaron live. Suddenly, a warm, moist tear ran down my cheek, and it was soon followed by what seemed to be millions more.

> There were no words that could describe the pain inside of us, and there was nothing that we could do to change the day's events. I had never seen my family as susceptible as they were the day the news of Aaron arrived at our home. Life as I knew it changed. I no longer thought of it as a solid structure that was almost impossible to break. The only thing left for me to do was to take Aaron's death and mold it into a positive force. I wanted to make him and me proud. So I gathered up all of my dreams and hopes for the future and leaped into life with him by my side.

This story has stayed with me for years. And the best thing about it is that it's helped Lori to remember and honor her brother, "to take Aaron's death and mold it into a positive force." This is the real power of stories—to make sense of our experience. It is the power to heal.

## RECIPE FOR REVIEW Telling a Story

### The Ingredients of a Story

See Figure 4.1

### Consistent Tense

Use the same **tense** (past, present, future) for events in sequence. Use the appropriate tense for those events that might be out of that sequence (see also Chapter 23).

### Point of View

While **point of view** is a complex aspect of storytelling that you may wish to explore further, the most basic need is to choose a point of view (first, second, or third person) that suits your purpose, and use pronouns consistently to fit that point of view (see also Chapter 24).

## A TASTE FOR READING

1. José Antonio Burciaga tells quite a few small stories in his essay "Tortillas" in Appendix VII. List as many as you can find. Then choose one and explain how it is used in the essay.

2. Read the excerpt from Shoba Narayan's *Monsoon Diary* in Appendix VII. Summarize the story that precedes the recipe. What does the story add to the recipe? What does the recipe add to the story?

## Ideas for Writing

1. Tell a story about a friend or family member that illustrates something important about him or her.
2. Write a paragraph or short essay in which you summarize your favorite story as a child. Then explain why it was your favorite.
3. Tell the story of the worst or funniest experience you've ever had at work or school, or eating in a restaurant.
4. Tell the story of an argument you had recently with a friend or family member—from your point of view. Then, rewrite the story from *the other person's* point of view. What did you learn about the incident?

chapter 5

# Working with Descriptive Details

Good storytellers provide details about how things look, sound, smell, taste, and feel. They are sensual. Our response to a good storyteller might be, "Oh yes, I *see* that, I *see* what you mean." Good cooks are also highly sensual. They offer us a plate of peppers, vibrant red, yellow, orange, and green. We hear the crisp fragments of a potato chip break between our teeth; we feel the rubbery skin of an apple give way to a burst of juice. The more we can use vivid descriptive details in our writing, the more our readers will understand and enjoy it. In fact, as you might imagine, description is an especially important part of the hospitality industry, from menus and restaurant reviews to proposals for prospective investors.

Effective descriptions often derive from the evidence of the five senses and should be very specific. If we wrote "that meat was amazing," we would actually communicate very little. But what about "the cube of tenderloin had a crispy brown outer layer with inner stripes of light brown and finally pink"? There is no need to use fancy or dramatic language; often a simple, almost scientific, vocabulary is more effective. You're trying to create a picture for the reader, not knock her out with your big words.

Descriptions can also suffer from a reliance on old, tired words and phrases. Were you "hungry as a horse" when you entered the restaurant? Did the dish "melt in your mouth"? When someone first said these things, they probably seemed very exciting. However, they are used so often by so many people that they have become clichés (see Chapter 11). If anyone can fill in the blanks (cool as a _______, red as a _______, pretty as a _______), the phrase is so old that it can't really tell us anything new.

**EXERCISE 5.1 Enjoying Description**

Find a story or article you've enjoyed and look carefully for descriptive passages. Copy one of them onto a sheet of notebook paper. What drew you to this particular passage? Does it contain details from the five senses? Does it "paint a picture"? Explain.

## SENSORY DETAILS

The five senses—sight, smell, taste, touch, and hearing—are an excellent source of descriptive details. We don't need to use every one of the senses for every description, and we don't need to go overboard, but a well-chosen phrase about a sound or an aroma can pull the reader immediately into the world of our thoughts. Listen to the *sound* of the dinner rush in the sentence below:

**STUDENT WRITING** Anthony Guarino

The clock strikes six, and your first ticket hits the board. There is an eerie silence until the sound of cold raw fish hitting a scorching hot grill fills the kitchen.

Interestingly, contemporary Americans rely heavily on their sense of sight.[10] When asked to describe an experience, they will concentrate on what they *saw*. Yet the other four senses are also at work, receiving data and sending it to the brain for analysis. And one of them may be far more effective in communicating your thought or feeling than sight is. The sense of smell, for example, can bring back a memory with extraordinary speed and intensity—the aroma of your mother's spaghetti sauce transports you to the dinner table of your childhood, or the unmistakable scent of a certain aftershave propels you into the breathless embrace of an old boyfriend.

In the kitchen we're awash in all five senses. But which do we rely on most heavily during the cooking process? For many people, it's the sense of taste—but not for everyone. Monica Bhide, a "food writer by night,"[11] visited Chef Jonathan Krinn in his restaurant in Washington, D.C. Bhide relates that Krinn's interest is in French American food, her own in Indian. While making a sauce, Krinn shocked Bhide by tasting a spoonful of the liquid. "Don't you know you're not allowed to taste while you cook?"

But Krinn is shocked in his turn. If you didn't taste the sauce, "How would you know when to season?"

Unlike most of us, Bhide learned to cook not by taste but by "sight, smell, sound and texture." Her father taught her to listen to the sound of roasting spices: "coriander whimpers, cumin smolders, mustard sizzles and cinnamon roars." He taught her to watch the colors as the spices cooked and to notice the smells. "Roast, sizzle, temper, broil, boil, bake, simmer, sauté, fry—we had to do it all by watching and listening." The reason lies in the Hindu culture. "The first offering of the food was for the gods. If you tasted while you cooked, it made the food impure." When you think about description, try not to rely on a single sense; especially, try not to rely on sight alone when the other senses can make a powerful contribution.

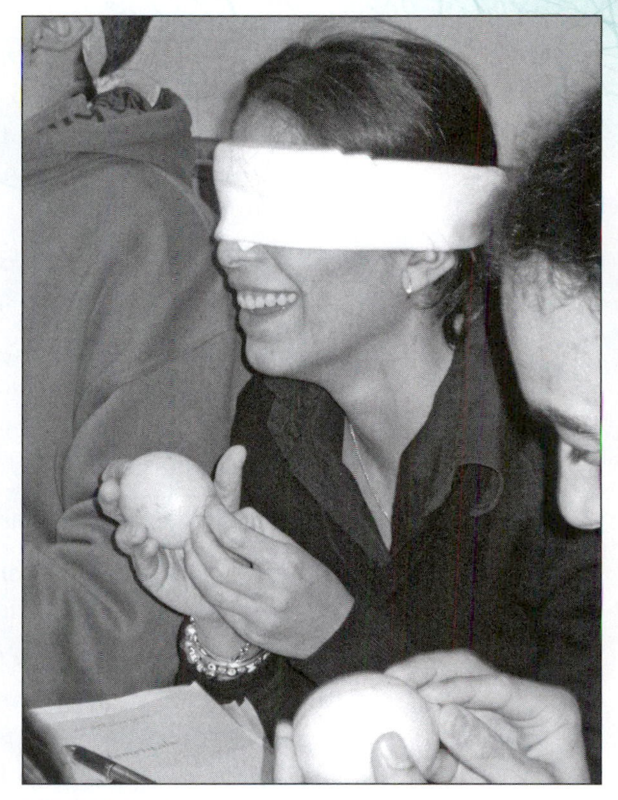

**FIGURE 5-1** After "getting to know" her lemon in Exercise 5.2, this student is delighted to find it again.

## EXERCISE 5.2 Describing a Lemon

First, take out a sheet of paper and describe a lemon in two to three sentences. Next, get a real lemon—but don't look at it! Close your eyes, and get to know the lemon using the other four senses.[12] See Figure 5.1. Take several minutes to do this. Finally, put the lemon aside (don't look at it!), and write a description of that particular lemon. How does this description differ from your first one?

## SPECIFIC DETAILS

When writing a description, try to avoid both vague or general words as well as complicated or fancy words. Instead, give the reader specific details—whole scenes can be conjured by a car's make and model or the name of a candy bar.

> I sat in the front seat of an old Dodge Ram pickup truck and nibbled on a Snickers bar.

When you know the name, use it.

Compare the following two descriptions, the first using the fun (think of Tony the Tiger) but very vague expression "great," the second using simple but specific words.

> The chocolate ice cream was great.

> I slipped the rich chocolatey spoonful between my lips, tasting the dark velvet, the bright ice.

That ice cream probably is "great." But don't *tell* the reader that; *show* her. In general, try to avoid the all-purpose, somewhat vague terms listed in Figure 5.2. Instead, use specific details to show the reader what you see and hear and taste so that it is the *reader* who exclaims, "Oh, that's beautiful" or "That sounds amazing."

### EXERCISE 5.3 Using Specific Details

Rewrite the general descriptions below to include specific details. The idea is to *show* the reader what you're feeling so that the reader says, "Oh, that's beautiful," or "That sounds amazing."

1. The sunset was beautiful.
2. I heard a terrible sound.
3. The kitchen smelled great.
4. _________ (fill in the blank) was a bad movie.
5. _________ (fill in the blank) is an amazing song.

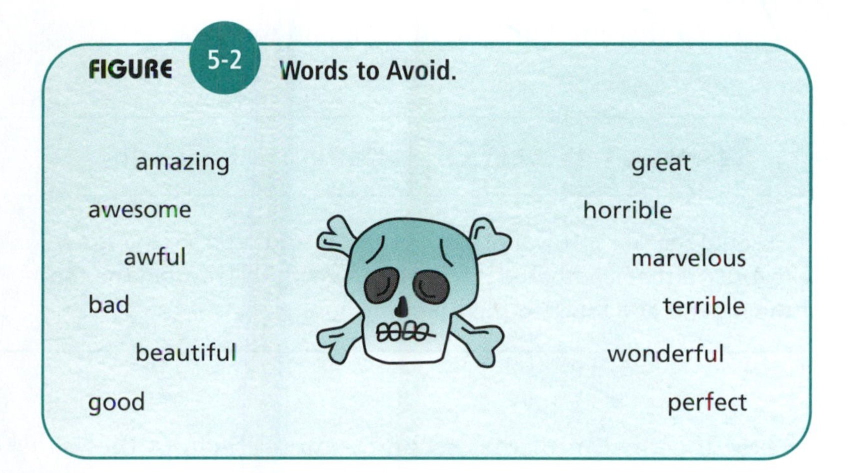


FIGURE 5-2 Words to Avoid.

## DESCRIPTIONS OF FOOD

Although they are filled with sensory experience, food items are some of the most difficult things in the world to describe. Try to approach a food as if for the first time. Look at it with new eyes. Savor the tastes and smells and textures. Use specific, sensory words. Avoid clichés. As you read the first example below, notice that the writer uses clear, straightforward language to create a vivid picture.

**STUDENT WRITING** Roth Perelman

I looked to my left to find the strawberries piled high in their crates, their plump, intense, red bodies spotted with black seeds like freckles, tapering down to a rounded bottom, topped with a leafy green halo, ripe and delicious, ready to be eaten.

Texture—the mouthfeel—is also an important part of our experience with food. In this next example, the writer reminisces about his favorite childhood dish.

**STUDENT WRITING** Dylan Chace

My aunt would spread the peanut butter on the toast right after it came out of the toaster, so the peanut butter was melted when she brought it to me. I can still remember the warm silkiness of the peanut butter as it hit my mouth.

And how about this combination of sight and touch?

Nathan E. Bearfield

I could see the soft ivory-colored rice protruding from the straw-yellow sauce that smothered it. . . . Steam slowly rolled from the top of the mound and touched the tip of my nose.

Notice how the rhythm of that second sentence echoes the *rolling* of the steam it describes.

The sense of smell is crucial to a full enjoyment of food descriptions and, of course, of food itself. Although we can perceive the four basic tastes—salty, sweet, sour, and bitter (Figure 5.3)—with our tongues alone, we need our noses to sense the full range of flavor. Have you ever noticed that your appreciation for the flavors of food and drink is significantly diminished during a head cold? Experiment. Hold your nose and breathe through your mouth. Then eat something. Now let go of your nose and try another bite. That's the difference between *taste* and *flavor.* Look at the passage below.

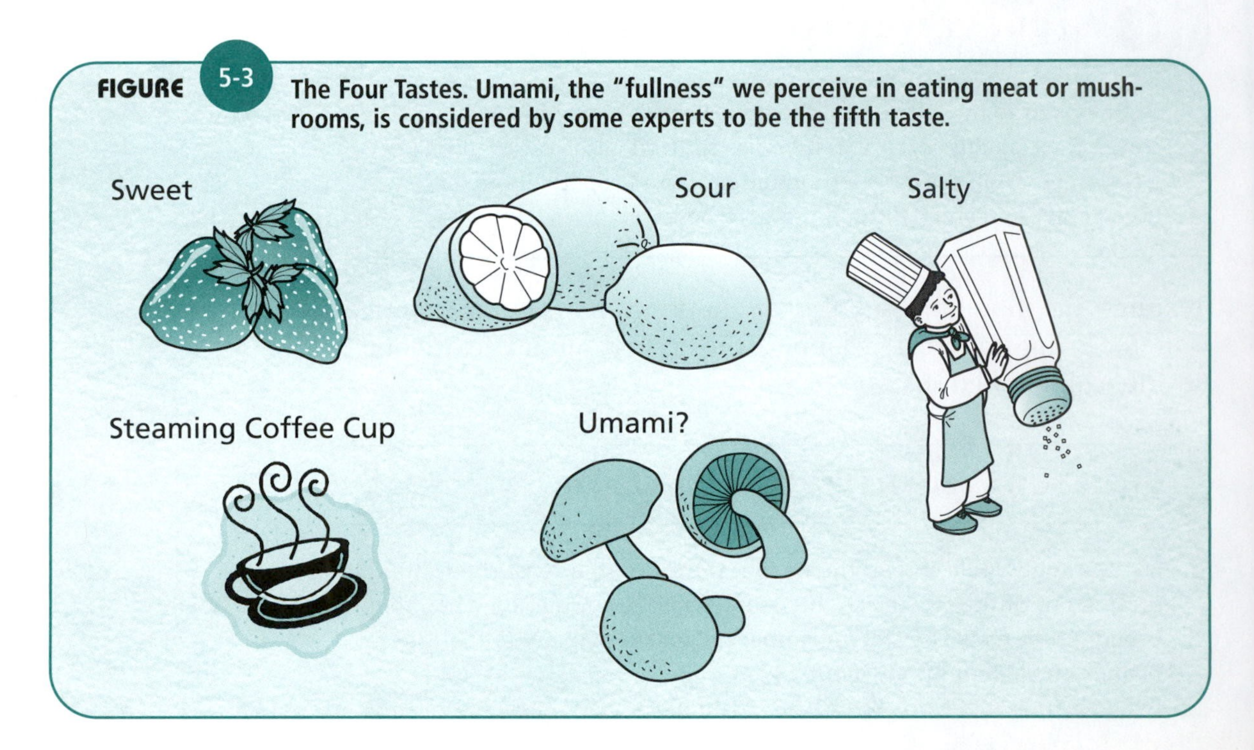


**FIGURE 5-3** **The Four Tastes. Umami, the "fullness" we perceive in eating meat or mushrooms, is considered by some experts to be the fifth taste.**

## STUDENT WRITING | Elizabeth Best

The aroma of the peach is very complex, with many notes intertwining in the nostrils. You can pick up a floral note, almost like a honeysuckle. Next you may pick up a sweet, almost musky scent. Perhaps the strongest aroma is honey. It smells like a very rich, wild honey, like that of the wild clover. Almond is present with the underlying aroma of vanilla. A faint scent of nutmeg can be sensed, lingering for a second.

**FIGURE 5-4 Words for Flavors and Aromas.**

tasty, flavorful, palatable, savory

pungent, hot, spicy, peppery, tart, sour bitter, salt, gingery
Dijon mustard, curry, hot like jalapeño peppery, fiery

pungent like onion garlicky, cayenne mild, bland

sour, acerbic, acrid tart, vinegary, green

aromatic, fragrant perfumed

incense, musky, potpourri, roses

sweet, sugary honeyed, candied
syrupy, caramel smoky, mushroomy

fruity, apples oranges, lemony,
banana, raspberry

nutty, chocolate mint, tarragon,
basil, rosemary, thyme

## EXERCISE 5.4 Writing Descriptions of Food

Select a food item such as a piece of fruit or a hearty soup. Carefully study the appearance, aroma, texture, sound (as you eat it), and flavor of the food. Make notes at each step. Review the selection of words for flavors and aromas in Figure 5.4. Then write a paragraph describing your food item using specific, sensory details.

## DESCRIBING A SPACE

In describing a space, it is important to help the reader *see* the relationships between the various objects. Writers do this by using specific descriptive details connected with spatial transitions. These words and phrases explain where one thing is in relation to another. In the example below, the transitional expressions are underlined.

**STUDENT WRITING** Matthew K. Greene

A three-part sink lined one whole side wall of the kitchen, with containers and storage units placed neatly on a shelf above. On the opposite wall there were two refrigerators, one walk-in with food stored for the whole week, and a regular fridge with items that were used throughout the day. In between the two were more shelves with extra bags, boxes, and fortune cookies. Along the back of the entire length of the kitchen were four huge woks where all dishes were prepared. Next to the woks were a fryer, a small prep table, and a rice cooker that got extremely warm.

As you read the complete essay below, notice how the details and the words that connect them create a mental picture of Rice Kitchen.

**STUDENT WRITING** Matthew K. Greene

### Rice Kitchen

For the past year and a half before I came to college, I worked in a Chinese restaurant called Rice Kitchen. The whole restaurant was rather small because we mainly did delivery and take out. The kitchen was even smaller during the work day or seemed that way because there were usually about nine people working in a small area. We had one person on the fryer, one person prepping vegetables, and two head chefs working the woks. There would be two delivery guys, one bagger, someone working the front desk, and my boss mediating and sometimes helping. That's nine people speaking three different languages (American, Spanish, and Chinese), all working in close quarters trying to get the job done without stepping on anyone's toes.

Now, because the kitchen was so small, it needed to be very organized, and everything needed to be in its place. A three-part sink lined one whole side wall of the kitchen, with containers and storage units placed neatly on a shelf above. On the opposite wall there were two refrigerators, one walk-in with food stored for the whole week, and a regular fridge with items that were used throughout the day. In between the two were more shelves with extra bags, boxes, and fortune cookies. Along the back of the entire length of the kitchen were four huge woks where all dishes were prepared. Next to the woks were a fryer, a small prep table, and a rice cooker that got extremely warm. Two feet back away from the woks was another small prep table, as well as a rack with every vegetable we needed and a flip fridge with all the pork, beef, and chicken.

All these items were packed into a small area, but everything was well organized and in its place. Before the end of the night, everyone would help clean so the kitchen was spotless for the next day of work.

The first paragraph paints a clear, even endearing portrait of the restaurant's tiny kitchen, which seemed "even smaller during the work day" because it was so crowded. The writer specifies both the jobs of the nine people at work in the small space and the three languages being spoken. Good description is produced with this type of detail rather than with a string of flowery adjectives.

## EXERCISE 5.5 Working with Descriptive Details

Describe a space you have worked in, the layout, the sounds and smells. Use specific sensory details that help the reader see and hear (and smell and taste, perhaps) what you do.

The following essay also describes a place. In this example, however, the focus is not on the effective organization of the space but rather on the feeling that the space evokes.

## STUDENT WRITING Jaren Niimi

### Hapuna Beach

When I look outside, all I see are bare trees and powdery white snow. I dream of my favorite place, Hapuna Beach, which is a few minutes from my house in Hawaii. I have been there so many times that memories are very clear in my mind.

There are fine grains of white sand that fill the cracks between my toes. Left to right, the rolling surf breaks in the crystal blue water of the bay. I can grab my board and catch wave after wave, riding deep in the hollow barrel that feels like something from another world. It is just me and the wave with its breathtaking thrill ride that Mother Nature created.

However, on the days when waves are none to one, I can grab my snorkel and fins to explore the underwater world. There is lively marine life all around me with turtles gliding through the water just a few feet away. There is a variety of fish, such as the red and green parrotfish that feed on the dome-shaped coral heads thriving in what seems like an aquarium. The serene atmosphere of the underwater world is very relaxing, unlike anything else that I know. We often stroll down the long white sandy beach talking about junior high school when we had no worries. The water glides over our feet as it washes ashore.

At the end of the day we all sit on the sand watching the sunset. It is amazing that in a matter of minutes the sky changes to so many colors. The blue sky turns into beautiful pastel colors of red and orange as the sun slowly fades away under the horizon of the ocean. The night falls as we relax on the sand under the stars, hoping to see one fall and make a wish that may come true.

I can remember these things in my mind like it was yesterday. My favorite place to be—when I think about it, a smile comes to my face. For now I am in a place very far away where the weather is cold and there is snow on the ground.

---

This is a lovely piece. From the first sentence we are drawn into the writer's world, looking out the window at the "bare trees and powdery white snow," and then looking inside to his memory of the beach. The three body paragraphs each describe different experiences—surfing, snorkeling, wishing on a star. The present tense makes us feel that these scenes are unrolling before our eyes, the water gliding over our toes. The word "amazing" works here since it is followed by the specific details of color and the lapping cadence of the final clause: "as the sun slowly fades away under the horizon of the ocean." In the last paragraph, the writer takes us out of memory and back to the snow. The description is effective because of the choice of details and because of a vocabulary as warm as the sand and a rhythm as fluid as the ocean itself.

## USING FIGURATIVE LANGUAGE

Your descriptive writing can often be spiced up with **figurative language**, a group of methods by which you compare one item with another to create pictures or explore ideas. A **simile** makes the comparison directly with the use of *like* or *as*. For example, the aroma of the peach was "like a very rich, wild honey." Or, in the passage below, the *mole* looks "like a volcano":

**STUDENT WRITING** Randy Gonzalez

At the end it's a plate looking like a volcano hot and steamy just erupted red sauce all over the chicken running down like a river streaming so delightful and when I put it in my mouth it just melts the taste buds all over."

Notice how the sentence itself streams over the page like the red sauce!

In the next example, the writer uses a simile to express his passionate rejection of apple varieties other than his favorite.

**STUDENT WRITING** Jeong-woo Kim

If people are working for a pastry shop or studying for a pastry course, they always use one of my favorite apples, the green Granny Smith apple. They might be able to use different varieties, such as the Golden Delicious, but I think the best cooking or baking apple is the bright green Granny Smith because it does not become too soft. If people use another apple, it gets mushy and disgusting because the texture is not crisp as the Granny Smith. About the taste of other varieties, please don't ask me because they taste like baby food. Sometimes even a baby doesn't want to eat them. So what is the best apple to bake or cook with? It's the Granny Smith apple, the crispest apple in the world.

Sometimes we imply the comparison by simply stating that one thing is another, that is, we use a **metaphor**. In the passage below, a kitchen is compared to the fires of hell!

## STUDENT WRITING | Nathan E. Bearfield

The kitchen was a cardboard box inferno that employed three line cooks and a chef. I truly believed Lucifer himself would have asked for a glass of water there from time to time. The only movement on the line was a pivot from the steam table to twelve spider-webbed burners that roared full blast, singeing anything that dared to come close to them.

Preparing curried butternut squash soup at a safe distance from the "spider-webbed burners."

Notice also how the writer uses the word "spider-webbed" both to describe the burners' appearance and to suggest a sinister personality. They're alive and malevolent, "singeing anything that dared to come close to them." Nice.

Let's go back to the "volcano" for a moment. The writer concludes the essay with a metaphor: his family *is mole.*

**STUDENT WRITING** Randy Gonzalez

Our family is a plate full of ingredients, a lot of good flavors, so that if you took a bite you would want more of it.

---

Using figurative language can bring a new dimension to our *thinking* as well as to our writing. We have done something similar throughout this book in exploring cooking as a metaphor for writing: peeling the potato, seasoning and plating the dish. In the passage below, the writer uses figurative language to look at the character of Dave Boyle from the film *Mystic River.*

**STUDENT WRITING** Tomer Blechman

Dave cannot trust his mind any more; he feels torn up between his two friends from the past. The dilemma they represent is whether to follow his heart (to kill the pedophile), or to control his feelings and to identify what is good and bad. Dave is more complex: he is not balanced between the two but is moving from side to side, taking his two friends with him. He doesn't have roots. His roots are buried in the basement.

---

Dave is damaged and out of balance. Like a tree whose roots have been severed by some monstrous lumberjack, he has been drained of hope and joy, even of life. His connection to his childhood, to his *roots,* was severed when he was kidnapped and tortured by pedophiles. In addition, the word *buried* inevitably suggests death: Dave has been a living corpse following his ordeal in the basement.

## EXERCISE 5.6 Using Figures of Speech

Fill in the blanks in the comparisons below. Then choose one, and expand the idea into a descriptive paragraph.

1. The _________ smelled like a _________.
2. The _________ looked like a _________.

3. The ________ sounded like a ________.

4. The ________ tasted like a ________.

5. The ________ felt like a ________.

Description can also include details about your feelings. Again, they don't need to be flowery and exaggerated; try for simple, clear, direct. In fact, many writers will portray feelings somewhat indirectly through the character's actions or remarks.

> Jane turned her face away while a hot red flooded the delicate skin of her cheek.

Feelings can also be described through metaphors and similes. In the story below, the author writes that fear "gripped me like a giant hand."

**STUDENT WRITING** Dale Andrews

## Deployment

It was the last day of February last year. I was standing outside in formation. This would be one of our last briefs before deploying to Kuwait. I was a sergeant in the 101st Airborne Division at Fort Campbell, Kentucky. It was cool out. Morning had begun to fade away. It was starting to get warmer, and the fog had started to dissipate. I was tired. We were all tired, but we were all very wide awake.

I can remember some of the details so vividly. The new desert uniforms were stiff and scratchy. I could smell many different things—all the new equipment and the booze from a couple of the guys next to me. My mind faded in and out during the briefing. I really don't remember what it was about. I had only slept for about an hour the night before. The whole unit had to be in early this morning. There had been a lot to do. The process had to be followed, weapon issue, casualty cards to fill out, and troops to inspect. The list went on.

The brief ended. Some of the soldiers went inside the gym to visit with their families. Some of them stayed outside fixing their rucksacks and smoking cigarettes. Fear was heavy in the air. I saw it in everyone's eyes. The wives that were there tried to be strong. Most of them were crying and hugging their husbands.

We had only one more hour until we would go to the flight line. This would be the last time some of us would ever see our families. The older soldiers knew how dangerous this deployment would be, but they were supportive of the younger ones despite their own fears. I was one of those older soldiers. I tried my best not to show my fear.

It was time to leave. I hugged my girlfriend tight and promised that I would do my best to come back in one piece. Then all of the soldiers got on the bus. As I found my seat and felt the bus creep forward, the fear of what would be gripped me like a giant hand.

To this day it still makes me shudder when I think about that moment in time. There were scarier things yet to come, but that memory is the most vivid. I think it is because we are afraid of what will happen and what we don't know.

---

The style is clear and direct, with a military precision. The details tell the story: the 101st Airborne Division, deployment to Kuwait, stiff and scratchy desert uniforms, rucksacks, cigarettes, tears. Only near the end does emotion grip the narrator—and the reader—like a giant hand.

## RECIPE FOR REVIEW Working with Descriptive Details

### Sensory Details

1. Use data from the five senses to describe what you see, hear, smell, taste, and touch.
2. Try using different senses to find a fresh description.
3. Avoid clichés.

### Specific Details

1. Avoid vague or general words like *wonderful* or *amazing*. (See Figure 5.2.)
2. When you know the name, use it: *Ed Harris, Butterfingers, Vidalia onions.*

## Figurative Language

1. A **simile** compares two items using *like* or *as,* for example, the *mole* looks "like a volcano."

2. A **metaphor** implies the comparison by stating that one thing *is* another: "our family is a plate full of ingredients."

## A Taste for Reading

1. Read M. F. K. Fisher's "Borderland" in Appendix VII, paying particular attention to the descriptions of chocolate and of tangerines. What words does she use for appearance, flavor, and texture? Are these effective descriptions? Explain. Where does she use figurative language?

2. Read Galway Kinnell's "Blackberry Eating" in Appendix VII. How does he describe blackberries? Why are they "icy"? What do blackberries have in common with words?

## Ideas for Writing

1. Find a dozen or more words for feelings, such as *angry, frightened, overjoyed.* Then write a paragraph in which you describe one of these feelings without using any of the feeling words. Show, don't tell.

2. Look on the Internet for *flavor wheels or aroma wheels.*[13] What new descriptive words did you discover? Write a paragraph in which you use five or ten of these words.

3. Describe a dish that you have eaten at a restaurant recently. Consider its appearance, aroma, flavor, and texture.

4. Look up some of the other figures of speech, such as *personification* and *hyperbole.* Describe an ordinary household item or food using at least two of these figures, as well as a metaphor or simile. Which part of the description is most successful? Why?

chapter 6

# Using Examples

There's nothing like a good *example* to make communication more effective. "Eat foods rich in iron," advise the scientists. Pause while we all think of spinach. Longer pause while we think of eating spinach every day. Are there other examples? we ask hopefully. "Yes," they reply. "Soybean nuts, red meat, chicken and pork, fortified breakfast cereal, raisins and prunes."

"Brush your teeth at least twice a day," say the dentists, "and avoid problems." Like what? ask the free-thinking twelve-year-olds. "Well, for example, bad breath, tooth decay, gum disease, and potentially inflamed arteries!" That's much clearer, isn't it? And it's much more effective as advice.

"If you improve your written communication skills," your instructor intones, "you can have a successful career as a food writer." Sure, you reply, but what *is* "food writing"? Your instructor responds with a list of examples: "There are many different kinds of food writing, for example, recipes and cookbooks, menu descriptions, restaurant reviews, scientific articles, and textbooks."

Like stories and descriptions, examples are ways of thinking that we use automatically when speaking and that transfer effectively to writing. The preceding three paragraphs are a series of examples with examples! Examples make the idea concrete; they put a picture in the reader's mind. We use examples to illustrate or explain a point or sometimes to "prove" one.

**The Role of Examples.**

| Category | Role | Examples |
|---|---|---|
| Many different foods are rich in iron. | to illustrate | spinach, soybean nuts, , red meat chicken, pork, fortified breakfast cereal, raisins, prunes |
| Poor dental hygiene causes serious health problems. | to prove | bad breath, tooth decay, gum disease, inflammation of the arteries |
| There are many different types of food writing. | to explain | recipes, cookbooks, menus, restaurant reviews, scientific articles, cookbooks |

## USING SPECIFIC EXAMPLES

Examples often come to mind as the most natural way of *explaining* oneself. "What do you mean by that?" we might ask. "Give me an example." For instance, suppose you come across the term "environmentally-friendly systems" in an essay. Before the question "What's that?" has time to form in your mind, however, the writer below offers three examples of such a system: "organics, sustainable and biodynamic farming." And to explain the concept further, he provides an example.

**STUDENT WRITING** | Travis P. Becket

An example would be instead of using a pesticide to kill a natural insect predator of a crop, a different crop is grown near the primary crop that attracts a different insect that feeds off the predator insect. All of the plants are indigenous, as well as the insects they bring, and the ecosystem is not affected negatively.

Examples are especially useful in explaining concepts or points of view that might be unfamiliar to the reader. In the passage below, the writer uses specific examples to explain the relationship in his childhood between his state of mind and the sauce he chose for dipping his chicken fingers.

**STUDENT WRITING** Thomas Monahan

There were so many sauces to choose from, depending on my mood. If I was happy, it would be the sweet and sour. If I was angry and resentful, it had to be honey mustard. And, of course, there were those "I don't know" moods when I would get BBQ sauce.

Examples can give shape and substance to the idea in your mind. They can create an image that *illustrates* the point. In the passage below, the writer illustrates the ways he might use a knife if stranded in the wilderness.

**STUDENT WRITING** Danny Choung

The knife is the most important item you will need [to survive]. With a knife, you can protect yourself from any wild animals, such as mountain lions, coyotes, and wolves. You can also use a knife to cut branches to make shelter. The knife also comes in handy when you prepare food, such as skinning an animal or cutting meat. You could make other types of weapons with a knife. You could make a spear by sharpening the end of a stick.

Notice the examples within the examples. The knife would be used to protect himself from wild animals, to make a shelter, to hunt, and to prepare food. Within the category of wild animals, the writer offers further examples: mountain lions, coyotes, wolves (Figure 6.1).

  **Using Specific Examples.**

With a knife you can...

protect yourself from wild animals, such as mountain lions, coyotes, and wolves cut branches to make shelter

prepare food, for example, skinning an animal or cutting meat

make other types of weapons: you could make a spear by sharpening the end of a stick

Often a general statement is best illustrated through a specific example, as in the following paragraph about the movie *Chocolat.*

**STUDENT WRITING** Jina Chun

The most interesting scene in the movie is when Vianne asks people to spin the dish and then tells what kinds of sweets best suit each person. For example, there is a young boy who is interested in the dark arts. When Vianne spins the dish, he sees a skull. Vianne recommends to him a dark chocolate. She sees through people's souls and makes the best chocolates for them. Soon, people begin to discover the mouth-melting effects of her wonderful treat.

---

The little boy's dark chocolate helps us understand the general statement that Vianne "tells what kinds of sweets best suit each person."

And it is especially important that we do understand this since it is "the most interesting scene" for the writer.

Examples can also be used to win the reader over to one's point of view, especially when the reader is likely to resist. In stating that writing and cooking have similarities, for instance, the next writer invites the audience to appreciate this point by offering specific examples from each field.

**STUDENT WRITING** Joseph Pierro

Both writing and cooking allow you to get a feel for your reader or customer. If you're writing for a romance novel, you're not going to have very many jokes or pictures. If you were writing a children's story, you wouldn't make the reading very difficult. It is just the same with cooking. You don't go to an Italian restaurant and start cooking Japanese food; moreover, you wouldn't cook a roasted tenderloin for a vegetarian.

Romance novel writing is one "cuisine" in the realm of fiction, just as Italian and Japanese cooking are cuisines in the world of food. The writer uses this analogy to illustrate one of the similarities between writing and cooking: that it's important to address the needs of your audience. For example, "You wouldn't cook a roasted tenderloin for a vegetarian." The examples are parallel and precise, and thus effective.

In this next paragraph, the writer illustrates the similarity between cooking and writing with the example of "the method that is used to perform both."

**STUDENT WRITING** Chris DiMinno

One good example that shows how closely related cooking and writing are is the method that is used to perform both. For instance, in cooking a chef would start with planning the meal, getting his or her ideas together. A chef would take a good amount of time in thinking about what to prepare, what to serve it with, and how to present it. In writing, the process is very similar. A writer would start by free writing, or by doing some other form of idea mapping such as clustering, or simply by brainstorming ideas on a piece of paper. This would clear the writer's head as well as plan for the piece that is to be written.

Here the student writer describes part of the method, brainstorming, for both cooks and writers. He also uses transitional phrases used to introduce his examples: "*one good example*" and "*for instance*."

We can't stress enough that examples must be *specific*. Whether you're providing details of an event, summarizing statistical results, or using sensory details, be concrete. Name names. In the example below, the author names a specific person, a specific book, and two specific films.

**STUDENT WRITING** Matthew Berkowski

There is actually one more similarity between cooks and writers. There is the odd occasion where the cook is a writer. Take Anthony Bourdain, author of *Kitchen Confidential,* for example. He is a recognized chef, yet he wrote about the food industry from a chef's perspective. Other authors use food in a symbolic way to represent people's emotions, for example, in *Chocolat* and *Like Water for Chocolate*.

**EXERCISE 6.1** **Finding Specific Examples**

What does it mean to be a good friend, a good cook, or a good student? Brainstorm a list of examples that illustrate your ideas. Be sure they are vivid and specific.

## USING RELEVANT EXAMPLES

In much of our writing, there is an element of "proof." We want the reader to understand that our ideas are important and reasonable, even if he doesn't actually agree with them. Examples can be of critical importance here. By choosing an example that relates to the main idea, we can be more persuasive. An off-topic example, on the other hand, tends to cast doubt on our main idea. Suppose we tried to insist that knives are a useful survival tool because so many different types are manufactured. Is the number of types really relevant to the knife's usefulness? Probably not. But see how the examples in the paragraph below *are* relevant.

**STUDENT WRITING** | Nicholas Castellano

The knife can be used for both hunting and building on the island. For instance, I can use the knife to sharpen a straight stick to use as a spear. The spear can be used to catch fish in the water or to stab unsuspecting birds or other small animals. With the knife I can cut my way through thick vegetation. I can also cut down large leaves and branches to build a shelter. The shelter will keep me safe from wild animals and bad weather. The knife will be necessary for my survival.

The knife is useful, we discover, because it helps provide food and shelter, assistance that is unrelated to the variety of knives available in camping stores. We *see* the uses of the knife from the sharpened stick with its sudden "stab" to the shelter of leaves and branches. When the paragraph wraps up with the firm statement "The knife will be necessary for my survival," we can readily agree. The examples have been so specific and relevant that the final sentence seems simply to state the obvious.

In this next paragraph, the writer uses an example to explain his opinion that chefs ought not to preach a particular political message to their customers, although they may very well live out a particular ideology.

**STUDENT WRITING** | Payson S. Cushman

The best example of a politically engaged chef is Alice Waters of Chez Panisse. Her movement to foster the farm-restaurant connection, as well as her programs to provide school lunches and teach children the concept of sustainable agriculture, illustrates the political potential of chefs in America today. However, there is a direct separation of her political message and the dining experience at her restaurant. Despite the fact that all, or as much as possible, of the food served comes from the smaller, usually organic farms, the diner can easily separate the message from the experience.

Illustrative examples are common in writing about literature and film, as in the passage below that illustrates the hard work Martin puts into his Sunday dinners in *Tortilla Soup*.

**STUDENT WRITING** Na Yeon Kim

He peeled spines on cactus, ground pan-fried yellow and red paprika slices in a stone mortar, sautéed banana slices, and simmered vegetable soup with squash flower.

This Southwestern Chicken and Chili Soup with its assortment of toppings could have been served at one of Martin's own Sunday dinners.

In the next paragraph, examples are used to explain Quoyle's "brokenness" in *The Shipping News*.

**STUDENT WRITING** David Wilson

Quoyle begins his brokenness from birth. During his youth he was tossed off the dock to learn how to swim. This shows how ruthless his father was. His father always told him that he was never going to be good at anything. Quoyle, of course, believed what his father told him. He works a series of dead-end jobs as a dishwasher, a cashier at a movie theater, and an inksetter, a job where he did more napping than working.

And in the paragraph that follows, the writer gives an example of a scene in *Mystic River* that illustrates Sean Devine's honesty.

**STUDENT WRITING** René S. León

Kevin Bacon plays the character Sean Devine, who during the movie showed many qualities. The ones that drew my attention the most were his honesty, professionalism, and a calm and detached personality. At the beginning, Jimmy Markham asked Devine to joyride in a car around the corner. He refused to do it. That scene reinforced my belief in his honesty. He could have chosen to do it, but instead he decided not to. I was impressed that as a kid, he knew what was wrong and what was right.

When essays are weak or unclear, it is often because the writer (unlike those quoted above) does not illustrate his ideas with specific, relevant examples.

**EXERCISE 6.2** **Finding Relevant Examples**

Look back at your brainstorming from Exercise 6.1. Which of the examples seems the most relevant? Why? Try to find two more relevant examples that might help the reader to understand your point. Write a short essay, using the three examples.

## ORGANIZING YOUR EXAMPLES

Once we've found specific, relevant examples, we must decide on the most effective order in which to present them. In the essay that follows, the writer introduces the following main idea:

> My inner circle has no limits; furthermore, money, time, and materials have no meaning.

He explains what he means by that, and invites our agreement, through the examples of the men within this inner circle.

### STUDENT WRITING | Glenn Pueblo Abutin

In the beginning of high school I befriended three people. They were Eugene, Eric, and Tim. Together we had good times and worked out the bad times. *Barcada* is a word in Tagalog that means brotherhood. It is defined as "My friendship is earned, not given," said my friend Tim. Everybody has friends; however, we have an outer circle and an inner circle. I would consider a classmate, roommate, coworker, and a well-known client to be a part of my outer circle. I treat them as equals; however, the outer circle is limited to a certain degree. My inner circle has no limits; furthermore, money, time, and materials have no meaning.

Attending the Governor's Magnet School for the Arts, I met Eugene. He is very loyal and forgiving. There were many occasions I would make fun at him and he would get really upset; nevertheless, he would be my friend. One time he let me stay with him rent free when I was laid off.

I met Eric at the same time when I met Gene. Eric is different from my other friends. He is very opinionated. If you have an idea or a plan, he challenges it until you prove it correct. I taught Eric how to play the blues. He would play his guitar, and I played my saxophone. We exchanged ideas and expanded our playing ability.

I met Tim through association with Gene and Eric. Tim is the wisest of the three. He invited me to the U.K., and he opened my eyes to the world. He explained that *utang na loob*, which in Tagalog means "depth of gratitude towards your parents," is not restricted to them. During my layoff in Louisiana, Tim decided to help me move back to Virginia. He drove from New Jersey to Louisiana to help me move.

To have good friends it takes time, trust, an open mind, and a caring heart. I have *barcada*, and my *utang na loob* is everlasting to them.

Each friend exemplifies the idea of "earned" friendship. The first is loyal and forgiving, and he lets the writer stay with him "rent free." The second challenges the writer to defend his thinking and to develop his musical ability. The third is the "wisest of the three," explaining key concepts of "friendship" and "gratitude," as well as driving from New Jersey to Louisiana to help the writer move. *This* is the "inner circle," and the examples are convincing.

Notice the way in which the essay builds to the last example (Figure 6.2). While Gene is understandably valued for his loyalty and support, the second friend, Eric—who is "different from my other friends"—pushes for growth in thinking and playing. This constant challenging of one's ideas is not always a quality that we value in a friend, however, and it makes sense that the paragraph about Eric follows the one about the more conventionally supportive Gene. The third friend is Tim, and, in the introduction to the essay, the writer gives us the information that Tim was the one to define *barcada*. Tim is "the wisest of the three" friends and "opened my eyes to the world." For these reasons, it seems appropriate to conclude with him.

The essay is about friendship so close it is more like "brotherhood," and the use of the Tagalog word here, *barcada*, adds flavor and texture to the story. When the second Tagalog phrase, *utang na loob*, appears in the fourth paragraph, the writing is further enriched. Not only do we glimpse a culture where "depth of gratitude towards your parents" is valued, but we catch an echo of the author's voice. The final sentence of the essay ties both phrases together as it reiterates the theme of the inner circle:

> I have *barcada*, and my *utang na loob* is everlasting to them.

## EXERCISE 6.3 Organizing Your Examples

Look again at the short essay you wrote in Exercise 6.2. Put the examples into a different order, and rewrite the paper using this new sequence. Which version do you prefer? Why?

FIGURE 6-2 Organizing Your Examples.

## USING EXTENDED EXAMPLES

Sometimes a complex concept is best explained through an extended example, that is, one that must be developed through several paragraphs or pages. In the research paper on organic farming and sustainable agriculture quoted at the beginning of this chapter, for instance, the writer explores the possibility of changing from conventional to sustainable farming through a series of examples from Burkina Faso in Africa to Madhya Pradesh in India to a research project at the University of California–Davis. Each example takes three to four paragraphs to develop, at the end of which the reader is better able to understand the concept.



**EXERCISE 6.4** **Extended Examples**

Read Travis P. Becket's research paper on organic farming in Appendix VI, paying close attention to his use of extended examples. Describe one such example and how it works to make his ideas clearer.

Thinking is often rich and complicated, and our writing should naturally reflect that complexity. We may develop our ideas through different methods within the same essay; in the same way, we may use several different cooking methods to prepare a single dish. Whether we use examples to explain our ideas, or season our writing with phrases from another language, or both, our ultimate goal is to communicate our thoughts effectively.

## RECIPE FOR REVIEW Using Examples

**Examples** are events, stories, facts, or other specific information that is used to illustrate, explain, or prove a general point. Examples should be *relevant* and *representative.* Ask yourself whether the example really proves your point.

When you use examples to develop your essay, decide on an *effective order*, perhaps least to most important. Consider also how *many* examples you need to do the job. Too few will leave the reader unconvinced; too many will leave him overwhelmed or bored.

An **extended example** develops a complex point through the use of one specific event, story, or person. Check that the example truly illustrates, explains, or proves your point. Be sure to include specific information.

## A Taste for Reading

1. Explain how Eric Schlosser uses examples in the first and last paragraphs of the excerpt from *Fast Food Nation* in Appendix VII.
2. In what way can M. F. K. Fisher's "Borderland" (Appendix VII) be described as an "extended example"?

## Ideas for Writing

1. What is your favorite type of music, film, or cuisine? Why? Use specific examples to develop your answer.
2. Choose one of the following topics, and use specific examples to develop your answer:

   How would your life be different if you were five years old? Eighty years old? A different race? A different gender? A different species?

3. What is a problem in the food service industry? Write a paragraph or essay in which you explore specific examples of the problem.
4. Choose a general statement or maxim such as "Beauty is only skin deep" or "Don't judge a book by its cover." Write an essay in which you develop a single specific example that illustrates the general statement.

chapter 7

# Compare and Contrast

In the last three chapters, we explored several different ways of approaching a topic. We began with telling a story, one of the most basic ways of thinking. Next we talked about adding descriptive details—evidence of the senses and of our emotions. Finally, we looked at the use of examples to develop and explain a topic. In this chapter we'll examine another very common and effective mode of thinking: **comparison** or, as it is often expressed, **compare and contrast**.

We use this mode frequently in the process of making a decision. For example, imagine that you are at the farmer's market scanning the different varieties of squash or apples. How will you decide which one to buy? You will probably begin by *comparing* the varieties in terms of appearance, flavor, and texture. You will think about how you plan to use the squash or the apple and how much each variety costs. Some types may be organically grown; others may not. Will you go with a familiar Golden Delicious or try a less common variety, such as Mutsu or Elstar? As a result of your study of the similarities and differences between the varieties, you are better equipped to choose among them.

Imagine that you are at a farmer's market scanning the different varieties of squash.

Or perhaps you need to buy a car. There will be many aspects to that decision, including cost, size, maintenance, and mileage. Perhaps you'd like to try one of the new hybrids. Will you rent or lease? And what about the color? Particularly with such an expensive purchase, it is very useful to break down the similarities and differences among the options so that you can make a wise decision.

## THE MATRIX

Many, many stories and films are developed through comparison and contrast. For example, the story of *The Matrix* grows out of the differences between the "the matrix," the illusion of reality generated by computer software that has taken over the planet, and the "real world," a chill wasteland below the earth's surface. Two dining scenes near the middle of the film capture the heart of this contrast.

The first scene takes place in a fine restaurant in the matrix. The tables are lit with candles and set with white linen and gleaming silver. Harp music plays in the background. The customers enjoy a tender steak and a glass of deep red wine. The honey-tongued villain, Agent Smith (who is actually a software program), is wining and dining one of the human rebels, Cypher, who has agreed to provide information that would allow Smith and his colleagues to suppress the human rebellion. In return, Cypher wants to escape the danger and discomfort of the real world and be reinserted into a luxurious lifestyle in the matrix. His conversation with Agent Smith is about betrayal.

The second scene follows immediately and provides a powerful contrast. Unlike the restaurant with its comfortable seating and rich, glowing colors, the ship's tiny mess hall is all bare metal and harsh, fluorescent lighting. The meal is a colorless concoction that looks like watery oatmeal—a far cry from the steak Cypher enjoyed in the previous scene. The conversation is largely informative, as the rebels explain the unappealing but nutritious dish to newcomer Neo. The diners work as a team, sharing facts and advice as well as food, and this meal could hardly be more different from the first one.

These two scenes illustrate some of the most important points about the compare and contrast method. It's clear and compelling—almost humorous in the dramatic difference between the succulent steak and the watery oatmeal. It can drive the action of the entire film (or essay) as the characters continuously talk about and travel between the matrix and the real world. Further, the contrast leads us to the film's central question: Is it better to live a more colorful, comfortable life as an enslaved human battery in the matrix, or would you rather endure the bleak and difficult, though free, existence of a rebel in the real world? While the traitor Cypher chooses comfort, Neo chooses freedom.

As you develop a topic through comparison and contrast, you may decide to organize your paper according to one of the simple formats described below. Use specific examples and details—the beautiful suits in the matrix, the torn sweats in the real world—to make your points clear and vivid. Finally, go beyond those details to explore the significance of the comparison.



## GETTING STARTED

Brainstorming similarities and differences can be energetic and fun. You can create a table or grid (Figure 7.1) and fill it in quite easily.

The items you're comparing probably have some things in common—if they didn't, it would be silly to compare them in the first

**FIGURE 7-1 Brainstorming differences between two scenes in The Matrix.**

Two Dining Scenes in *The Matrix*

| | Scene in the Matrix | Scene in the Real World |
|---|---|---|
| food | succulent steak<br>red wine | watery oatmeal |
| table setting | white linen<br>gleaming silver<br>candlelight<br>harp music | metal table<br>aluminum spoon<br>harsh fluorescent lighting<br>no music |
| clothing | elegant suits | torn sweats |
| focus of conversation | betrayal | survival |
| desire of human characters | comfort, even at the cost of slavery and betrayal | freedom, even at the price of danger and death |

place. You know the saying, "You can't compare apples and oranges." But we *can* and probably *should* compare different varieties of apple when we are trying to choose the best one for our needs. For example, suppose we are choosing between Red Delicious and Granny Smith apples. They are similar in that they are readily available throughout the year, can be eaten fresh or used in cooking, and are about the same price per pound. How would we choose between them, then? By looking at the *differences.*

> While the Granny Smith apple is round and smooth with a bright green color, the Red Delicious is oval in shape with bumps on one end and a dark red hue. The Granny Smith is crisp in texture with a tartness to its flavor. In contrast, the Red Delicious has a mealy feel and a sweet taste.

Notice that we want to examine the same characteristics of each apple—appearance, texture, flavor. It would not be helpful to say something like, "While the Granny Smith apple is round and smooth with a bright green color, the Red Delicious has a mealy feel and a sweet taste." A meaningful comparison studies the same points of each item.

Sometimes you will have to choose between similar items, as in the example of the apples. At other times, you will decide to compare your item with another, perhaps more familiar one, in order to increase the reader's understanding. For example, in order to explain a papaya to someone who has never tasted one, you might talk about the texture of the more familiar peach or the acidic bite of an orange. In any case, choose two or more items that have enough in common to make sense of a comparison.

## EXERCISE 7.1 Developing a Topic

First, choose two or three somewhat similar items to compare. They might be varieties of a fruit or vegetable (such as Gala and Braeburn apples), methods of cooking (steaming and grilling), characters in a movie (Luke Skywalker and Han Solo), or characters in different movies (Luke Skywalker and Frodo Baggins). Or you might choose two kinds of music or two musical artists. Second, make a chart of the differences between the items you're comparing. Finally, put this information into sentences.

Let's look back at our sentences about the apples. To introduce the comparison, we might give some information about the similarities between these two varieties.

> Both the Granny Smith and Red Delicious varieties of apple can be eaten fresh or used in cooking and are available throughout the year.

A concluding sentence might suggest a preference between them.

> Although the Red Delicious is a beautiful apple and looks well in a fruit bowl, many consumers prefer the versatile, flavorful Granny Smith for snacking and cooking.

Let's put them all together now:

> Both the Granny Smith and Red Delicious varieties of apple can be eaten fresh or used in cooking and are available throughout the year. While the Granny Smith apple is round and smooth with a bright green color, the Red Delicious is oval in shape with bumps on one end and a dark red hue. The Granny Smith is crisp in texture with a tartness to its flavor. In contrast, the Red Delicious has a mealy feel and a sweet taste. Although the Red Delicious is a beautiful apple and looks well in a fruit bowl, many consumers prefer the versatile, flavorful Granny Smith for snacking and cooking.

**EXERCISE 7.2** **Adding the Framework**

Look back at the information you developed in Exercise 7.1. Write an introductory sentence that outlines the similarities of your items. Then write a concluding sentence that gives the reason behind the comparison, for example, a decision reached or new information learned.

## ADDING DETAILS

As with all of our writing, it is the details that make the ideas real and fresh. In the culinary world, our subjects themselves are often unusually real and fresh. For example, one August my class tasted peaches and apricots, both stone fruits, locally grown and dead ripe. We made a chart (see Figure 7.2) and carefully recorded our observations of their appearance, aroma, flavor, and texture. In the passages that follow, notice how the student writers use descriptive details to bring the comparison to life.

**FIGURE 7-2** Tasting of peaches and apricots.

A Tasting of Local Apricots and Peaches
Both are stone fruits, related to almond

| | Apricot | Peach |
|---|---|---|
| **Aroma** | honey<br>pumpkin | floral/honeysuckle<br>fresh mountain air |
| **Texture** | drier | creamier |
| **Flavor** | earthy<br>sweet/tart<br>wilder<br>floral<br>citrus<br>spicy (nutmeg)<br>nutty<br>melon | bright<br>sweeter<br>tamer<br>richer,not quite as lean<br>nutty (pecan; close to skin) |

**SERVING SUGGESTIONS**

Use vanilla with both apricots and peaches to draw out spicy, aromatic quality

Use cracked black pepper with peach; pepper is floral and spicy, sharp and astringent to contrast with peach

## STUDENT WRITING Lori Vrazel Kallinikos

The outward appearance of the peach displays brilliant reds, oranges, and yellows that resemble the sky at sunset. The appearance of the peach seems to beckon those who view it to come over to take a closer look. . . . The apricot does not possess all of the vibrant colors of the peach but instead delivers one outstanding color of its own. The apricot looks like the sun as it rises in the morning.

In the passage above, the writer uses a contrasting image to highlight the difference in color: the peach "resemble[s] the sky at sunset" while the apricot "looks like the sun as it rises in the morning." It would have been far less effective to use two *different* images here; for example, the peach "resemble[s] the sky at sunset" while the apricot looks like a fresh egg yolk.

The next writer contrasts the *aroma* of the peach with that of the apricot.

**STUDENT WRITING** Elizabeth Best

The aroma of the peach is very complex, with many notes intertwining in the nostrils. You can pick up a floral note, almost like a honeysuckle. Next you may pick up a sweet, almost musky scent. Perhaps the strongest aroma is honey. It smells like a very rich, wild honey, like that of the wild clover. Almond is present with the underlying aroma of vanilla. A faint scent of nutmeg can be sensed, lingering for a second. However, the aroma of the apricot is deeper, reminiscent of a hay field. The apricot also smelled "greener" than the peach and lighter. It did not smell as rich and full-bodied as the peach.

This writer does an excellent job of exploring the aromas—we sense her genuine effort to communicate her experience despite the fact that odors and flavors may be two of the most difficult things to describe. She also uses an effective transition, "however," to move from the peach to the apricot, and she continues to make the contrast clear with phrases such as "greener than the peach" and not "as rich and full-bodied."

**STUDENT WRITING** Elizabeth Best

The texture of the peach is very juicy. The juices run down the hands and face as you bite into it. It is sticky, reminiscent of a sorbet. In contrast, the apricot is noticeably drier. It has a very creamy feel in the mouth. I like to compare it with custard, because of the rich, full mouthfeel.

This passage offers several fresh and vivid descriptions of the fruits' textures: the juices "run down the hands and face," the peach is "sticky" like a "sorbet." The comparison to the two desserts—sorbet and custard—works well. Again, it would not be as effective to relate the peach's texture to that of a sorbet but the apricot's to cold cream! Notice also the use of the transition "in contrast." Other effective transitions that show contrast are *although, but, however, instead of, on the other hand,* and *unlike.*

The next passage uses a story format, as well as descriptive details, to describe the taste of the peach. The present tense may make both the events and the sensory information more real and vivid to the reader.

**STUDENT WRITING** Kimberlie Endicott

After I smell the peach, I move on to taste it. I bite into the skin that has a nutty flavor, and as my teeth bite closer toward the pit, I begin to taste a bit more tartness. As the juices enter my mouth, I taste a small amount of acid. As I begin to chew, I receive a more almond flavor.

**EXERCISE 7.3 Adding Details**

Take ten minutes and brainstorm about the topic you chose in Exercise 7.1. Try to include specific events, sensory details, and examples that will explain the differences with greater depth and impact. Write a new paragraph, combining the information from this exercise with your work in Exercises 7.1 and 7.2.

## ADDING DEPTH

A good comparison is not only about what we observe, however, but about what we can *imagine* or what the comparison *means*. In her paper about the peach and apricot tasting, one student added a paragraph on the uses of the two fruits.

**STUDENT WRITING** Elizabeth Best

When using peaches and apricots in baking, you should treat them as differently as you would apples and oranges since they will contribute different components to the product. If you want a bolder flavor, use the peach and add some sliced almonds and vanilla to enhance it. If you want a subtler flavor, use the apricot. Adding some fresh ginger will play up its "green" components and make for an interesting complexity of flavors. Think of its velvety feel and how that will affect the overall consistency of the dish. A pudding or ice cream would show off the apricot's creamy texture.

Note how the writer has moved beyond the measurements taken in the classroom, the measurements of aroma, texture, and flavor, in order to explore the *uses* of the two stone fruits. With these observations, her knowledge of baking and pastry, and her imagination, the writer has found the *meaning* of the comparison—and it is this search for meaning that is sometimes neglected or omitted in a compare and contrast paper. Perhaps we could think of it like a sandwich. A simple sandwich of bread and cheese is easy to make and offers the relatively simple contrast of the two flavors and textures. A more complex and interesting sandwich, however, will explore different textures of the bread, various kinds of cheese, and an assortment of condiments.

The point of compare and contrast is to *explore* the idea, widely, deeply. Don't settle even for a simple decision such as "I like the peach better," or "I like the apricot better." Ask questions about *why* or *how*. Look at the paragraph below.

**STUDENT WRITING** Scott Morozin

As I tasted the two fruits, I had a chance to try and savor a flavor regardless of whether I enjoyed it or not. I forced myself to dissect each quality individually and assess each one. The result was remarkable. The peach is a crisp, refreshing summer fruit, which I can imagine eating on a hot day. The apricot, though just as wet and moist as the peach, has a lower flavor, a sort of darkness in its taste. I can imagine eating apricots on a cloudy day or maybe even after a rainstorm.

**EXERCISE 7.4** **Adding Depth**

Read over your information from Exercise 7.3. What have you learned? Why is the comparison interesting or important? How can this information be used? Write a paragraph about your thoughts.

## ORGANIZING THE COMPARE AND CONTRAST ESSAY

As in all your writing, the ideas you discover in a compare and contrast essay may end up organizing themselves—in time, space, importance. There may be special challenges, however, in terms of helping the reader keep track of which item you're talking about. Is it the peach that has a touch of nutmeg, or the apricot? One way to keep the ideas clear is to structure the essay around the comparison itself. Two such structures are often taught, the first of which involves writing all about one of the items you're comparing, then all about the other. For example, some students who had tasted the peaches and apricots wrote all about the peach first—the aroma, texture, and flavor, say—and then all about the apricot—again, aroma, texture, and flavor. This type of organization is sometimes called a "block" format (Figure 7.3). Other students focused on the characteristics, writing about the aroma of the peach and the apricot, the texture of the peach and the apricot, and the flavor of the peach and the apricot. This might be called an "alternating" or "point-by-point" format (Figure 7.4).

Remember our paragraph about the apples? It was structured in this second method—alternating. We focused on the three characteristics—appearance, texture, and flavor—and noted the differences between the apples.

> While the Granny Smith apple is round and smooth with a bright green color, the Red Delicious is oval in shape with bumps on one end and a dark red hue. The Granny Smith is crisp in texture with a tartness to its flavor. In contrast, the Red Delicious has a mealy feel and a sweet taste.

The same information might be structured differently, following the *block* format.

> The Granny Smith apple is round and smooth with a bright green color. It is crisp in texture with a tartness to its flavor. In contrast, the Red Delicious is oval in shape with bumps on one end and a dark red hue. It has a mealy feel and a sweet taste.

In either case, we could use the same introductory and concluding sentences.

### EXERCISE 7.5 Organizing the Comparison

Look back at your paragraph from Exercise 7.3. Rewrite it in two ways, first organizing it by subject, second by characteristic. Which do you prefer? Why?

FIGURE 7-3 Block Format.

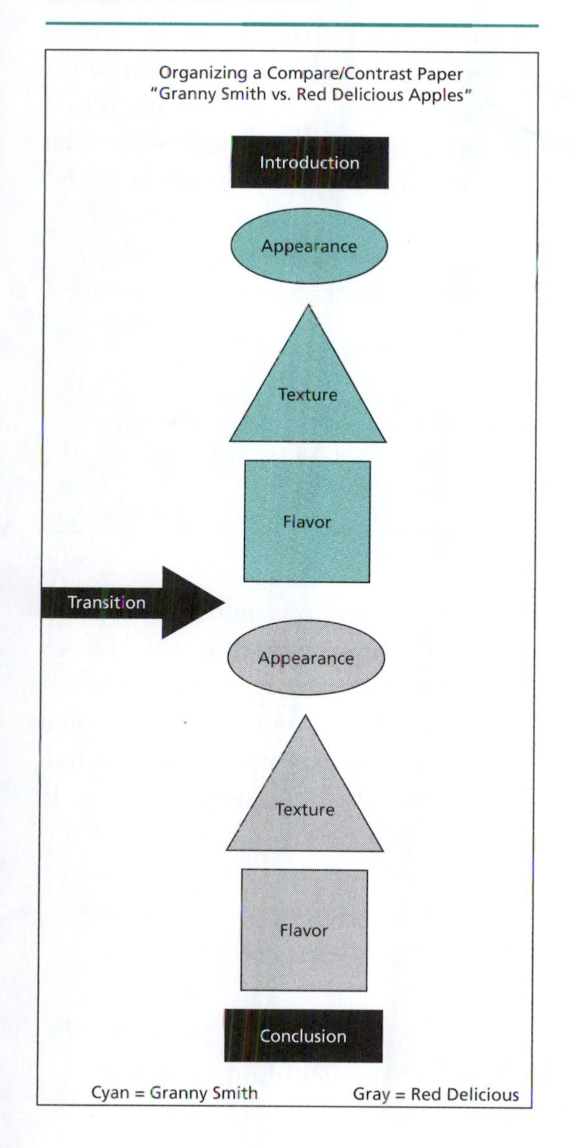


FIGURE 7-4 Alternating Format.

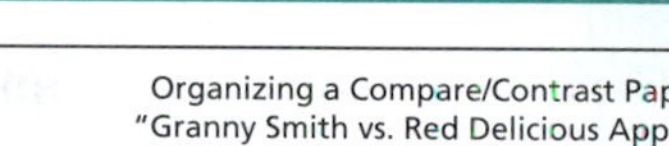

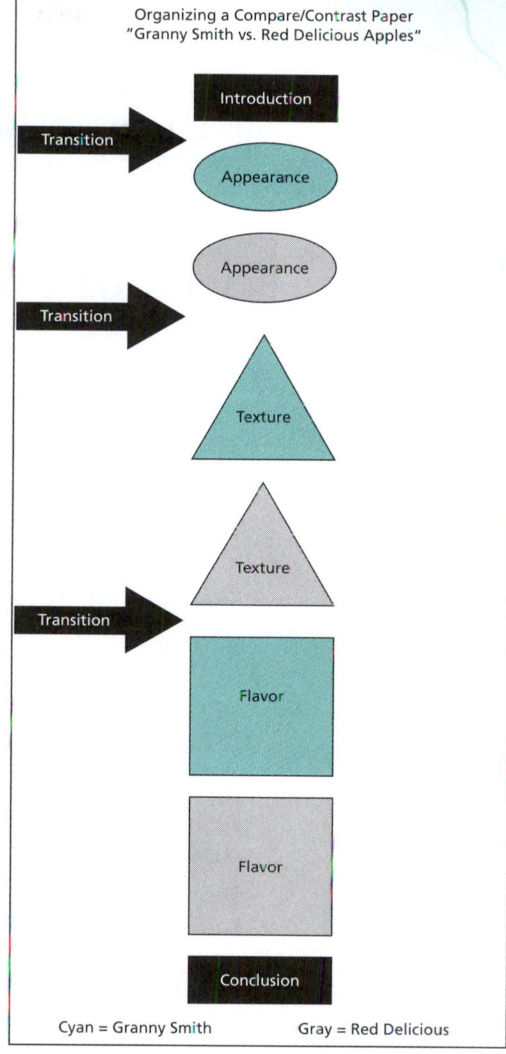




## BLOCK FORMAT: COOKIE TASTING

Once we've explored the comparison and decided on an organizational plan, we can write the rough draft. Casey Mondry, for example, wrote about tasting two cookies (see below), and he chose the block format, writing first about one cookie and then the other. He began with the appearance, then moved to the flavor and texture, and ended with an evaluation of each cookie. His detailed and thoughtful rough draft explored two ideas: first, that the appearance of the cookies matched their flavor and second, that he didn't really care for one of them. Both ideas were fully developed with a careful discussion of his reaction to the two cookies and such effective descriptive details as that the cookie had "earthquake-type cracks" and that it tasted "somewhat sugary, with the light flavor of molasses from the brown sugar and a little saltiness."

### EXERCISE 7.6 Drafting the Essay

Draft an essay based on the information you developed in the previous exercises. Pay attention to organization. Explore the significance of the comparison.

In revising his essay, Casey first checked the *content*, and it seemed complete. The overall *organization* of the essay was also effective—a clear block format. However, he did look more closely at the introduction and conclusion, add some background information, and sharpen the focus. In the introduction to his rough draft, for example, he had jumped into the paper feet first: "I found the two cookies, mudslide and chocolate chip, to be quite a bit different." As Casey revised this introduction, he gave some background on how he came to be tasting these two cookies in the first place, and he offered a brief preview of his conclusion, namely, that he preferred one over the other.

### STUDENT WRITING Casey Mondry

#### Revised Introduction

During writing class today we tasted two cookies from the same bakery, the mudslide and the chocolate chip. I thought that the look of both was appealing and told a lot about how they would taste. Although I liked both, I found that I preferred one to the other as an everyday treat.

He also added and revised some transitional expressions. In his rough draft, for example, he began the paragraph on the second cookie in this way: "The chocolate chip cookie was more my type, having a nice golden brown outside that was lighter in the center and darker as it got to the rim of the cookie." The revised sentence uses a stronger transitional phrase "on the other hand" and smoothes out the wording in the second half.

> The chocolate chip cookie, *on the other hand*, was more my type, having a nice golden brown outside that was lighter in the center and *darker at the rim*.



Finally, he trimmed the extra or repetitive words throughout the paper. In the rough draft he wrote: "I enjoyed how the walnut complemented the dark chocolate flavor, although this cookie had too heavy of a flavor for my taste." Notice the improvements in his revision:

> I enjoyed *the way* the walnut complemented the *lingering* dark chocolate flavor, although this cookie *was too heavy for me*.

As you read the full revision printed below, notice the vivid descriptions of each cookie and the thoughtful discussion.

**STUDENT WRITING** Casey Mondry

## Cookie Tasting

During writing class today we tasted two cookies from the same bakery, the mudslide and the chocolate chip. I thought that the look of both was appealing and told a lot about how they would taste. Although I liked both, I found that I preferred one to the other as an everyday treat.

The mudslide had a very rich dark brown color with earthquake-type cracks and the tips of a few nuts peeking out. When I split the cookie in half, some irregularly shaped, dime-sized crumbs broke off. The inside was dense with areas that looked gooey, and there were whole walnuts that gave a light tan contrast to the rich dark brown color. The density and color made the cookie look heavy or rich. When I bit into the mudslide, the rich bittersweet chocolate flavor came out very quickly. Parts of the cookie were moist and creamy like ganache and contrasted wonderfully with the drier, crisper outer edges. I enjoyed the way the walnut complemented the lingering dark chocolate flavor, although this cookie was too heavy for my taste. While its complexity went very well with my coffee, it was difficult to cleanse my palate with water, due to the mudslide's rich after taste and the gooey parts that stuck in my teeth.

The chocolate chip cookie, on the other hand, was more my type, having a nice golden brown outside that was lighter in the center and darker at the rim. The top was smooth with specks of the chocolate chips poking through. The bottom was porous and looked like it would be perfect for soaking up milk, one of my favorite ways to eat a chocolate chip cookie. However, I chose to have water; that way I could eat the cookie plain and evaluate it without intrusions. Breaking into the cookie, I could tell it was going to be dry because of how it crumbled. The inside looked aerated or fluffy and lightly tan. When I took my first bite, I didn't get a chocolate chip but instead tasted the rest of the cookie. I found the cookie somewhat sugary, with a light flavor of molasses from the brown sugar and a little saltiness. The cookie crumbled and dissolved in my mouth if I let it sit on my tongue long enough, and in doing so a hint of vanilla came out. I thought it was very nice to have just a few milky sweet chocolate chips in it. I find chocolate overwhelms most of the other flavors in cookies, perhaps because its flavor stays longer. I also enjoyed the finish of the cookie. It was clean and didn't stay too long. That way when the next bite came it was just about like retasting the entire cookie.

With both cookies, the look fit the taste well. The light golden brown and sparse chocolate chips in the chocolate chip cookie suited its subtle but somewhat chocolate taste. The rich color and complex texture of the mudslide also matched its flavor. I liked the chocolate chip for its subtlety, and I didn't care for the overpowering mudslide.

## EXERCISE 7.7 Revising the Essay

As you read over your draft from the previous exercise, think first about the main idea. Is it clear? Does the paper remain focused on the main idea? Do you have enough information to support, explain, or illustrate the main idea? Next, evaluate the organization. Do you still like the format you chose, block or alternating? Does the introduction give sufficient background information and lead the reader to the focus of the paper? Does the conclusion explain the point of the comparison? Have you used transitional phrases that make the comparison clear?

## ALTERNATING FORMAT: TWO WIVES

A compare and contrast essay can also be structured through a focus on the points of comparison, looking alternately at, for example, the aroma of the peach and then the aroma of the apricot. In the paper that follows, the writer compares two wives in the film *Mystic River* and uses the alternating or point-by-point format to highlight their differences in terms of loyalty to their husbands, ability to protect their families, and appearance at the end of the film.

**STUDENT WRITING** Soyang Myung

### Mystic River

The story in *Mystic River* unfolded calmly like a river flowing in gentle waves. In this largely silent movie, I saw how important the housewife was in the family. Also, I found differences between the two wives, Celeste and Annabeth.

Three childhood friends, Jimmy, Dave, and Sean, grew up in a small and shabby village in Boston. They used to spend their time playing hockey. One day, David was kidnapped by two strange men. A few days later, David escaped successfully from them. But their happy time in childhood ended. They didn't keep in touch with each other. Twenty-five years later, those three guys met again by a tragedy: Jimmy's daughter was killed. In that time, David's wife Celeste suspected her husband had killed Jimmy's daughter. As Jimmy pursued the question of the accident, Celeste confessed to Jimmy what she thought. Jimmy burned with revengeful thoughts and killed David, but it turned out David was not the murderer. Jimmy told his wife, Annabeth, everything that happened to him. When she heard about David, she might have been shocked. But she wasn't agitated. She showed her husband her strong loyalty for him. While Celeste's distrust of her husband destroyed her family, Annabeth's belief in her husband kept her family firmly together.

Celeste and Annabeth were both housewives, but they had a different belief about their husbands. While Celeste had a distrust, Annabeth had a strong loyalty for her husband. For instance, when Celeste's husband David came home with a bloody T-shirt, she was embarrassed. The next day, when it was discovered that Annabeth's daughter had been murdered, Celeste began to suspect her husband was the killer. She made a mistake. She confessed to Jimmy (Annabeth's husband) what she thought without any objective evidence. On the other hand, in the case of Annabeth, when she heard that her husband had killed David, she was not agitated. Even though she was stunned, she remained firm; furthermore, she consoled her husband. That showed us that their beliefs about their husbands were contrary to each other.

Secondly, I was able to see a difference between the two women's abilities to protect their families. Celeste lied about the truth to Jimmy. That meant that she was an egoist and didn't take care of her husband and her family, including her son. She just wanted to protect herself from David. In the long run, her irresponsible conduct destroyed her family. Because of her hasty judgment, her son Michael would live under a fatherless family, and she would live with a guilty conscience. On the other side, Annabeth tried to protect her family from tragedy. As she said "Your father is a king," she encouraged her husband and made her family hold fast. While Celeste exposed her family's problem, Annabeth concealed hers.

Finally, the two women's appearances indicated how they were different. In the movie, Celeste looked sad and full of doubt. Her face was filled with worries and clouded with anxiety. But Annabeth had a fair and a brilliant look. In the parade at the end of the movie, that was distinguished clearly. After the end of the murder case, Celeste avoided the eyes of other people, such as Annabeth and Jimmy, while Annabeth was looking at Celeste as if nothing had happened.

By those three points, I saw how different the two wives were. They were different with regard to their loyalty to their husbands, their ability to protect their families, and their appearance in life. Through seeing this movie, I realized how important the housewife's role and her belief about her husband are, because those things could affect not only individuals but also the entire family.

---

The introductory paragraph lets us know that the items being compared are two wives, Celeste and Annabeth. In the next paragraph, the writer summarizes the main story line and shows how it impacts these women in terms of their belief in their husbands, their ability to protect their families, and their appearance by the end of the film. The writer then goes through these points one by one, contrasting their implications for Celeste and Annabeth. In the end, the essay touches on the importance of this examination: "Those things could affect not only individuals but also the entire family." By going from one point to the next, the contrast is emphasized more heavily than it is by using a block format, as in the essay on the cookie tasting.

**EXERCISE 7.8** **Choosing a Format**

Rearrange the cookie tasting essay included earlier in this chapter in an alternating format. How does this change in emphasis affect the reader's interpretation of the information?

## EXPLORING SIMILARITIES

A compare and contrast essay might also examine the similarities between two or more items that are clearly different. This text, for example, explores the surprising and often instructive similarities between the clearly different processes of cooking and writing. Let's look again at the following paragraph about cooking and writing, which we first saw in the introduction.

**STUDENT WRITING** Robert A. Hannon

A cursory overview of cooking and writing could lead to the impression that the two subjects have little in common; however, a closer examination reveals many similarities. These similarities include the need to pay acute attention to detail, the certainty that only hard work will result in assured success, the need to train, study, practice, and do research, and the importance of being able to work under pressure. Other, somewhat more subtle, similarities exist: both pursuits involve simplicity of tools and equipment; they both rely on a foundation of tradition passed along through literature, and they are both vitally dependent on criticism. Paradoxically, they both have rich, almost rigid traditions; however, to be successful, the professional cook and the professional writer must acknowledge these traditions while striving to be creative, thereby standing out in similarly competitive fields of endeavor.

"Even though writing and cooking have different mediums, the similarities are there; you just have to look closely." — Jonathan A. Campbell

Note how the first sentence points out that these two items—writing and cooking—are so different that we wouldn't ordinarily think of comparing them. Only a "closer examination" reveals the similarities, which culminate with the role of creativity in a successful career in either field. The paragraph also uses effective transitions such as "*both*" and "*similarly*." Other transitions that show similarities include *also, in the same way, just as, like, likewise,* and *neither.*

Whether you choose to look for similarities or differences—or both—remember to organize your points clearly and look for the more interesting observations and inferences that lie beneath the surface.

## RECIPE FOR REVIEW Compare and Contrast

1. **Develop the topic** by choosing two or three *similar* items and exploring the *differences* between them (for example, two types of chocolate cookies), or by choosing two *dissimilar* items and exploring the *similarities* between them (for example, writing and cooking).

2. **Organize the paper** either by writing all about one of the items and then all about the other (**block format**), or by looking at each characteristic and writing about how each item measures up (**alternating or point-by-point format**). Study the following formal outlines of our short paragraphs on Granny Smith and Red Delicious apples found earlier in this chapter.

**Block Format**

I. Introduction—Two Apples
II. Red Delicious
 A. Appearance
 B. Texture
 C. Flavor
III. Granny Smith
 A. Appearance
 B. Texture
 C. Flavor
IV. Conclusion—Granny Smith is more flavorful, versatile

**Alternating Format**

I. Introduction—Two Apples
II. Appearance
 A. Red Delicious
 B. Granny Smith
III. Texture
 A. Red Delicious
 B. Granny Smith
IV. Flavor
 A. Red Delicious
 B. Granny Smith
V. Conclusion—Granny Smith is more flavorful, versatile

3. **Use transitions** to clarify for the reader which item you are referring to in each sentence or passage (see Figure 7.5).

**FIGURE 7-5 Transitions for Compare and Contrast Writing.**

To indicate similarities: also, both, in the same way, just as, like, likewise, neither, similarly

To indicate differences: although, but, however, in contrast, instead of, on the other hand, unlike

## A Taste for Reading

1. Read Jeffrey Steingarten's "Hot Dog" in Appendix VII. How does the essay use the comparison between skiing and eating seafood? Do you like this approach? Explain.

2. David Mas Masumoto's "The Furin" from *Epitaph for a Peach* (Appendix VII) also uses elements of compare and contrast, though more subtly than Steingarten's essay. Describe the comparisons between farming techniques, between children and adults, between the author and bamboo.

3. Galway Kinnell's poem "Blackberry Eating" (Appendix VII) revolves around a comparison between blackberries and certain words. Explain the similarities he finds.

## Ideas for Writing

1. Conduct a tasting in class of two or more varieties of fresh fruit, cookies, chocolate, cheese, or whatever food you choose. Determine the characteristics to be compared (for example, aroma, flavor, texture) and use a chart to record your observations. In small groups or as individuals, put the information into a paragraph or short essay. Look for fresh, vivid descriptive details and for the meaning behind the comparison.

2. Experiment in the kitchen with two methods of cooking the same food (for example, steaming or grilling vegetables) or with varying the ingredients in a recipe (for example, sweetening muffins with sugar or fruit juice). Write an essay explaining what you did and what you learned. Include vivid sensory details.

3. Compare and contrast two movies (or stories or poems) on the same topic or within the same genre (murder trials, weddings, horror, comedy).

4. Explore the similarities between chefs and doctors.

# Seasoning
## Revising Your Writing

UNIT 3

**Chapter 8**

Organizing Your Thinking

**Chapter 9**

Writing Introductions and Conclusions

**Chapter 10**

Using Transitional Expressions

**Chapter 11**

Revising Word Choice

chapter 8

# Organizing Your Thinking

The work of organizing your thinking stretches throughout the writing process. Although in Chapter 2 it's identified as the second step of the five, you may actually be moving paragraphs and sentences around up until the moment you submit the paper. Reorganizing an essay goes hand in hand with revising its content. As you add or delete information, you may need to change the order of sentences or paragraphs. A clear and logical sequence makes your ideas clear to the reader, and that is your primary objective.

During a meal, organization takes different forms. First, there's an order of events,[14] both in the kitchen and the dining room. In the kitchen, food preparation has begun before the customer arrives and continues until cleanup at the end of the night. In the dining room, customers are seated, offered drinks and appetizers, served their main course and often dessert, and finally given the check. Second, both the kitchen and the dining room are organized in terms of space so that meals can be prepared and customers served in the most efficient way. The very word "line" suggests an ordering of activity in a particular space. Finally, the menu is put together according to categories: appetizers, soups and salads, sandwiches. In the same way, an essay may be composed of different organizational patterns or categories.

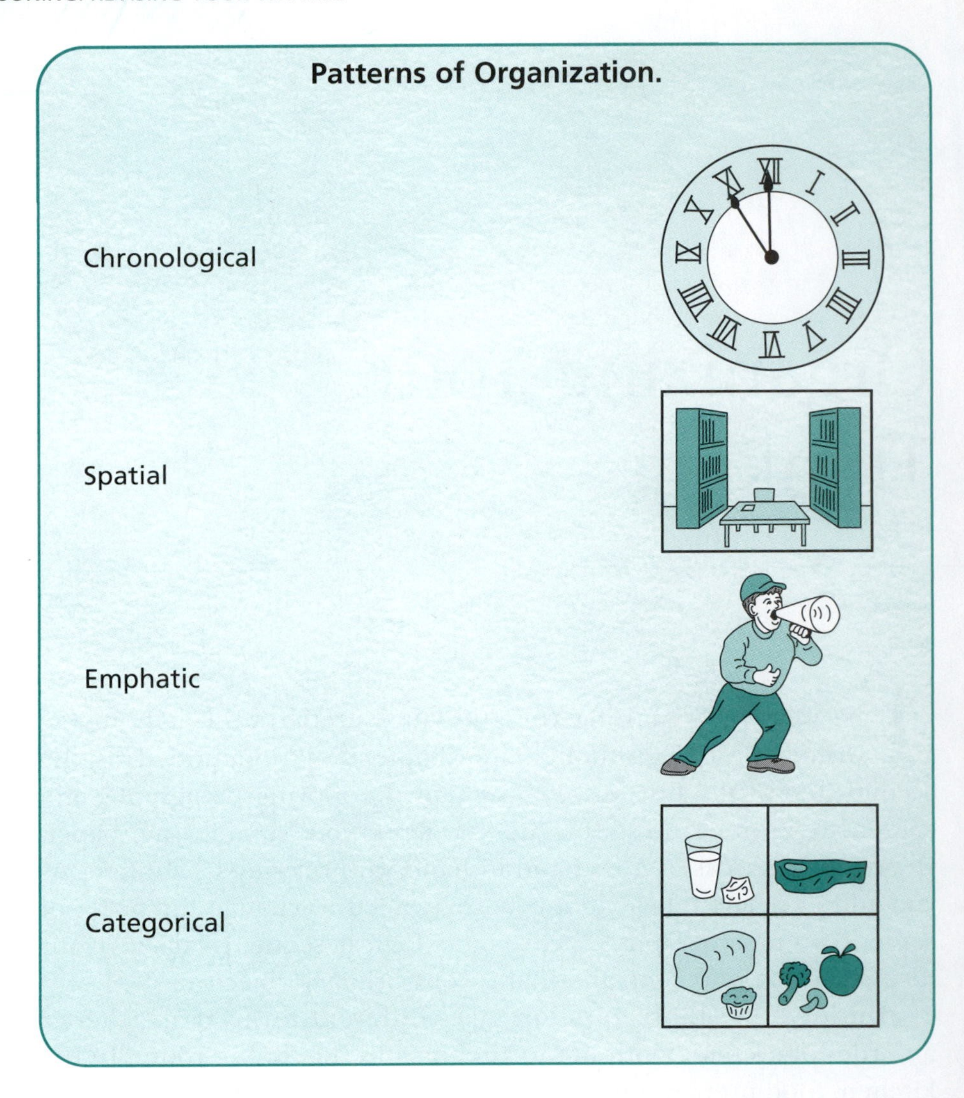


## PATTERNS OF ORGANIZATION

All essays, as we have said before, are organized generally into three parts: the introduction, the body, and the conclusion. Beyond that, there are no hard and fast rules about how to arrange your ideas, no single formula that must be followed. For each paper you're looking for a sequence that fits the content. For example, information might be arranged in **chronological order**, that is, according to the order of events. First this happened, then that happened, and now this. Storytelling, also called **narration**, generally follows this pattern. Look at this paragraph from the story about the primal cut prank in Chapter 4.

**STUDENT WRITING** | Brendan Cowley

After breaking down the primal, I remove the head and stick it upside down in the freezer. Twelve hours should congeal the blood into a thick pudding. The plan is to put the head in the beer cooler. The beer cooler is Carol's cooler, and she gets mad if anyone messes with it. Every day around 8:00 a.m. she goes down and takes inventory. I put the head upright on a sheet tray on top of a keg. I open the mouth and pull out the tongue so that is flopping around like a wet noodle. The trap is set; it's a waiting game now.



Sometimes an entire paper will be organized chronologically, sometimes just a paragraph or two.

Another pattern of organization has to do with the physical layout, the **spatial order**. This pattern works well with description. For example, in describing the features of a room, the writer might clearly indicate where each item is located in relation to others—to the left or right, above or beneath. If we were talking about a kitchen and said simply, "You'll find the bag of carrots on the counter," that would not be as clear as "You'll find the bag of carrots on the counter *next to the sink to the left of the door.*" Look at this paragraph from the description of Rice Kitchen in Chapter 5.

**STUDENT WRITING** | Matthew K. Greene

Now because the kitchen was so small, it needed to be very organized, and everything needed to be in its place. A three-part sink lined one whole side wall of the kitchen, with containers and storage units placed neatly on a shelf above. On the opposite wall there were two refrigerators, one walk-in with food stored for the whole week, and a regular fridge with items that were used throughout the day. In between the two were more shelves with extra bags, boxes, and fortune cookies. Along the back of the entire length of the kitchen were four huge woks where all dishes were prepared. Next to the woks were a fryer, a small prep table, and a rice cooker that got extremely warm. Two feet back away from the woks was another small prep table, as well as a rack with every vegetable we needed and a flip fridge with all the pork, beef, and chicken.

The information in an essay might also be organized according to its importance, that is, in **emphatic order**. In explaining a set of reasons, for example, or a cluster of effects, we might choose to begin with the most important. Or, we might choose to emphasize the most important point by placing it at the end of the paragraph or essay.

**STUDENT WRITING** Michael P. Otting

The experience in the meat room was thrilling. It makes me want to do well in my other classes so I can get to meat-fabricating class. This tour also gave me some new routes to different parts of the school. But above all else, the tour showed me that it is important to pay more attention to detail in everything I do.

Notice how the phrase *But above all else* lets us know that this is the most important reason. Such transitional expressions guide the reader clearly from one idea to the next, thus improving the overall effectiveness of communication.

Finally, you might organize your ideas in **categorical order**. This means that the paper follows the sequence and/or groupings that are already part of the subject you are writing about. For example, if you were writing about creating a wine list, the essay might naturally follow the order of the wine list itself—from the lightest whites to the heaviest red wines.

The organization of a paragraph or essay can also be driven by one of the **rhetorical modes**, such as **example** (Chapter 6) or **compare and contrast** (Chapter 7). The paragraph below contrasts a peach and an apricot.

**STUDENT WRITING** Elizabeth Best

The texture of the peach is very juicy. The juices run down the hands and face as you bite into it. It is sticky, reminiscent of a sorbet. In contrast, the apricot is noticeably drier. It has a very creamy feel in the mouth. I like to compare it with custard, because of the rich, full mouthfeel.

**Organizing a Wine List.**

WINE LIST

Sparkling Wines
Champagne

White Wines
Pinot Grigio
Sauvignon Blanc
Chardonnay

Blush Wines
White Zinfandel

Red Wines
Pinot Noir
Merlot
Cabernet Sauvignon

Dessert Wines
Icewine

3

The juxtaposition of peach and apricot highlights the differences in texture.

As we will go on to see in Unit 4, writing may also be organized according to the sequence of steps in a **process** (Chapter 12), as in the following example.

**STUDENT WRITING** Robert A. Hannon

To make Hollandaise, finely chopped shallots, white wine and cracked black peppercorns are reduced until they are almost dry; then they are mixed with a little water and added to a bowl. Next, egg yolks are added and heated over a pot of steaming water and whisked until their temperature reaches about 145° Fahrenheit.

Writing may also be organized through **cause and effect** (Chapter 13).

**STUDENT WRITING** Robert A. Hannon

At this point the eggs will have roughly tripled in volume and will fall off the whisk in long strands. If the eggs are undercooked, they will fall off the whisk like water, and if they are overcooked, they will curdle in the metal bowl, evidenced by small semi-solid chunks of soon-to-be-cooked eggs.

---

In the passage above, the *effects* of undercooking or overcooking the eggs are laid out methodically.

## EXERCISE 8.1 Identifying Patterns of Organization

Read each of the five paragraphs below. Then decide whether the primary organizing principle is that of narration, description, example, compare and contrast, process, or cause and effect. Write the principle on the line provided.

**1.** ______________

When there is no wind, the lakes are still. All the surrounding landscape reflects off the water and continues on throughout the sky as if it were a great mirror. At night the light from the moon shines on the water and reflects laser beam-like light in all directions.

Nicolas J. Goergen

**2.** ______________

I don't know why I felt this burning passion for fried chicken fingers. I guess I loved the flavor that fried chicken provides. That nice crunchy outer layer of flour or batter mixed with perfectly cooked chicken would always hit the spot! I also think that I was afraid of all the different choices that menus had. I had no idea what anything was, or what was in it. Since I was between four and ten, I did not want to take any chances. I did not want any of those tricky vegetables to end up on my plate.

Thomas Monahan

**3.** ______________

When Torcila, our neighbor, saw us at the window, we said "*Hola becina,*" and she let us in. *Tom and Jerry* was on, making us laugh. Not realizing the time, we were sitting for more than an hour. The door was knocked again and again. It was my mother. The first thing she said to us was "*Buenos tardes caballeros, ya esta el almuersa*" (Good afternoon, gentlemen. Is lunch ready yet?). We ran out of the room like rabbits when they see a dog.

René S. León

**4.** ______________

Instead of addressing the problem [of obesity] themselves, Americans often try to place the blame on the fast food corporations for making them fat. While this issue is obviously not the fault of chefs all over the country, there are alternatives that they could popularize to help keep our nation healthy. An example of this is Fast Good, a chain that sells healthful, non-genetically modified food that is available just as quickly as pulling through the McDonald's drive-thru.

Justine A. Frantz

**5.** ______________

After the chicken pieces are fried, they are put into a deep baking dish and smothered with heavy cream, garlic, and mushrooms. To loosen the sauce, chicken broth is added, and the dish is baked for an hour. My mouth is watering just writing about it! After that hour, the sauce should be boiling. Serve it hot so the chicken stays moist and delicious.

Vincent Amato

**6.** ______________

Both of the officers flip their points of view about African-Americans during the movie [*Crash*]. The one that started off being good and did not think of any racism turns out to be someone that shoots a black person. The one that starts out as a racist person and insults African-Americans comes to realize that it does not matter what race you are; everyone is equal [in a burning car].

Matthew Rutter

**Combining Organizational Patterns in the preceding writing samples**

"breaking down the primal" (Cowley) = chronological order + narration + description

"everything in its place" (Greene) = cause/effect + spatial order

"the texture of the peach" (Best) = description + compare/contrast

Perhaps the most important thing to understand about these patterns of organization is that you are likely to use more than one in any given essay. These patterns are not intended to be rigid guidelines; they're not like a fill-in-the-blank exercise. Essays are like meals; you may need more than one organizational pattern just as you may need more than one cooking method. Your goal is to develop and present your ideas as effectively as possible, through any one method or through a combination of several methods. The organization of your essay will be as individual as your thinking.

## FIRST STEPS IN ORGANIZING

The work of organizing takes place at many different points during the writing process. As you begin to think about a paper, you may find that the ideas fall into place rather easily. A story, for example, tends to follow chronological order, so its organization is relatively straightforward. However, the writer may choose to play with the sequence of events to emphasize certain points or to create a specific experience for the reader. Think about the shifting time lines in the films *Pulp Fiction* or *Memento*, for example. In any case, whenever you can, it is useful to map out a rough idea of your sequence of ideas. Even though this sequence may change many times over the course of revision, a map can be at least a temporary guide and framework for each draft and may assist in clarifying your thinking.

Let's look at an example. Suppose we were planning to write a paragraph on movies about food. Our first move, of course, is brainstorming, and our list might include the following: *Big Night, Chocolat, Dinner Rush, Tortilla Soup*, and *What's Cooking?* A paragraph about these movies might list them in the order they were released, that is, chronologically.

> *Big Night*. 1996. Directed by Stanley Tucci and Campbell Scott.
>
> *Chocolat*. 2000. Directed by Lasse Hallström.
>
> *What's Cooking?* 2000. Directed by Gurinder Chadha; starring Alfre Woodard, Joan Chen, Kyra Sedgwick, A. Martinez.
>
> *Dinner Rush*. 2001. With Danny Aiello and Sandra Bernhard.
>
> *Tortilla Soup*. 2001. With Hector Elizondo, Elizabeth Peña, and Raquel Welch; remake of Eat Drink Man Woman directed by Ang Lee.

With this rough organizational outline, we can draft something like the following:

> Set in the 1950s, *Big Night* is about a restaurant owned by two brothers, played by Stanley Tucci and Tony Shalhoub (now the Emmy-winning star of *Monk*). *Chocolat* takes place "once upon a time" in France and follows the story of a small village whose strict traditions are threatened when Juliette Binoche opens a chocolaterie (and is romanced by a gypsy, Johnny Depp). *What's Cooking,* set in present-day Los Angeles, concerns the Thanksgiving Day preparations of four ethnically diverse families. Placed in New York City, *Dinner Rush* follows several story lines, from the ambitious chef's attempts to impress a food critic to his father's brush with the Mafia. Finally, *Tortilla Soup,* also set in Los Angeles, tells the story of a chef who's lost his sense of smell and feels as though he's losing his three daughters as well.

This doesn't seem like a very interesting way to organize the material, however. Would the paragraph be less plodding if we grouped the movies by their settings, or if we grouped them by category? What are we trying to saying about these films? The paragraph should be arranged in a way that will help us to explain our main idea.

### EXERCISE 8.2 First Steps in Organizing

Reorganize and rewrite the paragraph above on food films. You may wish to add other films, such as *Tampopo* (1986, Japan), *Babette's Feast* (1987, Denmark), and *Eat Drink Man Woman* (1994, Taiwan; directed by Ang Lee).

## REVISING YOUR ORGANIZATION

What are some clues that your organizational scheme isn't working? The order of the sentences within each paragraph, as well as the order of paragraphs throughout the essay, should help make your points clear to the reader. One way to get a sense of whether the organization is working is to ask someone else to read it. If he doesn't follow your ideas, try to find out at what point the organization seems to break down.

If no obliging reader is at hand, you'll have to check the paper yourself. Jot down the main idea of each paragraph (see example below). Does this sequence of ideas seem logical, clear, and effective in communicating your meaning? Then examine the order of sentences *within* each paragraph. Does it make sense? It can be helpful to number the sentences, as in the following example.

**STUDENT WRITING** | Jesse Dowling

### The Meat Room (Rough Draft)

(1) When walking around the meat room, one gets a sense of history, from the walls riddled with names to the knowledge that is sealed inside, waiting to be grasped and wielded like a sword. (2) All the knowledge that is held in the dark and damp of the meat cellar—from the understanding of the equipment to the selections of deli meats to the cuts of meat themselves—it all comes together under one roof. (3) There was the smell; it was the scent of life and death. (4) Even though the air was tainted with the scent of blood, it symbolized all the people who would live off the lives of the animals who had given theirs. (5) It almost provided a sense of reverence. (6) From the moment I stepped onto the elevator, I got this feeling in the pit of my stomach—and it wasn't hunger. (7) It was a sense of life and death. (8) First it was the elevator, all the names. (9) It showed the past lives of the handful of people who had had the chance to experience the school. (10) So much culture on that wall—for the first time since I had arrived, I felt at home. (11) And the beauty of it is that this gem is not hoarded like Grandma's bundt cake recipe but passed down the generations, perpetually growing with time. (12) It is undying, not only eternal but reborn with each new student that has come to feast on the wisdom and wealth of information that the meat room has to offer.

A well-organized station in the meat room.

Some of the sentences clearly belong together, and the groups of ideas can be listed as follows:

| | |
|---|---|
| 1 | intro—knowledge waiting in meat room |
| 2 | various kinds of knowledge |
| 3–5 | smell of blood—both life and death |
| 6–10 | elevator—sense of history, belonging |
| 11 | knowledge of the meat room is passed to others |
| 12 | conclusion—knowledge is essentially eternal |

Although cutting and pasting are most well known nowadays as functions of the word processor, the old-fashioned method is still extraordinarily effective. When an essay's organization seems rocky or unclear, get a second copy of your draft and a pair of scissors. Then cut the copy up into individual sentences, groups of sentences, or paragraphs. You want to put similar *ideas* together, regardless of their original form. Once you've cut up the essay, physically move the pieces around and read them in different orders. This exercise forces you to "think outside the paper," literally. One of the most powerful roadblocks to good writing is a premature, rigid attachment to the first draft. And one of the most effective tools for removing such a roadblock is a pair of scissors!

In the revised version of "The Meat Room" printed below, the author introduces the idea of knowledge in the first sentence and chooses a chronological pattern of organization, moving next to the elevator, his first contact with the meat room. Then comes the smell, which—again according to the original sequence of events—became apparent only as the author began his descent in the elevator. Finally, the types and significance of the knowledge found in the meat room are explored, and the piece ends with the thought that knowledge is eternal. As you read the revised paragraph, think about whether—and why—this new organization is more effective.

**STUDENT WRITING** Jesse Dowling

## The Meat Room (Revised Organization)

When walking around the meat room, one gets a sense of history, from the walls riddled with names to the knowledge that is sealed inside, waiting to be grasped and wielded like a sword. From the moment I stepped onto the elevator, I got this feeling in the pit of my stomach—and it wasn't hunger. It was a sense of life and death. First it was the elevator, all the names. It showed the past lives of the handful of people who had had the chance to experience the school. So much culture on that wall—for the first time since I had arrived, I felt at home. Then there was the smell; it was the scent of life and death. Even though the air was tainted with the scent of blood, it symbolized all the people who would live off the lives of the animals who had given theirs. It almost provided a sense of reverence. Finally, all the knowledge that is held in the dark and damp of the meat cellar—from the understanding of the equipment to the selections of deli meats to the cuts of meat themselves—it all comes together under one roof. And the beauty of it is that this gem is not hoarded like Grandma's bundt cake recipe but passed down the generations, perpetually growing with time. It is undying, not only eternal but reborn with each new student that has come to feast on the wisdom and wealth of information that the meat room has to offer.

Notice that several transitional expressions have been added or revised to tie the paragraph together in the new sequence, for example, "*then*" and "*finally*."

**EXERCISE 8.3** **Revising the Organization**

Look again at the paragraph from Exercise 8.2. Try another sequence of ideas, and rewrite the paragraph. Which of the two do you prefer? Why?

It's often through revising and rewriting that your best thinking emerges. If possible, let time pass before you re-read and revise the rough draft. You don't need to remain committed to your *first* ideas, nor should you cling to the text of your first draft. Let your mind continue to range freely over the topic. Pursue interesting lines of thought. Look for better details and examples. And don't hesitate to experiment with completely different organizational plans.

## RECIPE FOR REVIEW Organizing Your Thinking

### Patterns of Organization

Although it can be useful to understand and employ these patterns, do not feel you should restrict yourself to a single principle or mode of organization.

1. Organizing principles: chronological (order of time), spatial (order of space), emphatic (order of importance), and categorical (order of type).
2. Rhetorical modes: narration, description, exemplification, compare and contrast, process, cause and effect.

### First Steps in Organizing

1. Once you've brainstormed a list of supporting ideas and details for your topic, put related points together and jot down an order.
2. Use this "map" as you write the rough draft of your essay.

### Revising Your Organization

1. Check your draft for a logical sequence of ideas within each paragraph and over the essay as a whole.

2. List or outline the main ideas; try rearranging them to get a clearer sequence.
3. Try cutting and pasting—with scissors and tape!
4. Rewrite the essay in the new sequence, adding or revising sentences and transitional expressions as needed.

## Cutting and Pasting

1. Obtain a fresh copy of your draft (photocopied or printed).
2. Cut up the draft by sentences, groups of sentences, or paragraphs.
3. Physically rearrange the "ideas" into a clearer sequence.
4. Rewrite the essay in the new sequence, adding or revising sentences and transitional expressions as needed.

## A Taste for Reading

1. What types of organizational patterns do you find in Jeffrey Steingarten's "Hot Dog" (Appendix VII)?
2. Create an outline of José Antonio Burciaga's "Tortillas" (Appendix VII), as we did for the short piece on the meat room in this chapter. Perhaps begin by listing the main idea of each paragraph; then describe the organizational patterns *within* each paragraph.

## Ideas for Writing

1. Pick a single topic and write two paragraphs about it that are organized in two entirely different ways. (Remember the two paragraphs on hollandaise sauce? One was process, the other cause and effect.) Which paragraph is more successful? Why?
2. Think about the way you typically write a paper. At what point do you reflect specifically about the organization? Do

you write down an outline or map? Have you ever tried cutting and pasting with actual scissors and tape? Organize your answers into a short essay.

3. Explain the organizational principle(s) behind such films as *Pulp Fiction* or *Memento.*

4. Are you an organized person in general? In the kitchen? What role does organization, or the lack of it, play in your life? Develop your answers into a short essay.

chapter 9

# Writing Introductions and Conclusions

The introduction to an essay is like the first scent to catch your attention as you step into a restaurant. The aroma may invite you in—or you may decide to dine elsewhere! Now, if you're reviewing the restaurant, you'll probably stay even if your first impression is not promising. Similarly, your instructor will read your essay whether or not the introduction is inviting. In most cases, however, customers and readers are free to try another venue if the first impression is poor. Therefore we want to make that first impression, whether it's an enticing aroma or a captivating introduction, a good one.

Now, imagine that the customer has finished the meal. We want to ensure that she *leaves* the restaurant with a good impression and that she passes that good impression on to her friends. In the concluding moments of the meal, we want to remind the customer that we delivered on our initial promise—that she enjoyed the meal. By offering coffee and dessert, we encourage the customer to form a positive opinion that she can then relay to others. Likewise, the conclusion of an essay is a chance for us to highlight our main point and show the reader that we kept the promise made in the introduction. For both cooks and writers, introductions and conclusions play important roles and present unique challenges.

We can think of the essay's introduction as a funnel that is wider at the top and narrower at the bottom.

## THE SHAPE OF THE INTRODUCTION

As we've said, the introduction to an essay is like the appetizer before the main course—or perhaps even like the delicious smells that emanate from the kitchen before any food is seen. Imagine a dog in a cartoon floating rapturously through the air as its nose follows the trail of a fabulous fragrance. In a similar way, the introduction to an essay draws the reader to the story that follows. The introduction has a second important task as well: to tell the reader what the writing is about. In terms of shape, we can think of the introduction as a funnel that is wider at the top and narrower at the bottom. The first sentence or two may offer a broad statement about the topic, while the next few may tell a story about it, ask a question, or provide background information. Finally, the last sentence narrows the focus of the paper, sometimes with a statement of the main idea.

In addition to getting the reader interested, the introduction states the topic of the essay and gives important background information. Like a funnel, the introduction opens wide to draw the reader in with a statement that is broadly appealing or a story that is uniquely interesting. With each new sentence, the introduction narrows its focus until it concludes with the specific subject of the essay. Look at the example below.

**STUDENT WRITING** Kenneth Zask

As a child, everyone has a dish that he or she loves so much it becomes an essential to survive. It's usually "fun" food, like French fries and chicken fingers, which can be picked up and eaten at any time. Yet every once in a while, a child comes along that enjoys eating entrees and side dishes other than the usual simple deep-fried meals. My favorite childhood dish was my grandmother's red cabbage.

The introduction begins with the general statement that "everyone" has a much loved favorite food. The second sentence outlines typical childhood preferences. The third lets us know that something different is coming. The fourth sentence narrows the focus to the writer's favorite dish—"my grandmother's red cabbage." The paragraph funnels the reader into the body of the essay, which will focus on that cabbage.

Here's another example of an introduction that **moves from the general to the specific**.

**STUDENT WRITING** Gerald Houston

We often seem to hold on to our past. We carry it on our shoulders, using it as an excuse, a hideaway, or as a compass for our future. Beating ourselves up for the past has become second nature to us. I should have. I would have. I wish I had. Although there's nothing that can change the past, by leaving our past behind we can now focus on the future, which can be the new foundation of our lives. In *The Shipping News*, the Quoyle family had trouble letting go of a past that hindered their growth. But they eventually learned that the past was merely a stepping stone. They learned that life still goes on.

Again, the introduction begins with a general statement—"we often seem to hold on to our past"—explains it, and ends with a specific focus. It is the Quoyle family who is holding on to "a past that hindered their growth."

In addition to narrowing the focus of an essay, the introduction may also **map out the main points**.

**STUDENT WRITING** James Virus

Today nearly everybody goes to college. It is almost a must-have for every job. The people that do go to college see that it is very different from high school. There have been many differences I have seen, but the three that stand out the most are the amount and difficulty of the work, the responsibility of getting up on my own, and the friends I have made.

3

The last line of this introduction makes clear that the next three sections will discuss the difficulty of college work, the responsibility of caring for oneself, and the making of new friends. This introduction offers a very specific promise to the reader in terms of both content and organization, and you may find such a road map to be an effective component of many academic essays.

While some introductions begin with a general statement, others start with a **specific image or quotation** that relates to the topic of the essay. Movies, of course, are all about the image. The first scene of *The Matrix*, for example, shows lines of mysterious figures unfolding across a computer screen. Initially uncertain what this image might mean, the viewer comes to understand that it represents the inner world of the matrix, the computer software that has relegated humanity to the status of a power plant.

Movies may also catch the audience's attention and introduce their themes through a quotation. The film *About a Boy*, for example, begins with the main character, Will, watching a game show in which contestants must identify the author of the famous quote "No man is an island." Our hero happily spouts the wrong answer (Jon Bon Jovi instead of John Donne) and explains the meaning this quote has for him. Will believes that "Jon" is wrong and that he himself is indeed an island, comfortably isolated from human commitment while completely entertained by his collections of CDs, DVDs, household appliances, and disposable dates. Throughout the movie Will refers to and refines his interpretation of this quote, which comes to embody the central theme in the movie. Quotations can also be effective in the introduction to an essay. Look at the example below.

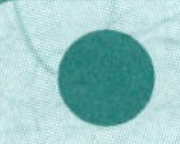

**STUDENT WRITING** David Wilson

The film *The Shipping News* is about a man named Quoyle who is relocated to Newfoundland. He goes through a bunch of poor choices to get into his current situation. His marriage to a girl that just jumped into his car was untrustworthy. The sudden death of his parents was surely unfortunate. Quoyle says in the beginning of the film: "My failure to dog paddle was only the first of my many failures."* At this point he is a "broken" man.

---

*Text from *The Shipping News* courtesy of Miramax Film Corp.

The quotation marks catch the reader's attention and bring life to the paragraph, while the content of the quote points toward the central theme of the film: After so many failures, can this "broken" man be healed?

Sometimes the introduction **tells a short story** that is related to the body of the essay. For example, the film *Miss Congeniality* begins on a playground in New Jersey where a tough little girl defends a boy from a bully. The boy doesn't appreciate her, however, and as she impulsively punches him in the nose, the title of the film drops ironically across her face: Miss Congeniality. Twenty years later, the little girl has grown up to become an FBI agent, and the movie continues to explore her twin impulses to protect—and to punch. The story of the little girl and the bully has both caught our attention and laid the groundwork for the film's main ideas.



**EXERCISE 9.1** **Comparing Introductions**

Pick a topic, such as your favorite food or the idea that success is 10% talent and 90% hard work, and write an introduction that moves from the general to the specific. Then write another introduction to the same topic in which you tell a short story. Does one of the introductions seem more successful than the other? Explain.

## MORE INTRODUCTIONS

Many students like to begin an essay with a **question to the reader**, such as "When was the last time you went on a picnic?"or "Do you like to snowboard?" There are some risks involved with this type of opening, however. First, the question often requires such a simple answer—perhaps "last summer" or simply "yes"—that the reader may not be interested enough to continue reading. Second, the reader may be more interested in the *writer's* point of view and thus not welcome a shift toward her own. In the end, this type of question may ring false since the reader's response cannot really be shared with the writer.

With that said, it is also true that some questions can make thought-provoking and effective introductions to your topic. A *good* question will be intriguing and open-ended, arousing the reader's curiosity and requiring a more complex and thoughtful answer. "What would make for an exceptionally unusual and delightful picnic menu?" or "Why might someone prefer snowboarding to skiing?" Note that questions do not need to use the dramatic and potentially intrusive word *you*. The example that follows is from an essay about the movie *Tortilla Soup*.

**STUDENT WRITING** Na Yeon Kim

What is a tortilla soup? As I'm not familiar with Mexican food, except for some representative dishes such as fajita, burrito and quesadilla, "tortilla soup" was a big question mark in my mind. My curiosity about the title first caught my interest; then I fell into the movie, which had such attractive and fruitful cooking scenes from the beginning to the end. The introductory cooking procedures were the most impressive scenes to me because they were powerful and vivid descriptions of Mexican food.

A related type of introduction presents a **problem to be solved**, such as improving your study habits or making a low-fat cheesecake. In the film *Galaxy Quest*, a hilarious spoof of the 1960s cult favorite *Star Trek*, the first scene reveals a number of problems facing the main characters, the cast of a television series that was cancelled seventeen years earlier. For example, Tim Allen's character respects neither his co-workers nor his fans, and he has failed to earn the love of the woman he admires (Sigourney Weaver). Alan Rickman is frustrated by the non-Shakespearean nature of his role (which he reprises at "Questerian" get-togethers), while Guy is frustrated that he had only a bit part in a single episode of the show—and no last name! All of the characters lack regular employment. When real-life aliens appear on the scene, they set off a series of events that lead eventually to solutions to these problems.

In the essay quoted below, the writer explores ways in which chefs can help with food-related problems. The introduction begins with the idea that "every occupation entails some sort of civil duties" and leads to a focus on those problems it might be a chef's duty to address.

**STUDENT WRITING** Justine A. Frantz

Every occupation entails some sort of civil duties. Some of these duties are quite obvious, such as a soldier's duty to defend his or her own country; however, other jobs include dealing with less obvious societal issues. Ordinary people may not think that chefs have much to do with society other than feeding the general public, but there are more responsibilities in being a chef than meet the eye. Food-related problems are rapidly increasing world-wide, and who better to help conquer these issues, such as obesity, eating disorders, and world hunger, than people who have a strong passion for and understanding of food.

3

Although introductions can take many different forms, they share a desire to catch a reader's attention and move her toward the subject of the paper.

These instructors outline some of a chef's responsibilities in relation to meat, one of the more expensive and potentially dangerous items on the plate.

## WRITING AND REWRITING THE INTRODUCTION

Since the introduction is the first part of the essay, it seems reasonable that it should be the first part we write, and that it should be all clean and tidy before we address the next part of the essay, the body. In fact, though, the opposite may be true. Although we might be tempted to work on the introduction extensively before straightening out the body of the essay, the introduction is often the last part of the essay to reach its final form. It makes sense that once we have found exactly where we're going, we can more effectively write an introduction that points the reader in that direction.

In writing the first draft of the essay, it is important not to get stuck on the introduction. Either breeze through and get to the body, or—if you're really stumped—jump over the introduction completely for the moment and write whatever part of the essay you can. If you labor too long over some part of the first draft, you often use up precious time. Once you have something down on paper—even if it seems weak—you'll be able to rewrite and reorganize. If you can't get anything down for an introduction, it works better to move on to another part of the essay, at least temporarily. Once you're satisfied with the body of the paper, go back and experiment with some of the different kinds of introductions discussed earlier. The chances are good that you'll find one that works.

Look at this rough draft of an introduction to a personal narrative.

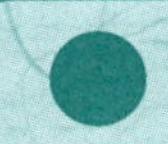

**STUDENT WRITING** Charles A. Dunn

I remember that one moment that changed my life. The day began like so many others before it. This one, however, was somewhat needful after juggling work and school all week.

This paragraph introduces the topic—the moment that changed the writer's life—but not much else beyond the stress suggested by "juggling." The story itself is quite dramatic, with serious and far-reaching consequences. After exploring that story, the writer revised his introduction, as in the following example.

**STUDENT WRITING** Charles A. Dunn

I vividly remember that one moment that changed my once mischievous life. It was two days before Christmas, so the atmosphere was cheerful. This day, however, ended up somber.

In the first sentence, the addition of three words brings a vitality and charm to this revised introduction. The memory is "vivid" and the writer "once mischievous." This new paragraph also sets the time of year at "two days before Christmas" and builds some suspense. Why, the reader wonders, did the "cheerful" day end up "somber"? She's hooked.

### EXERCISE 9.2 Rewriting the Introduction

Pick one of the sample introductions in the first section of this chapter and rewrite it in a different way. For example, if the introduction told a story, rewrite it as a question or a problem. Then discuss the result.

## THE CONCLUSION

As we have said before, the conclusion is like the coffee and dessert that follow the meal. It does not contain the majority of the words and information needed to make the essay's point; that is the role of the body. Instead, the conclusion provides time and space for the reader to think about what has been said by summarizing or highlighting the main idea. If the introduction promised to inform the reader about different varieties of apples, for example, the conclusion would review that information, reminding the reader that the promise had been kept.

The other important function of the last paragraph is to let the reader know that the essay is ending. Remember our cartoon character from Figure 3.1? We don't want him to float gracefully into the story only to fall off a cliff at the end! Unlike the season finales of many television shows, essays do not end with a cliffhanger. Instead, they provide that cartoon character with a parachute and let him down gently.

Movies, too, generally wrap the story up at the end. Some films end with pictures of what happens to the characters, while other films summarize the characters' fate in lines of text, for example, *Remember the Titans* and *Apollo 13*.

Let's look at some examples of concluding paragraphs. This first one **summarizes the main points** of the essay.

### STUDENT WRITING René S. León

Sean Devine showed me his honesty by rejecting the invitation to the bar from a female officer. He showed his professionalism by not allowing his personal problem with Lauren to interfere with his job. He showed his calm and detached personality when he announced that the body in a hole in the park was his friend's daughter. He knew how to handle himself, and that is the thing that makes him my favorite character.

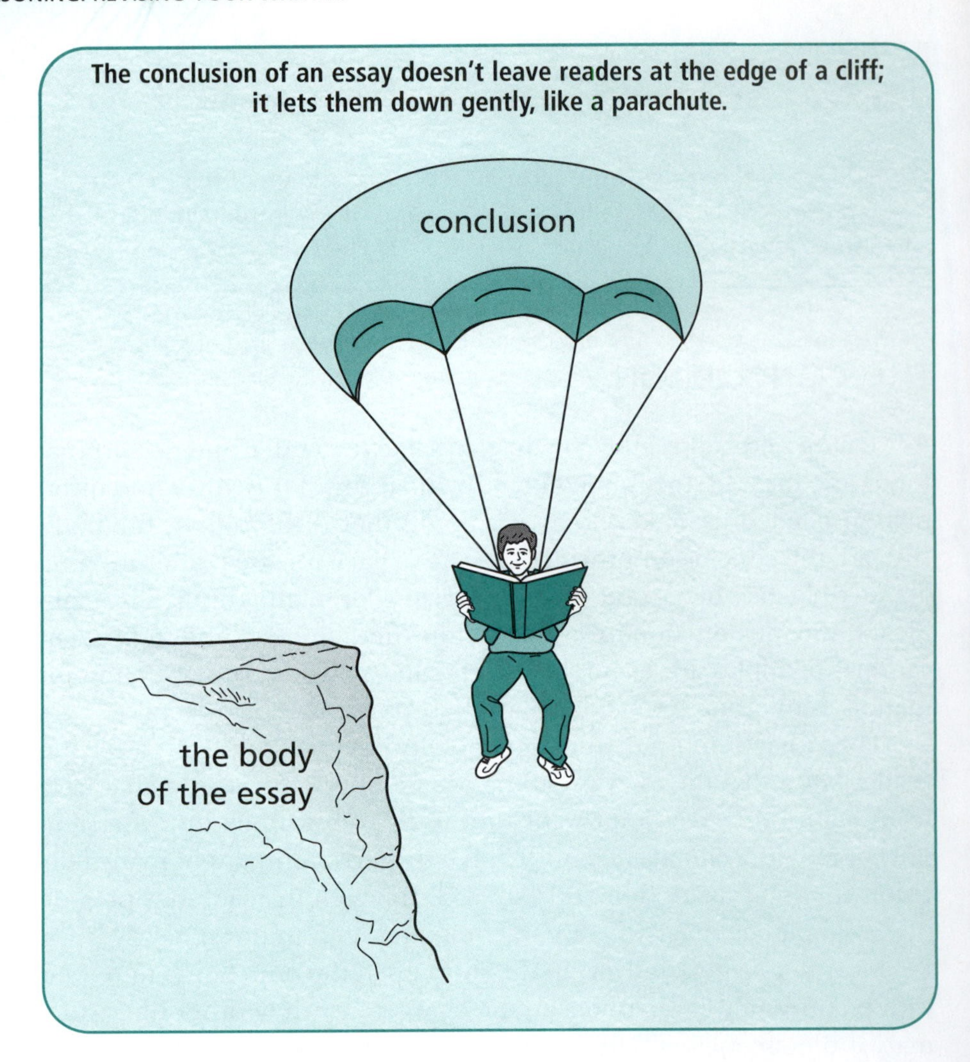


**The conclusion of an essay doesn't leave readers at the edge of a cliff; it lets them down gently, like a parachute.**

The next concluding paragraph summarizes the main points and **ends with a strong image**: "all of this *opened my eyes* to the art of Japanese food."

Adam Miller

I have always loved strange food, but sushi is my all time favorite, right down to the texture of the rice and the freshness of the fish. It is what got me interested in the culinary industry. The color, the design, all of this opened my eyes to the art of Japanese food.

In the following example, the writer sums up the main point of the movie *Crash,* then goes on to **state a new thought**.

**STUDENT WRITING** Mee-kyoung Kim

All these cases reflect a natural quality of human beings. Prejudice, selfishness and disenfranchisement make us lose our sense of judgment. Prejudice produces rage, and rage gives birth to tragedy. If nobody stops it, this vicious cycle goes on and on. This story goes beyond the people in the movie; they are us. We are all victims as long as we operate out of prejudice, selfishness and disenfranchisement. As the movie shows, those with one-track minds do not help us to live together. In the end, I came to know that love is the only thing to reduce the crashes and help us to live together in harmony.

In addition to highlighting the idea that "prejudice, selfishness and disenfranchisement make us lose our sense of judgment," the conclusion suggests an approach to ending that "vicious cycle": "love is the only thing to reduce the crashes."

Conclusions very often **explain why the topic is important** or **emphasize the importance of a particular point** in the essay, as in the example below.

**STUDENT WRITING** Michael P. Otting

The experience in the meat room was thrilling. It makes me want to do well in my other classes so I can get to meat-fabricating class. This tour also gave me some new routes to different parts of the school. But above all else, the tour showed me that it is important to pay more attention to detail in everything I do.

Sometimes the last page of a book or the last scene of a movie suddenly changes direction and **ends in a completely unexpected way**. Remember the famous twist in *The Sixth Sense?* It took most viewers completely by surprise. The twist was particularly successful, but it had more than shock value. It made sense of the plot, and it explained the "haunting" mood of the film. Twists can be effective in essays as well, especially when combined with a sense of humor, as in the example below.

**STUDENT WRITING** | Cheyenne Simpkins

Although I would be alone on the island, I wouldn't miss my complicated life. In fact, I would relish the solitude provided by the island. With a regular food supply of the mix plates, Mom's sweet potato pies, and the sunflower seeds, I could stay indefinitely—ironically becoming the only person to get fat while stranded on a desert island.

Many writers peter out when it's time to write the conclusion; they're simply exhausted! Don't give up, though. While the first impression made by the introduction is important, the conclusion provides an opportunity to leave a *lasting* one.

**EXERCISE 9.3** **Endings**

Choose one of the introductions you wrote in Exercise 9.1, and write two possible conclusions for it. Which one do you prefer? Why?

## TYING THE INTRODUCTION AND CONCLUSION TOGETHER

The conclusion in some way returns to the place where the essay began—but with a new perspective. Think first about the table in a fine restaurant. We sit down, looking at the immaculate tablecloth, the perfectly arranged silver, the spotless glasses. When we reach the end of the main course, the crumbs have been swept away, glasses have been refilled, and the table has been set for dessert. We're back to the place where we began, with a difference. Our stomachs are quietly full, our taste buds satiated. We have a glow of content as we anticipate the final summing up of the meal—coffee, brandy, dessert, cheese plate. As we begin that last course, we do so with the knowledge of what has gone before and with an awareness that the experience is coming to an end.

In the same way, the end of the essay looks back toward the beginning, performing a graceful arc within the reader's imagination.

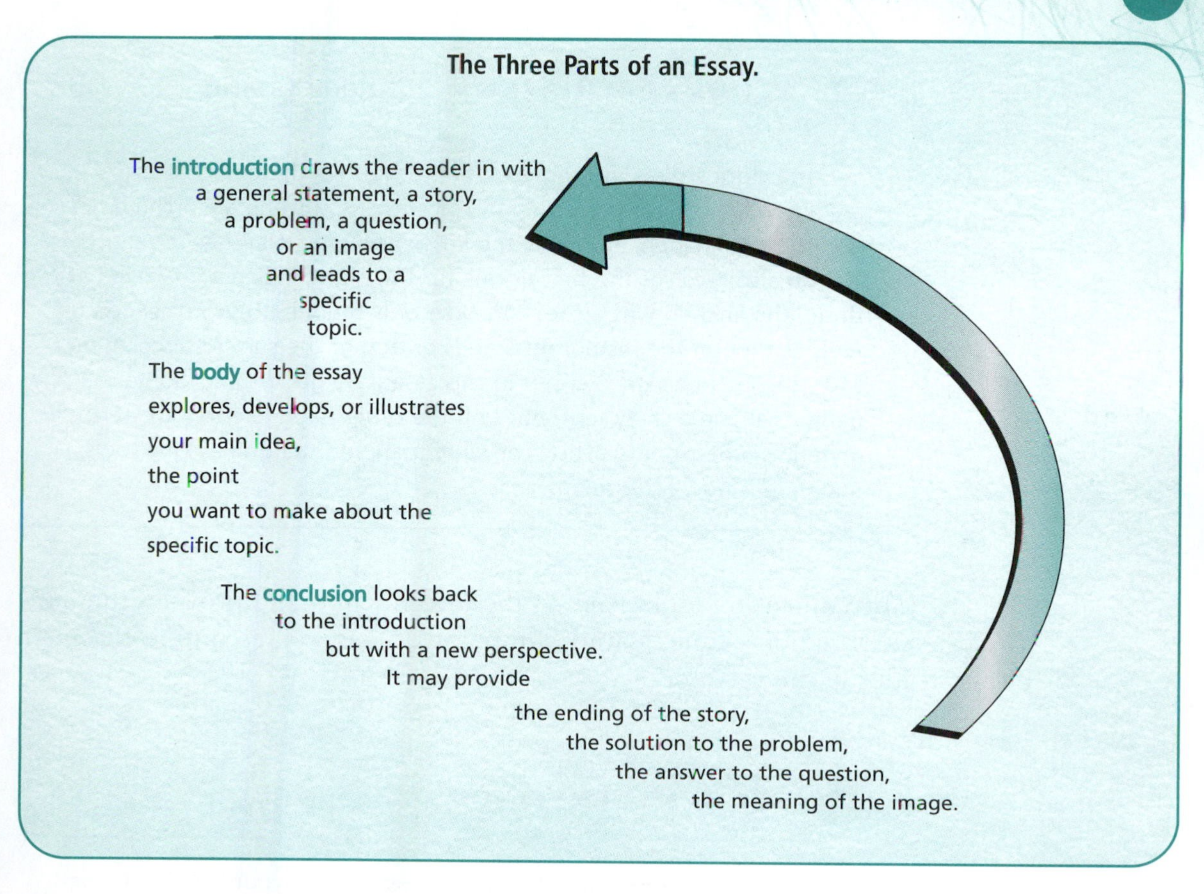


Conclusions are often especially effective when they clearly reflect some material in the introduction. Do not simply repeat the thesis statement, however! If the introduction told a story, let the conclusion allude to it. If the paper began with a quotation, it's nice to include or continue it, or in some way to refer to it again. A question in the first paragraph might be answered in the last, or a problem introduced in the beginning might be solved in the end. There are no hard and fast rules; just remember that the reader will be pleased to be reminded of where she has come from when she gets to the end of your paper.

In this first example, the introduction propels the reader into the action of a "scorching hot" kitchen in the middle of service. The second-person point of view and the present tense increase the sense of excitement.

**STUDENT WRITING** Anthony Guarino

The clock strikes six, and your first ticket hits the board. There is an eerie silence until the sound of cold raw fish hitting a scorching hot grill fills the kitchen. Suddenly the kitchen fills with light as the sauté station also receives the first tickets for the night. When you arrived at three, the kitchen was a cool 80°. Now only ten minutes into service, it feels as though the restaurant stands on top of the sun. At a scorching 110°, the kitchen's on fire, but in this case it's a good thing. Some people call me a crazy fool, but I on the other hand like to look at myself as a passionate artist. The sauté pan and the grill are my tools, and here is how I use them.

The conclusion looks back to the introduction and provides the same information—time, sound, temperature—as well as hitting the main idea at the very end.

**STUDENT WRITING** Anthony Guarino

It's one in the morning, and you've just burned your last ticket. The silence sweeps the kitchen again. The sauté station is cool. Everyone has started to clean up, thinking "Thank God, it's over"—not because they don't love it, just because now they get the sweet rewards. "Good job, guys, we turned them over three times." Grilling and sautéing make everything so fast that time flies, but yet the sense of accomplishment is still so immense. This is why these are my favorite tools.

Sometimes the conclusion gently **twists the image created in the introduction**. In this next example, the health concerns raised at the beginning of the essay vanish in a smile at the end.

**STUDENT WRITING** Kanyalanee Jirattigalachote

When I was a child, I was slender because I did not eat a lot of food, and my mother worried about my health. I looked like a weak girl, compared with my sisters. I weighed just 19 kilograms [about 42 lbs.] when I

**The conclusion mirrors the introduction. . .**

| INTRODUCTION | CONCLUSION |
| --- | --- |
| the clock strikes six | it's one in the morning |
| there is an eerie silence until the sound of cold raw fish hitting a scorching hot grill | the silence sweeps the kitchen again |
| the kitchen's on fire | the sauté station is cool |

the sauté pan and the grill are my favorite tools

was in second grade. Therefore, my mother wanted me to eat a lot of food to gain more weight. She tried to find my favorite dish to make me eat a lot. Finally, she found that I loved her Tom-Yum noodles. . .

### Conclusion

. . . If I had not had my mother's Tom-Yum noodles, I could not grow up to be a fat girl like today!

---

Finally, the conclusion can *drive home the emotional impact* of a topic, as in the following example.

 **STUDENT WRITING** | Tammi Bertram

### Introduction

When I was eight years old, my parents got divorced. My sister and I chose to live with our father and only visit our mother. In the meantime my father had to instantly change from being not only Dad but "Mr. Mom." He then had to take on both roles of parents in the house. He had to wash clothes, clean, raise my sister and me, and worst of all, cook. He didn't know how to cook much, but every Tuesday he knew how to make the best macaroni and cheese. I will always remember this dish and its meaning. It's not always how you cook it; it might just be who you're eating it with that makes it tasty. . . .

### Conclusion

Despite the fact that my dad really couldn't make much of anything, he knew how to make the most memorable meal. He did it all by himself, without help from anyone. This made him grow without my mother, yet my sister and I grew by eating a simple meal with three people that loved each other, instead of four that didn't.

## EXERCISE 9.4 Tying the Introduction and Conclusion Together

Choose one of the sample introductions earlier in this chapter. Think about what the body of that essay might say. Then write a concluding paragraph that ties it all together.

The conclusion of an essay or a meal asks the audience a question: Was it worth it? At a restaurant, the value is partly financial since the customers are presented with a check. As they look over the figures, they will have a reaction, perhaps a sense of satisfaction that the meal was worth the money. An essay's conclusion also asks the audience to evaluate the experience. Do we nod our heads in understanding? Or shake them in frustration? Was the essay worth the time it took to read? By looking back at the introduction, the conclusion strengthens our sense that the whole paper hangs together—a very satisfying feeling. Remember the Hapuna Beach essay in Chapter 5? The introduction to that piece led us from the wintry landscape of the real world to the tropical paradise in the writer's memory. The conclusion was a mirror image—moving us from the sand back to the snow. It was somewhat of a sad ending emotionally, but a very satisfying one structurally and thematically.

Movies are especially well designed to create that satisfaction through the use of recurring images and musical themes. *Mystic River*, for example, begins with a view of the Boston neighborhood along the river's edge in which the main characters grow up. Two hours later, after telling a complicated story of abuse and revenge, the movie returns to the river. And now the river has a symbolic meaning in keeping with its name—it is into the river that Sean Penn has thrown the men he killed. Like the River Styx in Greek mythology, the Mystic River lies between life and death. In the last scene, the simple piano tune heard first at the start of the film begins again. The camera moves closer to the water, skimming for a moment along the surface, then plunges beneath it into darkness.

## RECIPE FOR REVIEW Introductions and Conclusions

### Introductions

1. The purpose of the introduction is to catch the reader's attention and explain the subject of the paper. The introduction may also map out the main supporting points.
2. Although it appears first in your papers, the introduction is often the last part to be written. Only when you know exactly where you're going in the body of the essay can you guide the reader in the right direction.
3. Introductions may use general statements that narrow to a specific focus; tell a short story that is related to the body of the essay; begin with a quotation, a question, an interesting fact, or a strong opinion; or outline a problem to be solved.

### Conclusions

1. The purpose of the conclusion is to highlight the main idea of the essay and to let the reader know that this is the last paragraph.
2. Sometimes the conclusion, like the introduction, cannot be written until several revisions have helped you think through

the topic. At other times the conclusion may be the first part of the essay you're sure of.

3. The conclusion may raise a question or add information about the main idea, answer a question, or outline a solution to a problem raised in the introduction. Further, if the introduction contains a story, image, or allusion that relates to the main point, it is satisfying to the reader if the conclusion mentions it again.

**Tying the Introduction and Conclusion Together.**

If you begin with an image,

go back to it at the end . .

but with a new perspective.

## A Taste for Reading

1. Which of the nine pieces in Appendix VII has the most effective introduction, in your opinion? Why?
2. How are the introduction and conclusion of David Mas Masumoto's "The Furin" (Appendix VII) tied together? What idea or theme is emphasized by this connection?
3. Read the conclusions of the pieces by M. F. K. Fisher, Galway Kinnell, and Jeffrey Steingarten in Appendix VII. How does each piece's conclusion relate to its introduction?

## Ideas for Writing

1. Choose a movie (or book) that you're especially fond of. Describe the first scene. Is it effective in drawing you in? Explain. Then look at the last scene. Is it effective? Does it echo the introduction in some way?
2. Choose a restaurant in which you've eaten recently and evaluate your first impression.
3. Think about the endings of popular songs—most of them simply fade out, right? Then compare the endings of classical pieces, such as Bach's "Toccata and Fugue in D Minor," which end in a single crashing chord. Describe the different effects of these two kinds of endings.
4. Write a paragraph or brief essay in which you explain what happens in Lucille Clifton's "Cutting Greens" (Appendix VII). What phrase in the first three lines is echoed in the last two? How does the experience in the middle of the poem change the narrator's attitude?

chapter

10

# Using Transitional Expressions

One of the major themes of this book has been the importance of communicating effectively with the reader, the importance of making your ideas clear, organized, focused. In this unit we have already talked about ways of organizing your ideas and ways of beginning and ending your essay. **Transitional expressions** help in communication by showing how your ideas are connected in terms of time, space, similarity or difference, and so on. Think Transit Authority. A transition is like a bus. The reader hops in and is driven to the appropriate destination. Like a passenger on the bus, the reader doesn't have to do the work of walking, nor does he run the risk of getting lost. He's delivered right to the door of the next idea.

Transitions have another important similarity to the public transportation system: When they're working, we don't even notice them. But when they're *not* working, or when they're missing altogether, we're left frustrated and fuming.

## IDENTIFYING TRANSITIONS

Transitional expressions are words and phrases that show the reader where and how to move from one idea to the next. For example, transitions such as *first, next,* or *after the water boils* might be included in a recipe to make clear the order in which the steps are to be performed. Transitions might also show a cause and effect relationship between ideas, such as *consequently, therefore,* or *as a result.*

Transit/ion.

While inexperienced writers often miss the importance of transitions, readers find an absent transition as obvious as a missing fork; they find the wrong transition as jarring as a piece of shell in the egg salad. Think about a restaurant meal. Perhaps you've begun with a cup of soup, a hearty pumpkin seasoned with nutmeg and ginger. You're eagerly anticipating the next course—a tender filet mignon with béarnaise sauce—but the wait extends to thirty, forty minutes. When it finally arrives, the steak is delicious, but the *transition* between the two courses has created a problem. Or think about the role of transitions in songs, perhaps in Queen's "Bohemian Rhapsody." A jump in tempo, a guitar riff, or a change in key can signal a new idea or a new mood. Similarly, expressions such as *suddenly* or *on the other hand* or *the next thing that happened* are used in writing to mark the beginning of a new idea or a new mood, and these transitions explain *how* the new idea is related to the old.

Movies and television can also give us some information about the importance of transitions. On a talk show, for example, the transition to a commercial is signaled very clearly. Often the host will turn to the camera and say, "We'll be right back." Sometimes the transition is heightened by the camera zooming out. Only then does the series of commercials begin. Television shows themselves are often written to accommodate commercial breaks. Perhaps there is a dramatic pause, a close-up on a tear-stained face, as in a soap opera. Or perhaps, as in many television movies of the 1980s, the action builds toward a climax, pauses on the brink, and—cut to commercial. All of these devices work, that is, they let the audience know a shift is coming from the story to a scene outside it. In contrast, some contemporary shows do not have a clear transition from story to commercial, whether this is because the writers don't like to insert an unnatural pause or because the advertisers don't want to give the audience a chance to change channels. Whatever the reason for it, the lack of transition can cause a painful jolt.

This is why film editors exist—to make the transition from one scene to the next comfortable and clear. A fade-out/fade-in is smooth and pleasant, while a cut from the telephone in one house to the telephone in another house seems logical. The fact remains that transitions are themselves an ingredient of any story or essay, and what we do with them has a profound effect on the audience's understanding of and response to what we've written. Let's consider a final culinary example: When a colleague returned recently from an extended visit to restaurants and wineries in northern California, she reported that the best experience was not about the food, but about the sensitivity of the service: "The pacing of the *transitions* between courses was even and therefore pleasant."[15]

## TYPES OF TRANSITIONS

Transitional expressions explain how ideas are related in a number of ways. Some transitions have to do with **time**, for example, when an event takes place or when one event occurs in relation to another. You've used many transitions relating to time, especially when telling a story.

 **STUDENT WRITING** | Brendan Cowley

One day I step into the walk-in and notice a primal cut of beef that still has the head on it. As soon as I see this, the gears in my mind shake the dust off and start to turn.

---

The phrase "as soon as" indicates that the gear-turning occurred immediately after the writer saw the primal cut. Other transitions have to do with **space**.

In this well-run kitchen, the students make smooth transitions through the stages of the cooking process as they prepare for service.

**STUDENT WRITING** Matthew K. Greene

A three-part sink lined one whole side wall of the kitchen, with containers and storage units placed neatly on a shelf above. **On the opposite wall** there were two refrigerators, one walk-in with food stored for the whole week, and a regular fridge with items that were used throughout the day.

Some common transitions **introduce an example** that illustrates a concept or fact. The phrase "the best example" in the sentence below connects the idea of "a politically engaged chef" with a specific personality.

**STUDENT WRITING** Payson S. Cushman

**The best example** of a politically engaged chef is Alice Waters of Chez Panisse. Her movement to foster the farm-restaurant connection, as well as her programs to provide school lunches and teach children the concept of sustainable agriculture, illustrates the political potential of chefs in America today.

A final example of transitions concerns **moving from one item to a dissimilar one**.

**STUDENT WRITING** Elizabeth Best

The texture of the peach is very juicy. The juices run down the hands and face as you bite into it. It is sticky, reminiscent of a sorbet. **In contrast,** the apricot is noticeably drier. It has a very creamy feel in the mouth. I like to compare it with custard, because of the rich, full mouthfeel.

Study the list of common transitional expressions in Figure 10.1.

**FIGURE 10-1** Common Transitional Expressions.

| TYPE OF TRANSITION | EXAMPLES |
|---|---|
| time | after, before, currently, during, meanwhile, now, once, since, then, until |
| space | above, around, below, here, next to, on the other side, there |
| examples | for example, for instance |
| comparison/similarity | both, in the same way, likewise, similarly |
| contrast/difference | although, but, however, in contrast, on the other hand, yet |
| sequence | first, second, third, next |
| cause and effect | accordingly, as a result, because, consequently, for this reason, therefore |
| addition or emphasis | also, in addition, besides, further, in other words, moreover |

3

## EXERCISE 10.1 Identifying Transitions

As you read the following paragraph from an essay about Hitchcock's *Rear Window*, list all the transitional expressions on a separate piece of paper.

In the beginning of the movie Lisa is pushing Jeff to take the relationship a step farther. However, Jeff is feeling skeptical about the true intentions behind Lisa wanting to move forward in their relationship. He feels as though Lisa is this perfect woman who lives in a perfect world. I think Jeff doesn't feel as if he could fit into her way of life, nor do I think he would want to. This puts Lisa into a position in which she feels as if she has to constantly try to prove herself to him in one form or another. For example, she has a good restaurant deliver a wonderful meal and a bottle of wine, and all Jeff is able to say is that everything is perfect as usual. Throughout the meal it seems as though Jeff can only put down or challenge everything Lisa says. It almost seems as though he is setting up to break up with her.

Justin Henning

## USING TRANSITIONS

We use transitional words all the time—within a single sentence and between sentences. The examples in the previous section are of this kind. Try them yourself in Exercise 10.2. (See also the exercises in Chapter 21.)

### EXERCISE 10.2 Using Transitional Expressions

Choose appropriate transitional expressions from the list in Figure 10.1 to connect the ideas in each of the sentences below.

1. __________ you add the flour, be sure the butter and sugar are thoroughly blended.
2. Jeremy had a sunburned nose __________ he had stayed in the sun too long.
3. Rose was named after a flower. __________, Stella was named after a family friend.
4. __________ the meal was not quite ready, the children were sitting expectantly at the table.
5. __________, wash and peel the carrots. __________, dice them into a saucepan.

When transitions are missing from a piece of writing, we see more clearly the important role they play. Read the paragraph below, adapted from an essay by Soyang Myung.

**STUDENT WRITING** Soyang Myung

Celeste and Annabeth were housewives. They had a different belief about their husbands. Celeste had a distrust. Annabeth had a strong loyalty for her husband. Celeste's husband David came home with a bloody T-shirt. She was embarrassed. It was discovered that Annabeth's daughter had been murdered. Celeste began to suspect her husband was the killer. She made a mistake. She confessed to Jimmy (Annabeth's husband) what she thought without any objective evidence. In the case of Annabeth, she heard that her husband had killed David. She was not agitated. She was stunned. She remained firm. She consoled her husband. That showed us that their beliefs about their husbands were contrary to each other.

---

Without transitions, the reader has to stop and struggle a bit to follow the *connections* between ideas. It can be done, but it uses up precious time and patience on the reader's part. It's like having the customer cook his own dinner at a restaurant. Now read the same paragraph with the original transitional expressions restored (in bold type).

**STUDENT WRITING** Soyang Myung

Celeste and Annabeth were **both** housewives, **but** they had a different belief about their husbands. **While** Celeste had a distrust, Annabeth had a strong loyalty for her husband. **For instance, when** Celeste's husband David came home with a bloody T-shirt, she was embarrassed. **The next day, when** it was discovered that Annabeth's daughter had been murdered, Celeste began to suspect her husband was the killer. She made a mistake. She confessed to Jimmy (Annabeth's husband) what she thought without any objective evidence. **On the other hand**, in the case of Annabeth, *when* she heard that her husband had killed David, she was not agitated. **Even though** she was stunned, she remained firm; **furthermore**, she consoled her husband. That showed us that their beliefs about their husbands were contrary to each other.

---

Note how the transitions lead the reader smoothly from one idea to the next and tie the whole paragraph together. Further, the transitions add rhythm and variety to the sound of the sentences.

## EXERCISE 10.3 Adding Transitions to a Paragraph

Rewrite the paragraph below, adding five transitional expressions to make connections clear and tie the paragraph together.

The restaurant was in a medium-sized building with two floors and fits one hundred people. The feel of the restaurant was warm and exciting. The dining room had beautiful pictures on the walls and vibrant colors of green and orange on the ceiling. I walked in. I felt as though I was in Mexico. Music of Spanish guitars was playing throughout the building. It was a loud restaurant due to all of the customers. I was able to make conversation with the other patrons.

Taylor Jones

As useful as transitions are, there are times when they are overused or, more often, when the same one is repeated so frequently that it has the effect of clouding rather than illuminating the connections between ideas. Three of the worst offenders are *and, so,* and *then.* Read the paragraph below, again adapted from an essay by Soyang Myung.

**STUDENT WRITING** Soyang Myung

Celeste and Annabeth were both housewives, **and** they had a different belief about their husbands. **So** Celeste had a distrust, **and** Annabeth had a strong loyalty for her husband. **So** when Celeste's husband David came home with a bloody T-shirt, she was embarrassed. **So** the next day it was discovered that Annabeth's daughter had been murdered, **so then** Celeste began to suspect her husband was the killer, **and** she made a mistake. She confessed to Jimmy (Annabeth's husband) what she thought without any objective evidence. **So** in the case of Annabeth, when she heard that her husband had killed David, she was not agitated. She was stunned. **Then** she remained firm, **and** she consoled her husband **so** that showed us that their beliefs about their husbands were contrary to each other.

When we compare this version to the original paragraph reprinted earlier, we find that the simple and repetitive *so's* and *and's* lack the meaningful connections provided by *while* or *for instance.*

## UNIFYING AN ESSAY

Transitions play a similar role in full-length essays. They show connections *between* paragraphs, as well as within them, and keep the reader focused on the main idea. In the essay diagrammed in Figure 10.2 (see Chapter 15 for the full text), the theme of Quoyle's "brokenness" and

FIGURE 10-2 Transitions in David Wilson's essay on *The Shipping News*.

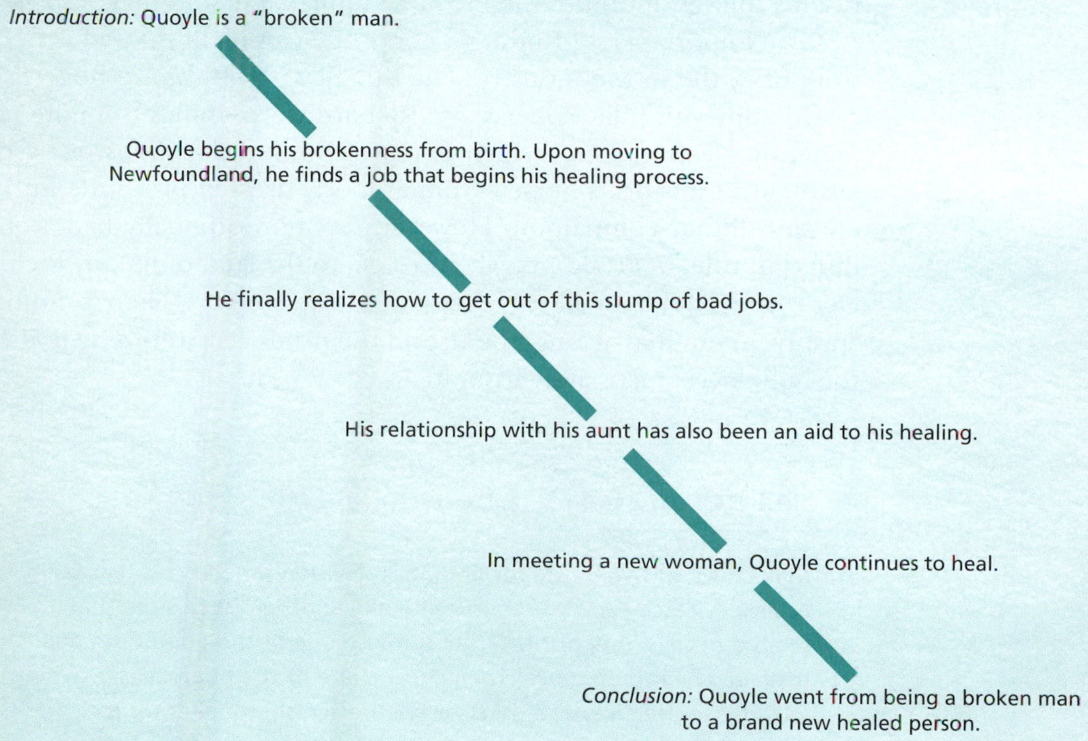

*Introduction:* Quoyle is a "broken" man.

Quoyle begins his brokenness from birth. Upon moving to Newfoundland, he finds a job that begins his healing process.

He finally realizes how to get out of this slump of bad jobs.

His relationship with his aunt has also been an aid to his healing.

In meeting a new woman, Quoyle continues to heal.

*Conclusion:* Quoyle went from being a broken man to a brand new healed person.



his subsequent healing, which is introduced in the first paragraph, flows through the essay like a stream.

Transitional expressions and the repetition of key words and phrases are two ways that writers connect ideas and tie their paragraphs together. When you're editing the final draft of your essay, read it carefully to judge whether you have the correct balance of transitions: enough to make the essay clear, smooth, and unified, but not so many as to make the writing seem stilted and artificial. You don't want to be a waiter that *hovers.*

## TITLES

The **title** of an essay is like your very first transition: it moves the reader from the outside world into the world of your ideas. The title usually refers to the topic of the paper and is designed to attract the reader's attention, often with a play on words or a vivid image. The title of the movie *Tortilla Soup*, for example, places the focus immediately on food, in fact, on Mexican food, as does the very first scene. Tortilla soup is also on the menu at a climactic dinner. By the end, tortilla soup represents the resolution of the movie's familial and cultural conflicts.

The importance of finding a good title can be illustrated with this story from the movie industry. The title of *Pretty Woman*, a hugely successful film with Julia Roberts and Richard Gere, comes from the popular Roy Orbison song, which is played while Julia shops on Rodeo Drive and transforms herself from a poorly dressed prostitute into an elegant dinner companion. However, this film originally had a quite different title, *3000 Dollars*, which refers to the sum of money Richard Gere offers Julia to be his "date" for the week.[16] There's nothing "pretty" about that arrangement, and the producers were wise to divert the audience's focus away from it.

**EXERCISE 10.4** **Choosing a Title**

Make a list of five movie (or book or television series) titles that seem especially catchy and correct. Then make another list of titles that seem especially dull or off target. Choose one of the good titles and explain why it is effective. Then choose one of the poor titles, explain why it is ineffective, and think of two better alternatives for it.

Single words often make effective titles, particularly when the word is part of the essay's main idea. Look at the title and introductory paragraph below.

**STUDENT WRITING** Roth Perelman

### Lost

Dreams, goals, or aspirations are something one strives to achieve throughout life. This is about my dream as a young man of nineteen and how I came to the realization that I would not be able to attain my dream without sacrificing the life I wanted. Few people who have met me after June 2003 know that I have not always aspired to be a chef. As long as I can remember, I wanted to be a Navy SEAL, "the best of the best" on the front lines fighting for my country.

As the essay continues, we learn that the writer's dream of a career with the SEALs was "lost" because of an injury. The title foreshadows the essay's main point.

In the next example, the title combines an image with a little word play.

**STUDENT WRITING** | Winston G. Caesar

### Garlic: The Sprouting of My Career

When I was young, I would love watching my mother cook. My job was to help get her ingredients for the family meal she was about to prepare. I was too young at the time to reach into the cupboard, so she would lift me up and I would grab the ingredients she needed. One day when I was helping my mother in our kitchen, she lifted me up and told me to reach for the garlic. I found a bulb of garlic and grabbed it, but to my surprise I saw another one that had started to sprout!

---

The author's interest in food and cooking "sprouted" along with the garlic bulb.

Another playful title is that of an early Wallace and Gromit film, *A Close Shave,* which refers both to the literal "shaving" or shearing of the sheep and, figuratively, to the characters' near escape from disaster. Titles may contain other elements of playfulness. In Chapter 13, you'll read part of an essay called "Oil and Water Don't Mix, or Do They?" by Matt Berkowski that explores the fragile nature of hollandaise sauce. These titles are clever and effective. In our academic writing, though, we probably wish to avoid such extreme examples from film as *Don't Be a Menace to South Central While Drinking Your Juice in the Hood* or *To Wong Foo, Thanks for Everything, Julie Newmar.*

Cookbook titles also vary from the plain and simple to the playful. An issue of *Food Arts* (December 2005) reviews these straightforward titles: *Cheese: A Connoisseur's Guide to the World's Best, The Herbal Kitchen: Cooking with Flavor and Fragrance,* and *Arnaud's Restaurant Cookbook: New Orleans Legendary Creole Cuisine.* In the same issue, *A Chef for All Seasons* borrows its title from a literary masterpiece, *A Man for All Seasons,* while *Eat This Book: Real Kitchen Recipes for Everyday Occasions* and *Crave: The Feast of the Five Senses* are simply fun. If you haven't thought much about titles in the past, consider the huge success of *The Joy of Cooking* and its spinoffs.

The best titles may become a symbol for the audience's experience of the story. The movie *Crash,* for example, reveals the origin of its title in the very first scene. By the end of the film, however, the word *crash* has been enriched by the audience's deepening understanding of the original quote. After the film is over, the title acts as a reminder both of the central *image* of the film—the car crash—and of the impact of its poignant stories on the viewer.

**EXERCISE 10.5** **Naming an Essay**

Create a title for the untitled story by Brendan Cowley in Chapter 4. Explain your choice.

## RECIPE FOR REVIEW Using Transitional Expressions

### Transitions

1. **Transitional expressions** are words and phrases used to show the connection between ideas.
2. Transitions may relate to time, space, examples, comparison and contrast, sequence, cause and effect, and emphasis. See Figure 10.1.

### Titles

1. **Titles** should relate to the topic of your essay and catch the reader's attention.
2. A good title will emphasize a particularly important or attractive point about your paper.

## A Taste for Reading

1. Read one of the pieces in Appendix VII, and discuss the use of transitional expressions.
2. Explain the meaning and discuss the effectiveness of the titles "Borderland" and "Hot Dog" in Appendix VII.

## Ideas for Writing

1. Make an outline or a diagram of the main ideas in one of the student essays in Chapters 3 through 5. Include transitional expressions.

2. Write a brief essay in which you describe the restaurant, bakery, or café you'd like to open. Choose a name for the property and defend the name's effectiveness.

3. Evaluate the titles of these shows on the Food Channel: *Pastry Daredevils, How to Boil Water, Semi-Homemade Cooking, 30-Minute Meals, Boy Meets Grill, Restaurant Makeover, Recipe for Success,* and *Good Eats.* Which are the most interesting? Why? Which are unappealing? Why?

4. Write a paragraph in which you explain the meaning and evaluate the effectiveness of one of the following movie titles: *Waiting, Sideways, Scarface, Tootsie.*

chapter 11

# Revising Word Choice

Once we've worked on revising the content and organization of a piece of writing, and on the transitions that guide a reader through it, we begin to look at the structure of individual sentences (see Units 6 through 8) and at the accuracy and effectiveness of individual words. The search for the best word is like the search for the best seasoning. Experimenting with different words and flavors is both constructive and entertaining, and the right decision can change an ordinary sentence or dish to an extraordinary one. Look at the sentence below:

> I saw a student trying very hard to remove excess fat from a pork tenderloin while another student tried to tie up a strip loin of beef.

Although this sentence is clear and correct, look what happens when another word is substituted for *tried* in the second line.

**STUDENT WRITING** Nelson Tsai

> I saw a student trying very hard to remove excess fat from a pork tenderloin while another student *battled* to tie up a strip loin of beef.

"Another student battled to tie up a strip loin of beef."

Using a strong verb such as *battled* adds important information, creates a sharp mental picture, and invigorates the text.

Choosing the right word is also about expressing your personality. Sometimes students misinterpret the purpose of a writing course as somehow to subtract their personality from their writing. In reality, the reverse is true: A writing course should make your written communication more effective in expressing both your thoughts and your personality. Choosing the best word is about choosing what suits the purpose and audience of your communication; it's about using words efficiently and not wasting the reader's time with extra ones; and it's about experimenting with words in a fresh and interesting way rather than trotting out tired old clichés like "it was cooked to perfection" or "it melted in my mouth" or "it was divinely decadent."

## CHOOSING THE RIGHT WORD

Looking for the best word can be fun, and with a thesaurus right there on the computer, it's easy to find all kinds of synonyms. Yet this search can also seem rather daunting, especially because English has nearly a million words to choose from! Was that steak you had last week *tender, juicy*, or absolutely *luscious*? Were you *astounded, flabbergasted*, or simply *surprised*? Would you describe yourself as *obstinate, resolute*, or merely *stubborn*? Check your computer's thesaurus under *stubborn* and you'll find quite a few alternatives: *obdurate, dogged, pigheaded, mulish, inflexible, willful*, and *intractable*. You'll find even more synonyms in a print thesaurus. The following words are also synonyms for *stubborn*, but with more positive associations: *persistent, resolute, unwavering, steadfast*.

We can think of words as having two levels of meaning. The **denotation** of a word is a simple definition. The italicized words that follow *stubborn* in the previous paragraph share the same denotation. However, they do not share the same **connotation**, the feelings or associations that make a word seem positive or negative, or simply neutral. *Obstinate*, for example, suggests that the stubbornness is unreasonable; it has a negative connotation. In contrast, *resolute* suggests that the stubbornness is brave, that someone is holding fast to a principle or a position even when it's dangerous to do so. It has a positive connotation. Or think about the difference between *aroma* and *stench.* An *aroma* sounds positively delicious, like the odor of a freshly baked pie or an expensive perfume, while a *stench* would more likely be used for dirty socks or rotting meat.

As you revise your writing, it is good to experiment with new words. Use the thesaurus to find interesting alternatives. However, then look them up in the dictionary to confirm that *both* the denotation and connotation fit your purpose. We don't want to confuse or offend our audience with a poor word choice. For example, you wouldn't make a good impression on a date's family by exclaiming over the "stench" coming from the kitchen! Try to select words whose connotations communicate your intended meaning accurately and effectively.



### EXERCISE 11.1 Denotation and Connotation

Look up the following words in a dictionary: *inflexible, obstinate, persistent, pigheaded, resolute*. Explain the different *connotations* of each word. Then, on a separate sheet of paper, rewrite each sentence, inserting the appropriate word.

1. The boy was ___________ in his refusal to go to bed despite being tired.
2. The activists were ___________ in continuing their peaceful protest, even when threatened with arrest.
3. The dedicated student was ___________ in finishing her homework.
4. In *Cool Hand Luke*, the ___________ title character is determined to eat fifty eggs, even though it nearly kills him.
5. Understanding teachers will not be entirely ___________ about due dates.

## Revising Slang, Jargon, Clichés, and Sexist Language

In order to communicate effectively with your audience, you must use words that the audience understands. Sometimes communication is blocked because we're essentially speaking a different language. For instance, it would be difficult to persuade a restaurant customer to order a particular item if the two of you didn't speak the same language or if you used such specialized technical terms that your customer didn't know what you meant. Further, it would be difficult to impress a customer with your chicken with peanuts stir fry if she had a peanut allergy. We want to use ingredients of food and of language that our audience will both understand and appreciate. In general, therefore, we want to avoid slang, jargon, and sexist language.

## Revising Slang Expressions

The term **slang** refers to words or phrases that have become popular within a certain group of people—teenagers, for example, or computer programmers, or restaurant workers—but which may not be recognized by all readers. Would the following terms be clearly understood by all age groups?

> I've got to hunker down and swot for that wicked hard test.

Probably many readers would need the following translation:

> I need to get to work and study for that very hard test.

Readers of different generations, even when they share the same profession, may not always share the same slang. You're probably familiar with some American diner slang from the middle of the last century, phrases such as *Adam and Eve on a raft* for two poached eggs on toast. In fact, some diner slang has entered the general vocabulary, such as *a cup of joe* for "coffee" or *86'd* for "out of." Yet would all food service workers today understand the following slang expressions? (Translations can be found at the end of the chapter.[17])

a. shingle with a shimmy and a shake

b. bossy in a bowl

c. Mike and Ike

d. dog and maggot

e. customer will take a chance

Slang expressions can be especially difficult for non-native speakers. One student from Korea was horrified to hear passersby talking about "da bomb" on a Los Angeles street. In this post-9/11 world, she immediately thought of a terrorist attack. Imagine her surprise and relief when an American friend explained that the passersby had simply been praising the attractiveness of the girl walking ahead of them! In any successful communication, we want to avoid using words that might be unclear or confusing to our readers. Therefore, as you revise your rough drafts, consider your audience, and replace slang expressions with vivid, concise, but more formal language.

**EXERCISE 11.2** **Revising Slang Expressions**

Rewrite the following sentences to avoid using slang expressions.

1. Carrie was jonesing for a cig.
2. We had to give her props on the bling.
3. Hey, dude, I need to borrow the porcelain.
4. The amputated foot in *Saw* made the audience want to hurl.
5. Girl, your *tude* is a real buzzkill.

## Avoiding Jargon

Another type of vocabulary that can be hard for some readers to understand is **jargon**, the technical words specific to particular jobs, professions, or specialties. If your readers aren't familiar with a particular job, they may have difficulty understanding its specialized vocabulary. Even when writing for people in the same profession, it is often best to use direct and ordinary language rather than to rely too heavily on jargon. Again, think about your audience. If you must use a technical term—if it's the only word for what you need to say—go ahead and use it, but define it for the reader. Figure 11.1 lists examples of jargon from a variety of fields.

Writers will sometimes use jargon intentionally to establish an atmosphere or a character. On *CSI*, for example, there's a lot of talk about *DB*'s, *DOA*'s, and *vic*'s. In works of fiction, these abbreviated expressions are appropriate; in an academic essay, however, you would most likely spell out "dead body," "dead on arrival," and "victim." Like slang expressions, jargon also may move into the general vocabulary. For example, we have "front-runner" from horse racing, "overdrive" from the automotive industry, and "touch base" from sports. Jargon can also easily become clichéd—so common as to lose most of it meaning. In the worst cases, jargon is used to impress or intimidate readers, or

**FIGURE 11-1 Examples of Jargon.**

| Specialty | Examples of Jargon |
|---|---|
| computer programming | Boolean operators |
| | HTML |
| | TCP/IP protocol |
| medicine | normal sinus rhythm |
| | sedimentation rate |
| | tension pneumothorax |
| wine | sommelier |
| | quaffer |
| | legs |
| culinary arts | AP weight |
| | salamander |
| | persillade |
| business | strategic relationships |
| | core competencies |
| | benchmarking |

to avoid an unwelcome truth. It's like an artificial smile rather than a warm handshake. The corporate world, for example, is fond of inflated language like "strategic relationships," "core competencies," and "benchmarking." Despite their impressive appearance, the precise meaning of such terms is not always clear.

Related to this kind of jargon is a type of word choice called **euphemism**, a watered-down term used supposedly to spare the reader's feelings when uncomfortable things like death or sex must be discussed. And, indeed, many people prefer to say someone "passed away" rather than "died" or to say "went to bed" rather than "had sexual intercourse." Yet euphemism too easily becomes a tool of *mis*communication. Think about the term "downsizing". It doesn't sound too threatening, maybe Grandma moving from the big, old house in which she raised five children to a condo in North Carolina. However, it really means *firing*—people losing their jobs and perhaps the ability to own any kind of home at all. This emphasis on what sounds "positive" rather than on what will accurately represent the facts is especially evident in the language of business, politics, and the military. While the phrase "surgical strike," for example, sounds clean, even life-saving with that reference to medicine, it actually refers to bombings in which many people died. The truth is best told in plain and simple language.

### EXERCISE 11.3 Avoiding Jargon

Rewrite the following sentences in clear, ordinary English. You may wish to consult a dictionary.[18]

1. The powder hound went off piste and ended up in the blood wagon.
2. The department implemented a new initiative designed to enhance investment in human capital.
3. Roger that. ETA ten minutes.
4. Jack described the wine as complex and supple with a long finish.
5. Derek received the following flame: Check the FAQ or RTFM.



## Avoiding Clichés

In the introduction to this chapter, we singled out the expressions "cooked to perfection" and "melted in my mouth" as **clichés** that are less effective in our writing than fresh, clear, and specific language. Although we often assume that clichés are easily understood by our readers, they may actually be both confusing and boring. Study the following example:

> The steak was cooked to perfection.

What does that phrase really mean? To you, "cooked to perfection" may mean bleeding onto the plate. To your dinner companion, it may mean so well done it would make a thud instead of a splooshy splat if you dropped it on the floor. Isn't the sentence below more specific?

> Grilled over a fragrant wood fire, the steak had a slightly crisp outside and a tender, deep pink inside.

You can often recognize a cliché by its fill-in-the-blank nature. Complete the expressions below:

Quiet as a __________

Hungry as a __________

Last but not __________

Suppose you wanted to describe a loyal friend *without* using a cliché. Instead of saying "Karen is a loyal friend: she stood by me *through thick and thin,*" you could give a specific, real-life example. "Karen is a loyal friend: she called me every night for two weeks after my boyfriend broke up with me." As with all your writing, be clear and direct. Avoid clichés; using such stale language is like serving store-bought muffins instead of fresh baked.

**EXERCISE 11.4** **Avoiding Clichés**

Rewrite the following sentences, replacing any clichés with fresh, specific language.

1. That awesome chocolate just melted in my mouth.
2. The line cook remained as cool as a cucumber throughout the dinner rush.
3. That old restaurant is going down the tubes.
4. The movie heroine just had to explore that dark basement, and you know what they say about curiosity killing the cat.
5. Jerry thought his new job would be better, but it was just out of the frying pan and into the fire.

## Revising Sexist Language

**Sexist language**, though clear enough, is no longer acceptable in formal written communication. Further, it may alienate the portion of your audience that has been left out. You wouldn't think of excluding women from your restaurant, so don't exclude them from your language. While in the past it was conventional to use *he* to refer to both men and women, writers now prefer to use *he or she*, or to use neutral plural forms such as *they* or *people*. In a longer work, such as this textbook, half of the chapters may use the pronoun *he* while the other half will use *she*. Traditional titles such as *chairman* can be rewritten neutrally as *chairperson* or, more simply, *chair*. Often the gender-specific suffix *-man* can be replaced by the neutral *-er*: *firefighter*, *police officer*, *worker*. Don't assume you know a person's sex just because you know his or her job description. Remember that men, like Ben Stiller's character in *Meet the Parents*, can be nurses, too.

### EXERCISE 11.5 Revising Sexist Language

Rewrite the following sentences to eliminate sexist or gender-specific language.

1. Each student needs to bring his book to mathematics class tomorrow.
2. The discovery of penicillin was a benefit to all mankind.
3. Every man should have the right to vote in his own country.
4. She was the chairman of the Academic Standards Committee.
5. Each new chef carves his initials on his knives.

Although slang, jargon, and sexist language are inappropriate in academic writing, they can be effective in other types. Word choice reflects background and personality; thus, slang, jargon, and sexist language can be used in fiction to reveal something important about a character. In journalism, too, vivid slang expressions make good quotes.

> "I was totally in the weeds last night, but chef just told me to man up," said one of the restaurant's line cooks.

As always, the purpose and audience of a piece of writing determine what type of vocabulary is most effective.

## ELIMINATING EXTRA WORDS

As we've said many times, when you're writing a rough draft, you're not thinking critically about exactly how you're expressing the ideas; you're just trying to get them down on paper. The idea of eliminating extra words need not even cross your mind until the revision process begins.

Why are extra words a problem? If your meaning is clear, what does it matter whether you used ten words or a hundred? Those one hundred words may easily contain repetitious or unnecessary phrases. Once these have been removed, your writing will be more efficient. Let's think about the service at a restaurant. The same meal could be served after ten minutes—or thirty. Which do you think the customer will prefer? In making your writing as concise and pointed as possible, you're respecting the reader's time, and perhaps in turn the reader will respect your point of view.

As you revise, omit words that are simply unnecessary. For example, the following wordy sentence needs to be revised:

**FIGURE 11-2 Revising Wordy Expressions.**

| Wordy | Concise |
|---|---|
| In my opinion, I think... | I think... |
| They proceeded to go... | They went... |
| Due to the fact that... | Because... |
| For the reason that... | |
| Because of the fact that... | |
| At the present time... | Now... |
| At this point in time... | |
| In the event that... | If |
| In the neighborhood of... | About... |
| In spite of the fact that... | Although... |



At this point in time, due to the fact that I am sleepy, I will proceed to get ready for bed. [wordy]

I'm sleepy now, so I'll get ready for bed. [concise]

Study the examples in Figure 11.2.

Wordiness is also caused by repeating the same term or by using synonymous terms.

Blake is a student at the local community college where he is the best student in his Economics class at college. [wordy]

This sentence repeats the word *student* unnecessarily. Note how a little revision eliminates an entire clause:

Blake is the best student in his Economics class at the local community college. [better]

Here's an example of a sentence brimming with synonymous and redundant words:

The color of the new tile was a very unique purple shade. [wordy]

*Color* and *shade* are so close in meaning that it makes sense to eliminate one. Further, *purple* actually *is* a color, so we don't need *color* or *shade.* Finally, can anything be *very* unique, since by definition *unique* means "one of a kind"?

The new tile was a unique purple. [concise]

Just as we revise our sentences to make them concise, we revise our paragraphs to avoid unnecessary or repeated sentences. While the repetition of key terms can be used effectively to unify an essay, wordiness is awkward and distracting. Read the rough draft below.

**STUDENT WRITING** Nicolas Quintero

Revenge is one of the most common human emotions. To some, revenge provides closure that regular actions don't provide. When you look up revenge in the dictionary, you will notice the definition says "to inflict punishment in return for injury or insult." In the movie *Mystic River*, revenge plays a large part in the story. In fact, the whole driving force behind the movie is revenge. Two out of the three main characters' whole purpose in the movie is revenge. The most obvious one will have to be Jimmy Markham, whose revenge was the cause of his daughter's death. But he is not the only one seeking revenge throughout the movie. David Boyle is seeking revenge toward his childhood monsters.

The paragraph contains good ideas but is encumbered with unnecessary and repetitive phrases. With a bit of revision, the statement of ideas becomes more clear and concise.

**STUDENT WRITING** Nicolas Quintero

Revenge, one of the most common human emotions, is the driving force behind the movie *Mystic River.* For two out of the three main characters, revenge provides closure. The most obvious is Jimmy Markham, whose search for revenge was the cause of his daughter's death. But he is not the only one. David Boyle is seeking revenge toward his childhood monsters.

If you find your rough drafts are wordy, don't despair. That's what revision is for. Don't forget that the rough draft is still peeling the potato; we'll cart the peels off to the compost heap later!

## EXERCISE 11.6 Eliminating Extra Words

Rewrite the following sentences in more concise language.

1. In my opinion, I think that it is difficult and hard to achieve a fluffy texture in a soufflé.
2. After the movie was over, we proceeded to go to the coffee shop for coffee after the movie.
3. Doreen served the punch in Styrofoam cups because of the fact that her glass cups and ceramic mugs were still packed away in boxes and stored in the attic.
4. The lively and energetic dog, who was a terrier, raced very quickly across the road in pursuit of a squirrel.
5. In today's society, there is a lot of technology that people use every day on a daily basis.



## RECIPE FOR REVIEW Revising Word Choice

### Choosing the Right Word

1. The **denotation** of a word is its basic definition. The **connotation** includes the feelings or associations that make a word seem positive or negative, or simply neutral.
2. Be sure that a word's connotation does not unintentionally confuse or offend your readers.

### Revising Slang, Jargon, Clichés, and Sexist Language

1. **Slang** refers to the informal words and phrases used by a relatively small group of people.
2. **Jargon** refers to the specialized terms used by people in a particular job, as well as to any wordy and pretentious language designed more to manipulate than to communicate.

3. A **cliché** is phrase that has been used so often that it has lost both precision and interest.

4. **Sexist language** refers to expressions that inappropriately specify one gender when both should be included, whether *he* to mean "people in general" or *she* to mean "any nurse."

5. As you revise your writing, replace slang, jargon, clichés, and sexist language with clear, vivid, and appropriate words and phrases.

## Translation of Diner Slang

a. shingle with a shimmy and a shake = buttered toast with jam

b. bossy in a bowl = beef stew

c. Mike and Ike = salt and pepper shakers

d. dog and maggot = cracker and cheese

e. customer will take a chance = hash

## Eliminating Extra Words

1. Refer to Figure 11.2 for ideas on avoiding or revising common wordy expressions.

2. As you revise, eliminate unnecessary or repeated words and phrases.

## A Taste for Reading

1. Choose one of the pieces in Appendix VII and examine the author's choice of words. Does he or she use slang, jargon, euphemisms, or clichés? Explain, using specific examples.

2. Read the first paragraph of M. F. K. Fisher's "Borderland" in Appendix VII. When the author uses the pronoun *he*, does she mean to restrict her comment to men only? If you had the power to edit this essay, would you remove the sexist language? Why or why not?

1. Discuss the advantages and disadvantages of using slang in an academic essay. Use specific examples to illustrate your point.
2. Visit the website *Weasel Words* **www.weaselwords.com.au**, which discusses the *dangers* of jargon. Why does the author believe jargon is dangerous? Do you agree? Explain.
3. Jargon can also be fun, however. How about "seagull manager" and "blamestorming"? Google "corporate jargon" on the internet, and see what you can find. Then write a short paper about the fun side of jargon, using plenty of specific examples.
4. Choose an adjective like *fat, thin, young,* or *old,* and look up ten synonyms in a thesaurus. Then look up the definition of each synonym in a dictionary. Finally, use each word in a sentence that illustrates its connotation as well as its denotation.

UNIT 4

# Cuisine

## Understanding Rhetorical Modes

**Chapter 12**

Process Analysis

**Chapter 13**

Cause and Effect

**Chapter 14**

Persuasive Writing

**Chapter 15**

Writing about Literature and Film

chapter 12

# Process Analysis

Explaining how to do something—giving directions—comes naturally to most of us. Perhaps you've taught a child to tie his shoes or make scrambled eggs. Or perhaps you've explained the layout of the kitchen to a new co-worker. In a face-to-face interaction, you can see at each moment whether your directions are being understood and followed correctly. You can make adjustments as you go. "No," you say to the little boy, "make the loop this way," or "See, the eggs are ready." In writing, though, as we've seen before, we do not have the opportunity to supervise the process directly, and we must put all the necessary information in the text.

## TYPES OF PROCESS WRITING

Process writing has three general types: **process analysis**, in which you explain how something is done; **process narrative**, in which you explain how you did something; and **directions** or recipes, in which you explain how to do something. All process writing includes a list of ingredients and equipment, an outline of the necessary steps in the correct sequence, and appropriate explanations and cautions. All process writing must use the vocabulary and level of detail appropriate to the audience and purpose of each piece. Process analyses and process narratives are written in essay format with an introduction and a conclusion. The sentences are not commands like "fold the egg whites into the batter" but statements: "The egg whites are then folded into the batter." The paragraph below is from a process *analysis.*

**STUDENT WRITING** Robert A. Hannon

To make hollandaise, finely chopped shallots, white wine and cracked black peppercorns are reduced until they are almost dry; then they are mixed with a little water and added to a bowl. Next, egg yolks are added and heated over a pot of steaming water and whisked until their temperature reaches about 145° Fahrenheit. At this point the eggs will have roughly tripled in volume and will fall off the whisk in long strands. If the eggs are undercooked, they will fall off the whisk like water, and if they are overcooked, they will curdle in the metal bowl, evidenced by small semi-solid chunks of soon-to-be-cooked eggs.

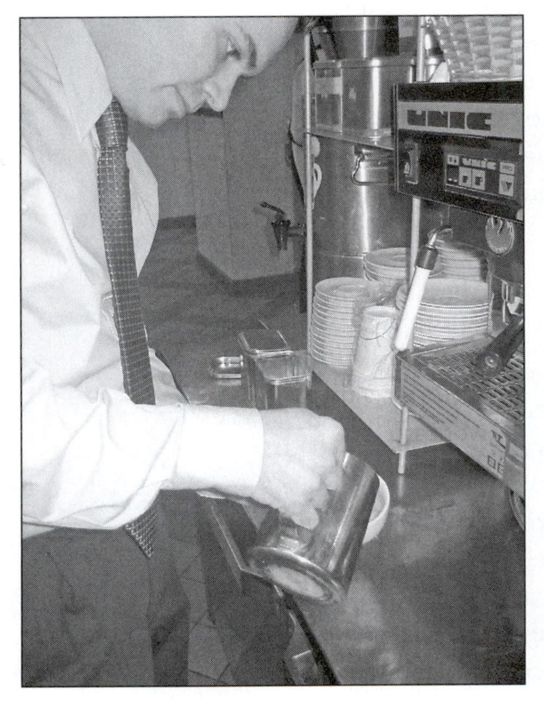

Preparing a latte.

One difference between a process analysis like the one above and a process *narrative* is that the analysis is written in the third person (*he* or *she, one*, etc.) or in the passive voice (as in the preceding paragraph), while the process narrative is often written in the first person (*I*). The analysis is also more likely to be in the present tense, while the narrative may be in the past.

A recipe is a set of directions or instructions for completing a process. Like a process analysis and a process narrative, a recipe lays out ingredients, defines procedures, moves forward in strict sequence, uses appropriate transitional expressions, and offers guideposts—ways to know if things are proceeding correctly. Unlike them, however, and unlike an essay, a recipe does not consist of a series of paragraphs, which are in turn made up of declarative sentences. Instead, a recipe is a series of *commands* whose subject is "you." Notice how the student paragraph about making hollandaise[20] sauce might be rewritten in recipe format:

**Ingredients**
finely chopped shallots
white wine
cracked black peppercorns
water
egg yolks

Reduce shallots, wine, and peppercorns until almost dry. Mix with a little water and place in bowl. Add egg yolks and, whisking continuously, heat over pot of steaming water until mixture reaches 145° Fahrenheit. At this point eggs have tripled in volume and will fall off whisk in long strands.

Since a recipe's purpose is to prepare the reader to reproduce the process, it will specify the amount of each ingredient. An essay need not do so. Further, in a recipe we may dispense with *the* or *a* before nouns—"place in bowl" rather than "place in *a* bowl"—while in an essay we use formal, complete sentences.

Finely chopped shallots, white wine, and cracked black peppercorns are reduced. [essay]

Reduce shallots, wine, and peppercorns. [recipe]

**EXERCISE 12.1** **Types of Process Writing**

Rewrite the process analysis on hollandaise sauce as a process narrative. Use the first person and the past tense.

## PURPOSE AND AUDIENCE

One of the most important aspects of process writing is understanding who your audience is and what terms and processes they already know. In the hollandaise paragraph, for example, would every audience understand how to "reduce" shallots, wine, and peppercorns? Probably not. Yet most would know how to "whisk" egg yolks. In writing for an audience of experienced cooks, you might explain the steps in less detail and feel free to use specialized culinary terms such as *mirepoix* or *en papillote*. For a general audience, however, you will want to define these terms or else explain the procedure without using them.

You must also decide how far to break down the process. As you write any given step, you are likely to assume that the reader already has a certain amount of knowledge or experience. In describing how to make scrambled eggs, for example, do you need to be as specific as "First, take the eggs out of the refrigerator"? That sounds funny, but the principle is an important one. In another essay, you might write "First, I saddled the horse." But does your audience know how to saddle a horse? Do you need to break down that part of the process further?

> First I lifted the saddle, making sure the stirrups were crossed over the pommel so that they didn't swing against the horse's belly and frighten him. Then I laid the saddle gently on his back, slightly up on his withers, and slid it into place. In this way his coat would lie smoothly under the saddle.

And if you write with this detail, do you need to define *pommel* and *withers*? Or is this *too much* detail? You can only answer these questions with a specific audience and purpose in mind. For example, for general readers, you need to define your terms. However, unless you are preparing them to saddle a horse themselves, you may not need to break the process down into small steps.

## EXERCISE 12.2 Purpose and Audience

Choose a process you are familiar with from home, school, or work (perhaps how to change the oil in your car, or how to make chocolate chip cookies), and write a short process analysis for a general audience. Think carefully about what terms to use, which ones to define, and how far to break the process down. Your purpose is to help the audience understand the process but not perform it.

**Types of Process Writing.**

| TYPE | PROCESS ANALYSIS | PROCESS NARRATIVE | RECIPES/INSTRUCTIONS |
|---|---|---|---|
| Purpose | to explain how a particular process works | to tell the story of how someone performed a particular process | to instruct the reader how to perform the process |
| Format | essay format | essay format | list of ingredients and steps |
| Point of View | third person: ***Shallots** are reduced.* | first person: ***I** reduced the shallots.* | second person: ***(You)** Reduce the shallots.* |

*(continues)*

| TYPE | PROCESS ANALYSIS | PROCESS NARRATIVE | RECIPES/INSTRUCTIONS |
|---|---|---|---|
| Verb Tense | Present or Past: *Shallots* ***are*** *reduced.* *Shallots* ***were*** *reduced.* | Past or present: *I* ***reduced*** *the shallots.* *I* ***reduce*** *the shallots.* | Present: ***Reduce*** *the shallots.* |
| Mood: statement v. command | statement: *Shallots are reduced.* | statement: *I reduced the shallots.* | command: *Reduce the shallots.* |
| Ingredients | Part of explanation | Part of story | Listed at the beginning |
| Sequence of Steps | Correct sequence | Correct sequence | Correct sequence |

## INGREDIENTS AND STEPS

All process writing includes some information about the ingredients and equipment used. In a set of directions like a recipe, the ingredients are often listed first so that the reader can assemble them in one place, so that he can prepare his *mise en place.* In a process analysis or narrative, the necessary ingredients or equipment may be discussed in an early paragraph or they may be mentioned only at the point at which they are brought into play. For example, note that this early paragraph in an essay on making breakfast for four speaks only generally about the required components.

**STUDENT WRITING** Nathan E. Bearfield

It all starts with a strong mental and physical *mise en place.* Your mental *mise en place* begins with an integral gathering of your thoughts. The order in which items are cooked, how fast they cook, and what the finished product should look like are "key" when cooking. Locating a sufficient supply of food for production should be the first step in your physical *mise en place*. Next should be the small amount of prep work for those ingredients that need more attention to detail. Finally, the process of cooking.

In a process narrative entitled "How to Grill a Birthday Steak," the introduction does mention one of the ingredients: the supremely important steak itself.

**STUDENT WRITING** Andrew S. Pohan

It was my sister's birthday, and I wanted to make it extra good for her. I knew I needed two days before the whole process would be complete: a day to shop for the ingredients along with an overnight marinade in the refrigerator, then a second day to cook and serve the steak. I thought about what I needed for the whole process and knew I needed to select just the right cut of meat for it . . . not too big, which would make it easy to fit in a large bowl for marinating.

In the narrative that follows, the equipment and ingredients are mentioned in the order of their appearance in the process.

**STUDENT WRITING** Jahnna Howell

The recipe began with one can of Campbell's tomato soup. The circular lid was quickly taken off. The juicy, royal-deep red contents of the can were meticulously scraped into a small pot sitting on the stove. Following the soup contents, one can of cool, crisp water was stirred in. As the tomato soup began to simmer on the stove top, the rice was almost done cooking. After a few minutes, my mother added a cup of steaming hot, fluffy, long-grain rice to the pot. I could smell the lovely aromas of tomato soup. My lips tingled: I was dying to taste it.

A student in baking and pastry arts studies her recipe for carrot cake.

The backbone of all three types of process writing is the sequence of steps, whether these steps are incorporated into paragraphs within an essay or listed by number in a set of directions. In a set of directions, the steps follow the ingredients and are listed as commands, as we saw in the earlier example about making hollandaise sauce. While the placement of the ingredients and the level of detail will vary depending on the audience and purpose of your writing, in all cases you will be careful to put the steps of the process in the correct sequence.

**EXERCISE 12.3** **Ingredients and Steps**

Add a separate list of ingredients and equipment to your paragraph from Exercise 12.2. Check that the steps are in the correct sequence.

## EXPLANATIONS AND CAUTIONS

Effective process writing often includes explanations and cautions. Particularly when you are writing a set of directions that you expect the reader to carry out, it is helpful for him to understand the *reason* for particular steps and to be *warned* of things that might go wrong. The writer below, in describing how he prepares his sauté station, explains why he holds items in a "nearby cooler."

**STUDENT WRITING** David Filippini

I fill the sauce bins and hold the rest in a nearby cooler, along with the cream and cheese, for easy access when I run low. This helps because I don't have to go all the way to the walk-in during a rush.

Here's an example from the breakfast essay.

**STUDENT WRITING** Nathan E. Bearfield

On another stove top burner, heat a non-stick 12-inch sauté pan. Your flame should not top medium. At this time locate some paper towels. They will help to absorb the fat when draining the bacon, thus making it less greasy. With a circular motion, move the hash browns in a counter clockwise motion to release them from the pan. At about this time—and do not attempt if stuck—flip the hash browns.

Another writer includes this important caution.

**STUDENT WRITING** Euijin Kim

Don't forget to cover with the pot lid while cooking. . . . The rib meat has to be soft enough to come off from the bone when people bite it. That's the key point of this food. If the meat does not come off from the bone, it cannot be Galbe.

When directions involve potential dangers like fire or electrocution, it is essential to caution the reader about them and explain how to avoid them. Further, for safety's sake, such warnings should come at the beginning of the set of instructions as well as before the dangerous step itself.

Sometimes process writing also includes variations on the steps or ingredients, as in this process narrative.

**STUDENT WRITING** Simon M. Imas

When I would make Dulce de Leche, I would put the whole can [of evaporated milk] in the pressure-cooker. I learned from experience that the longer someone lets it cook, the darker and more bitter it becomes. I would let it cook for about 45 minutes to get a deep brown appearance and slowly open it after waiting 30 minutes for it to cool a bit. The reason I do this step is because if you open it while it is still hot, a long stream of hot and sticky caramel will come shooting out. After opening the can, I was told to add whatever I wanted, but to make sure I didn't add anything that would mix flavors. My mother used to add walnuts, but I just add a couple of teaspoons of pure vanilla extract. Without the vanilla extract, the caramel would taste bitter.

The writer of the Galbe essay includes a personal recommendation.

**STUDENT WRITING** Euijin Kim

The first step to make Galbe is, soak it into cold water to get rid of the blood that remains inside the meat. This is a basic skill to steam meat foods. Then boil it in hot water in order to remove the extra fats. It is not a necessary work, however; it depends on one's taste. In my personal opinion, I recommend doing this step because Galbe is a fatty food and contains any amount of animal fat.

As you think through the steps of a process you are writing about, be sure to consider what might go wrong and how to prevent it. If you are writing about a process you know well, try to remember what it was like when you first performed it. What mistakes did you make? How do you avoid them now?

**EXERCISE 12.4** **Explanations and Cautions**

Re-read your paragraph from Exercise 12.3. Add an explanation of the reasoning behind two or three of the steps. Then try to anticipate what problems might arise in the process, such as the hash browns getting stuck or the pot not being tightly covered. Add this information to the paragraph.

## PROCESS AND PERSONALITY

Process writing is especially enjoyable when it reveals interesting information about the cultural background of the dish. For example, "Most Koreans believe that food tastes better when it cooks with touch of hands" (Jina Chun). Readers are also interested in the writer's individual experience and personality.

**STUDENT WRITING** Euijin Kim

In the first place, "Galbe" is not an easy food to cook. Even though I am a Korean, I have to pay all my attention to produce a wonderful Galbe.

In the next example, the writer's personality infuses the process analysis with humor and warmth.

**STUDENT WRITING** Vincent Amato

Chicken Supreme starts with frying the chicken in a special batter. Sorry, I'm not allowed to give specific ingredients! After the pieces are fried, they are put into a deep baking dish and smothered with heavy cream, garlic, and mushrooms. To loosen the sauce, chicken broth is added, and the dish is baked for an hour. My mouth is watering just writing about it!

Or consider the charming, personal conclusion of "How to Grill a Birthday Steak".

**STUDENT WRITING** Andrew S. Pohan

I watched Amanda sit there and eat while I talked to her about her day and about cooking in general. Usually when I am at the table with Amanda she is asking me how to make certain dishes, and I always try to explain as best I can. She had a smile on her face during the whole meal, which indicated to me that she was enjoying it. The conversation was quite entertaining as well. Amanda is very funny when it is just the two of us talking over dinner. After the meal Amanda gave me a big hug, said she really loved the meal, and told me she wants me to cook for her every birthday.

### EXERCISE 12.5 Process and Personality

Add some background information about the process you've been working with in this chapter. Conclude the piece with a discussion of what you've learned through the process, or why it's important to you. In other words, add a touch of personality.

Good process writing clearly outlines the necessary ingredients and the proper sequence of steps. Great process writing explains the reasoning behind the steps, offers tips to judge how the process is going, and warns of possible problems. Best-selling process writing offers a personal connection to an engaging author—the secret to the continued success of certain cookbooks.

**Classic Cookbooks.**

*James Beard's American Cookery by James Beard*

*How to Cook Everything by Mark Bittman*

*Jane Brody's Good Food Gourmet by Jane Brody*

*(continues)*

*The Way to Cook by Julia Child*

*An Omelette and a Glass of Wine by Elizabeth David*

*The Silver Palate Cookbook by Julee Rosso & Sheila Lukins*

*The French Menu Cookbook by Richard Olney & Paul Bertolli*

*Joy of Cooking by Irma Rombauer*

**QUESTION:** Why have these cookbooks been so successful?

4

## RECIPE FOR REVIEW Process Analysis

### Types of Process Writing

1. **Process analysis** explains how something is done.
2. **Process narrative** explains how you did something.
3. **Directions** or recipes explain how to do something.

All process writing . . .

- uses the vocabulary and level of detail appropriate to the audience and purpose of each piece and
- includes a list of ingredients, an outline of the necessary steps in the correct sequence, and appropriate explanations and cautions.

### Essay Versus Recipe Format

1. Process analyses and process narratives are written in essay format. They have an introduction and a conclusion, and sentences are written in the indicative mood (i.e., *not* a command).

2. Recipes are different in that they usually begin immediately with a list of ingredients. The steps themselves may be listed or appear in small "paragraphs," but they are not in essay format. The steps are written in the imperative mood (i.e., a command).

## A Taste for Reading

1. The excerpt from Shoba Narayan's *Monsoon Diary* in Appendix VII contains both a story and a recipe. What explanations and cautions are included in the recipe? Does the story make a good introduction to the recipe? Explain.

2. Study the process of preparing the tangerine slices in M. F. K. Fisher's "Borderland" (Appendix VII). What is the basic process? What explanations and cautions does she include? Do you like the way the story and the process are blended together? Explain.

## Ideas for Writing

1. Rewrite Exercise 12.5 as a set of directions for a general audience. List and define the ingredients, and decide how much detail to include about each step in the process.

2. Remember the movie *How to Lose a Guy in 10 Days*? Kate Hudson's character was the "How To" girl; she wrote a newspaper column in which she explained how to do things that her readers were interested in. Write a "How To" column of your own, perhaps on a humorous topic.

3. Find a recipe for the same item in two different cookbooks. Compare them in terms of vocabulary, level of detail, voice or personality. Is one easier to follow? Why?

4. In what ways is the writing process outlined in Chapter 2 like a recipe? How is it different from a recipe?

chapter 13

# Cause and Effect

Culinary students know all about cause and effect. When you beat the eggs, they froth up. When you add yeast, the dough rises. When you heat the water in the double boiler, the chocolate melts. The heat is the cause; the melting is the effect. Studies of cause and effect go beyond the kitchen, of course. Historians write about the causes of World War II while engineers study the causes of a bridge's collapse. Economists warn us about the effects of a stock market crash while legislators review the effects of a change in the drinking age.

Narration, or storytelling, shows *what* happened. Process analysis tells *how* it happened. **Cause and effect** explains *why* it happened. Like narration, process analysis, and the other rhetorical modes, cause and effect may be central to the subject and organization of an essay or it may be confined to a single paragraph within a larger piece. We'll look at examples of both in this chapter, though the principles remain the same.

## ANALYZING CAUSES

An essay or paragraph often focuses on the various causes of a single event. What were the causes of World War II? Why did that bridge collapse? What caused the hollandaise sauce to break? The passage that follows offers a partial answer to the third question.

The station is set to prepare Catalina vinaigrette.

**STUDENT WRITING** Matthew Berkowski

A hollandaise sauce may break because the fat added to the yolks might have been too hot or too cold, the fat may have been added too fast, or the hollandaise was held at too high of a temperature.

The temperature of the egg yolks is also important, as the next example makes clear.

**STUDENT WRITING** Robert A. Hannon

Egg yolks are added and heated over a pot of steaming water and whisked until their temperature reaches about 145° Fahrenheit. At this point the eggs will have roughly tripled in volume and will fall off the whisk in long strands. If the eggs are undercooked, they will fall off the whisk like water, and if they are overcooked, they will curdle in the metal bowl, evidenced by small semi-solid chunks of soon-to-be-cooked eggs.

Note that the writer describes the precise effects of undercooking and overcooking the eggs: the eggs will "fall off the whisk like water" or form "small semi-solid chunks." These details paint a vivid picture of the doomed hollandaise! Further, these causes and effects seem reasonable; we could test them by trying the hollandaise ourselves. Although there are times when we might not be sure what caused an event and so want to *suggest* causes that haven't been proved, it is best to avoid proposing *unreasonable* causes (the hollandaise broke because the chef wasn't wearing a side apron) or *simplistic* causes (the hollandaise broke because something went wrong).



## EXERCISE 13.1 Finding a Topic

Think of an event, condition, or situation that interests you. Now answer such questions as *What caused that event?* or *Why does that condition exist?*

Causes are often explained or presented in relation to time. In the example above, the causes of the broken hollandaise are *immediate* and occur a very short time before the event. Sometimes causes can be *remote*, that is, farther away in time. In the case of the hollandaise, the remote cause might have been that the chef did not get a good night's sleep and so forgot to warm the butter appropriately. Though part of the chain of events, the chef's poor sleep is not as close to the broken hollandaise as the cold butter.

A second way to think about causes (or effects, of course) is in terms of importance or strength, that is, to identify the *main* or most important and powerful cause of an event versus the *contributing* or less important and powerful causes. In the example above, while the chef's poor sleep *contributed* to the problem with the hollandaise, the *main* cause of the broken emulsion was the cold butter.

**Analyzing Causes: Why did the Sauce Break?**

| | | |
|---|---|---|
| time | immediate cause | the butter was added too quickly |
| | remote cause | the chef did not get a good night's sleep |
| importance | main cause | the butter was too cold |
| | contributing cause | the cooler's thermometer malfunctioned |

### EXERCISE 13.2 Identifying Main and Contributing Causes

Think about one of your good friends (or a favorite movie, actor, food, or restaurant). List all the reasons you can think of that you like this person, for example, a kind personality, wise advice, sense of humor, loyalty, similar interests, or other reasons. Now, among those reasons, which one is the most important or powerful? Why?

## ANALYZING EFFECTS

Some topics focus on the *effects* of an event. They answer the question "What happened *because* of this one event?" The example below describes the effects of a character's "irresponsible conduct" in the film *Mystic River.*

**STUDENT WRITING** Soyang Myung

Celeste lied about the truth to Jimmy. That meant that she was an egoist and didn't take care of her husband and her family, including her son. She just wanted to protect herself from David. In the long run, her irresponsible conduct destroyed her family. Because of her hasty judgment, her son Michael would live under a fatherless family, and she would live with a guilty conscience.

The event or *condition* here is Celeste's failure to take care of her family, her "irresponsible conduct," which results in her husband's death, her son's grief, and her own despairing guilt. Many essays on film and literature contain studies of cause and effect. Other essays on *Mystic River*, for example, explore the *effects* of the initial kidnapping scene on the three main characters. In this next paragraph, the writer explains how Dave Boyle uses images of vampires and werewolves to describe the effects of his childhood trauma.

**STUDENT WRITING** | Idan Bitton

Poor Dave cannot explain in words what he has been going through ever since he was a child, being molested by two men. Everything that is coming out of his mouth takes him closer to insanity. He compares the two molesting pedophiles to vampires. Vampires are creatures that have to suck human blood to survive, and Dave's molesters do, too. They sucked Dave's blood and life out of him. After the abuse, he lost himself, as if he had died. Dave is also scared that he is going to become one of them. He knows that once you get bitten you need to bite, like vampires do.

The effects of Dave's experience have been a psychological "death" and an agonizing fear that he himself will become a predator. Tragically, the resulting strangeness of Dave's behavior leads directly to his murder.

### EXERCISE 13.3 Analyzing Effects

Choose an event, situation, or condition, perhaps the same one as in Exercise 13.1. What are the effects of that event? Which are the most important? Why?

## CAUSAL CHAINS

We often recognize the presence of a *series* of causes and effects, that is, a **causal chain**, in which one consequence is the cause of another, which in turn causes another. Dave's childhood trauma caused him to act strangely, which attracted the attention of the detectives investigating a murder twenty-five years later. Yet because he fears becoming a pedophile himself, as a result of his early experience, he refuses to tell anyone what really happened. Since he doesn't tell the truth, the victim's father assumes he is guilty and kills him in revenge. And since Dave is dead, his son Mike is now doomed to repeat Dave's own sad, fatherless youth.

The broken hollandaise can also be part of a causal chain. Suppose the cook has a toothache which causes her to lose sleep, which in turn causes her to hold the hollandaise at too high a temperature. The high temperature causes the sauce to break moments after it is served to the customer, causing the customer to complain to the server, who in turn complains to the cook. Now three people have had a bad day!

**Causal Chain.**

the cook's toothache
causes her to lose sleep

which causes her to hold the
sauce at too high a temperature

which causes the sauce to break
before it reaches the customer

which causes the
customer to
complain to the
server, who
complains to the
cook ...

and all three have a bad day

## ORGANIZING A CAUSE AND EFFECT ESSAY

In organizing a cause and effect essay, you may wish to begin with the cause (or effect) nearest in time and continue in chronological order. The writer of the paragraph below introduces the topic—the drama of a broken hollandaise—and outlines three causes.

**STUDENT WRITING** | Matthew Berkowski

### Oil and Water Don't Mix, or Do They?

Just imagine you're at a very fine dining establishment; the garçon comes with your eggs benedict, and you notice that the hollandaise is broken. Broken hollandaise is not only visually demeaning to the customer, but it doesn't please the palate either. The chef should have recognized that the hollandaise was broken. But maybe the hollandaise broke on the way from the kitchen to the table. A hollandaise sauce may break because the fat added to the yolks might have been too hot or too cold, the fat may have been added too fast, or the hollandaise was held at too high of a temperature.

---

A problem may first arise with the *mise en place.* Is the fat too hot or too cold? A little later on in the process, the problem might be caused by the method: has the fat been added too quickly? Finally, the sauce might break *after* it's made if it is held at too high a temperature.

As the writer develops and explains each of these causes, he uses clear transitional sentences to keep the reader on track.

> *In order for a hollandaise sauce not to break,* the fat that is added to the egg yolk concoction cannot be too hot or too cold.
>
> *Another reason that the hollandaise may break* is that the fat might have been added too quickly.
>
> *Finally, hollandaise may break because* it was stored at too high or too low of a temperature.

Words like *reason* and *because* are useful transitions in a cause and effect essay. My favorite sentence is in the paragraph about *mise en place.* The fat cannot be too hot or too cold, we are warned.

> The fat, like the porridge in the Three Bears fairy tale, must be just right.

This essay is also interesting because it mentions an alternative to the tricky handmade emulsion.

> There is another method of making hollandaise that doesn't result in a sore arm or having to bother someone. Having a second person helping you makes this sauce produce pretty darn fast.

The writer concludes, however, that "making the hollandaise by yourself takes a little longer, but it strengthens your skills." He distinguishes between the more remote but important effect of improved skill and the immediate, practical result of a helping hand.

An invisible second person holds the ladle.

**EXERCISE 13.4** **Organizing a Cause and Effect Essay**

Choose one of the topics you brainstormed in Exercise 13.1, 13.2, or 13.3. Organize the causes and/or effects you identified into an interesting and logical sequence. Then explain why you chose this sequence.

## DEVELOPING AN ESSAY WITH CAUSE AND EFFECT

Like the other rhetorical modes, cause and effect is a way of thinking about or developing our ideas. It is likely to be only one of several modes used in writing an essay, just as braising may be only one of several cooking methods used in putting together a plate. In the following essay, the writer analyzes the evolution of her philosophy about food and cooking, a complex series of causes and effects. As you read, notice the various types of causes and effects but also the other methods by which ideas are developed.

**STUDENT WRITING** Hyun Ah Seong

My mother has a genius for making Kimchee, which is the famous and representative Korean dish served at almost every Korean meal. Though the recipe of Kimchee is common for every Korean, there are huge differences between each Kimchee according to the person who makes it. My mother's Kimchee is well known as fabulous among the close neighbors. Kimchee is made of fermented vegetables—such as cabbage or turnips—that are mixed with chili powder, some fish sauces, minced garlic, a little bit of granulated sugar, and various Korean seasonings. The Kimchee is pickled before being stored in tightly sealed pots or jars to ferment. In this process of fermentation, the Kimchee gets a unique taste that is spicy, pungent and sour, and a crispy texture. Especially the spicy and pungent flavor is aroused by chili powder.

In November—the season in which Kimchee is most often prepared—Koreans lay up large stocks of Kimchee for winter, because it is hard to find fresh vegetables then. People want to make Kimchee together with my mother because they want to get the delicious

Kimchee—the important commodity for their loving family. Preparing and making Kimchee with several close neighbors together is such a major event. While preparing all the ingredients and sharing my mom's hidden recipe, each of the neighbors often notices that the secret of my mom's Kimchee is not that much different from their own, so they catch a little surprise. The essential point of making a dish is not that far away from the dish's general process, which almost everyone knows. Just a small additive touch and some switches within the process make differences. For example, my mother uses several different kinds of chili powders together, not selecting only one kind of chili powder. The mixing of the hottest one and the mild but sweet chili powder boosts the flavor of Kimchee in a sophisticated way, and the result is different.

My mother also spends more time to find fresh ingredients. The red pepper for chili powder is famous in Chungyang, a southern area of Korea. She visits there in person and buys it for year-round use, though it takes four hours to get there. When I was young, I felt that she was too obsessive, even though she could make a delectable Kimchee. However, I have changed my attitude toward admiring my mother as I became a woman. Her efforts and passion to give a delicious dish to her family must have been accumulating in my mind, and now I see that making and completing a dish need a person's spiritual faith. I received inspiration from what she did; to live as a chef is a wonderful way of being absorbed in ingredients. I could create a totally different dish with them, and then I could give pleasure to people. I am more enthusiastic about the cooking process than about just eating a dish that is made by others and served. I especially like to think what I will cook for family or friends, what vegetable would be good for this season; then I imagine how the taste will be, what kinds of food would be appealing to them, and so on. I believe caring about people and preparing food are strongly linked to each other, and both of them create the thing we always crave, "love."

Like my mother who made an effort to look for better ingredients, the basic philosophy that I keep in mind as a prospective chef is to put love into my dishes. As the generations have changed and people's living patterns developed, food is not only providing nutrients but is also another method to work off people's frustration. Thus the custom of eating and sharing nice food has been developed through human history. Since I decided to be a chef, I feel that I am charged with a sense of duty as if I have become a therapist who gives a mental peace to people who need treatment. Sharing my food will be a prescription (recipe) for people who have a lack of fullness from their lives.

Chef and student display the end result—a successful emulsion.

The first sentence states the central idea—"My mother has a genius for making Kimchee"—and the introductory paragraph goes on to describe the ingredients of this popular Korean dish. Cooking, as we've said, is all about cause and effect, and the writer notes that the process of fermentation affects the flavor and texture of the vegetables. "Especially the spicy and pungent flavor is aroused by chili powder."

The writer then explores both the causes of her mother's "genius for making Kimchee" and its effects on her as a child and as a adult. Words like *because* and *result* emphasize the relationship between cause and effect.

> The mixing of the hottest one and the mild but sweet chili powder boosts the flavor of Kimchee in a sophisticated way, and the result is different.

The writer also identifies a more *remote* cause than the mixing of the chili powders: "My mother also spends more time to find fresh ingredients." She travels four hours to find a certain pepper, for example. As a girl, the writer found this "too obsessive." Only later did she realize that her mother's "efforts and passion to give a delicious dish to her family must have been accumulating in my mind." The result is a philosophy of cooking.

> Like my mother who made an effort to look for better ingredients, the basic philosophy that I keep in mind as a prospective chef is to put love into my dishes.

Perhaps the most important cause of her mother's genius for Kimchee is the love she puts into it.

When we think about cause and effect, we are asking *Why*? We are tapping into one of the human being's most important traits: curiosity. In exploring ideas through questions about causes and effects, as Hyun Ah does, we may discover our own essential thoughts and beliefs.

**EXERCISE 13.5** **Developing an Essay with Cause and Effect**

Write an essay based on the outline you developed in Exercise 13.4. What methods, other than cause and effect, did you use in developing your ideas?

## RECIPE FOR REVIEW Cause and Effect

### Definitions

1. Cause and effect is a way of approaching a topic that focuses on *why* something happened or *why* something is the way it is (**causes**) or on *what effects* an event will have or *what the consequences* of a particular event (**effects**) will be. Essays may focus on causes only, effects only, or a combination of the two.

2. Causes and effects can be organized in terms of *time* (immediate or remote) and *importance* (main and contributing).

3. Causes and effects sometimes form a sequence in which one event hinges on the next to form a **causal chain**.

## Cautions

You should be able to prove that two events are connected; don't mistake one event being followed by another for one event *causing* another. Be sure your causes or effects are not unreasonable or oversimplified.

## A Taste for Reading

1. Read the excerpt from Eric Schlosser's *Fast Food Nation* in Appendix VII. What has contributed to the enormous success of the fast food industry? What are the consequences of that success?
2. Read Jeffrey Steingarten's "Hot Dog" in Appendix VII. What are some of the causes and effects he describes in this essay? What is the effect of the title?

## Ideas for Writing

1. Think of a food you particularly like or particularly dislike. Brainstorm the reasons why, including more and less important reasons, past and present reasons. Perhaps taste the food as you brainstorm to increase the number of sensory details in your writing.
2. Pick a topic in cooking or baking, such as why emulsions form or why bread rises. Do a little research into the science behind the event, perhaps by reading the relevant section from *On Food and Cooking: The Science and Lore of the Kitchen* by Harold McGee. Organize the cause or causes in terms of importance or time, and write the essay.
3. Pick a favorite of some kind, for example, a favorite song, color, place, or actor. Then, in a short essay, explain *why* it's a favorite.
4. Choose a topic from contemporary culture, such as Why are reality television shows so popular? or What effect did *Super Size Me* have on McDonald's and on American culture as a whole? You may wish to research the topic (see Chapter 16). Then answer the question in an essay, citing any sources appropriately both within the text and on the Works Cited page (see Chapter 17).

chapter 14

# Persuasive Writing

A restaurant menu usually has at least two purposes: to *inform* customers about the available items and to *persuade* them to place an order. The persuasion may take the form of appealing menu descriptions, colorful photographs, or tips on healthy choices. A menu may also *entertain* customers with humor, drawings, or intriguing tidbits of information. Like menus, writing may also have several purposes. In Chapter 1, we talked about three of them: to inform, to entertain, and to persuade. Beyond simply informing or entertaining its readers, a persuasive piece hopes to influence the reader's opinion, even to cause an active response such as writing a letter to a government representative, giving money to a charity, or buying a particular product. Like a menu, persuasive writing uses a variety of techniques to elicit a response from the reader.

## INFORMATIVE VERSUS PERSUASIVE WRITING

Think about the difference between a straight news story and an editorial. While the news story restricts itself to informing the public, to presenting the facts of the story without a particular bias, the editorial presents an opinion and seeks to enlist the agreement of the reader. Let's look at an example. Suppose there's an outbreak of *E. coli,* which is eventually traced to bags of spinach. A news story might give details of the investigation and an update on the victims' health; the news story is about *informing* the reader. On the other hand, an editorial on this same topic might express an *opinion* about whether the government is doing enough to ensure the safety of our food supply. It might suggest that readers take specific action, such as boycotting certain brands or writing to their legislators.

Another example of the difference between **informative and persuasive writing** is the Food and Drug Administration's mandated Nutrition Facts label on a food package versus an advertisement for that same food. The label is intended to inform, to present factual material without editorial comment. The advertisement, in contrast, is designed to persuade the consumer to buy the product. While the Nutrition Facts label on the side of the box simply informs the reader that the product contains, for instance, whole grain rolled oats and zero grams of saturated fat, the advertisement on the front of the box proclaims the food's effectiveness in reducing cholesterol. Whether or not this claim is true, its *purpose* is persuasive rather than informative.

Persuasive writing is an ancient and respectable art. However, if persuasive techniques discard the truth and discourage debate, they can become **propaganda**. Like editorials and advertisements, propaganda attempts to influence the beliefs and actions of its readers, but "not through the give-and-take of argument and debate, but through the manipulation of symbols and our most basic human emotions."[21] Sometimes certain movies are criticized for being manipulative rather than truthful or realistic. Viewers clap and cheer as the music swells with triumph and the underdog sports team claws its way to a win. They sob along with the violins as a child succumbs to terminal illness. They jump two feet in the air and spill their popcorn as an unexpected explosion rocks the movie theater. Are these reactions produced by genuine emotion or by clever manipulation?

A good piece of persuasive writing encourages its readers to explore a question for themselves as well as to agree. Propaganda, on the other hand, discourages that kind of independent thinking (see Figure 14.1). In the *Propaganda Critic*, Aaron Delwiche writes that "propagandists love short-cuts—particularly those which short-circuit rational thought. They encourage this by agitating emotions, by exploiting insecurities, by capitalizing on the ambiguity of language, and by bending the rules of logic."[22] For example, an advertisement may seek to transfer an audience's respect for a certain symbol, such as a doctor's white coat, to a particular goal, such as the sale of toothpaste. While

**FIGURE 14-1 Debate vs. Propaganda.**

| Debate | Propaganda |
|---|---|
| Appeals to reason/logic | Avoids reason/logic |
| Appeals to ethics | Ignores ethics |
| Appeals to emotion | Creates fear |
| Invokes pity or compassion | Ignites hate |

these techniques may be successful, it is best to avoid them in persuasive writing.



### EXERCISE 14.1 Identifying Persuasive Writing

Read the editorial page of a newspaper and find two editorials or letters to the editor that interest you. What point does each try to make? How does each writer try to "prove" that point? Are their arguments successful? Explain.

## AUDIENCE AND PURPOSE

As we do with any writing, we want to understand who the readers of a persuasive essay are and, therefore, which arguments will be most likely to influence them. With much informative writing, our concerns with the reader have to do mainly with how much information to include and what type of vocabulary (general or specialized, formal or slang) to use. With persuasive writing, however, we need to look more deeply into the mind of the reader and try to discern his beliefs and values.

For example, take a topic like the drinking age. Should the drinking age be 18 or 21, or should it be abolished? Whatever position you take in your essay, imagine first that you are writing a letter to the editor of the student newspaper on a college campus. Second, imagine that you are writing a letter to local members of MADD (Mothers Against Drunk Driving). Finally, imagine that you are writing a letter to the owner of a local bar. How would you change your arguments and perhaps even your style to communicate most effectively with each particular audience? Is your purpose to raise awareness, stimulate debate, or move the audience to action? Does your audience share your views, oppose your views, or have yet to form an opinion? The answers to these questions will dramatically affect the content and style of your writing.

### EXERCISE 14.2 Audience and Purpose

Write a paragraph from each of the points of view in the preceding paragraph. Work from a clear idea of what your purpose is and what your audience's opinion is likely to be.

## TYPES OF PERSUASIVE ARGUMENTS

Persuasive writing is only necessary where there is more than one opinion about an issue, plan, or decision. Classically, persuasive writing has relied on three approaches The first, and perhaps the most popular in academic settings, is the *appeal to reason or logic.* This type of argument uses reasoning, supported by facts and examples, to draw logical conclusions. For example, you might argue that investing in a good education for all citizens will bring financial rewards to the government. Since those with a good education are able to earn a higher income, they are also able pay higher taxes. Writers may also use an *appeal to ethics.* Show the reader that your proposal meets the needs of something they value, such as equal opportunities for education and employment. Of course, this isn't the easiest thing to do because people don't always share the same values. With persuasive writing, it is extremely useful to know something about the audience ahead of time. Finally, writers may make an *appeal to emotion*—a favorite of speechwriters of all persuasions. An emotional appeal on behalf of equal access to education might describe the life of hardship and uncertainty to which many citizens would otherwise be doomed.

Persuasive essays may use one type of appeal or a combination of two or three. For example, in an argument against lowering the drinking age from 21 to 18, you might present statistics showing that when the drinking age was raised from 18 to 21, the number of fatal alcohol-related accidents decreased.[23] Therefore, you might conclude, it would be *illogical* to lower the drinking age again. You might also argue your position from an *ethical* standpoint: the state should not condone policies that have a high probability of injuring or killing its citizens; hence, it is as appropriate to have a drinking age of 21 as to have red lights at intersections or mandated seatbelt use. Finally, you could appeal to the *emotions* of your readers by telling the story of a carload of teenagers killed because the driver had been drinking, or worse, the inebriated driver walked away from the crash that killed a family of four.

Controversial topics by definition have more than one side. In looking at arguments on the other side of the drinking age debate, you might logically ask, "If 18-year-olds are considered responsible enough to vote, why shouldn't they also be considered responsible enough to drink?" In terms of ethics, you might argue that the state is wrong to mandate "morality." A favorite emotional argument is that if 18-year-olds are considered old enough to die for their country, then they should be old enough to buy a beer. A truly persuasive paper will acknowledge the arguments against its point of view and then disprove them. An example of this format of argument and response can be found at the *Alcohol Policies Project*, a website maintained by the Center for Science in the Public Interest.[24]



**EXERCISE 14.3** **Developing Your Arguments**

Choose a controversial topic, and brainstorm a list of arguments on both sides of the issue. For each side, try to include at least one from each of the three types: reason, ethics, and emotion.

## A FOOD-RELATED CONTROVERSY

For professional chefs, one controversial topic is the degree of responsibility they have to make a difference in industry-related problems, such as obesity and poor working conditions. If assigned a paper on this topic, our first step would be to brainstorm: In what way *could* professional chefs make a difference?

## STUDENT WRITING | Justine A. Frantz

The most common problem with food, particularly in the United States, is obesity. Americans' waistlines are seemingly ever increasing due to the lack of exercise and healthy food, along with an increase in the consumption of fatty fast food. Instead of addressing the problem themselves, Americans often try to place the blame on the fast food corporations for making them fat. While this issue is obviously not the fault of chefs all over the country, there are alternatives that they could popularize to help keep our nation healthy. An example of this is Fast Good, a chain that sells healthful, non-genetically modified food that is available just as quickly as pulling through the McDonald's drive-thru. If more chefs take this approach to preparing food, America will surely become a more figure-friendly nation.

While dismissing as unreasonable the claim that fast food corporations are making America fat, this writer suggests that the industry might address obesity by making healthy meals as convenient as fast food meals. In defense of her logic, she offers the example of a successful, real-life alternative to fast food, Fast Good.

Chefs and restaurant owners can make a difference not only in the kind of food they serve but in the amount. The importance of controlling portion size is widely recognized by doctors and nutritionists in helping to maintain a healthy weight.

## STUDENT WRITING | Ashley Roosa

The percentage of Americans that are obese is an unhealthily large number. I myself can be considered obese. It's not the easiest thing to deal with when aspiring to be a pastry chef. I constantly have to hold back from tasting everything I can due to the fact that I don't want to gain more weight. As chefs, there are many things that we can do to help this epidemic. I feel that what leads to obesity may not be what we eat, but how much of it we eat. Chocolate mousse cake or crème brûlée really isn't as bad for you as you may think. But when eaten in large quantities for periods at a time, that is when it becomes unhealthy. If chefs control the portion size of the food that they produce, they are indeed making this epidemic better. Portion control is the biggest thing a chef can do to control this epidemic.

Notice that this writer engages the reader's sympathy with a personal statement: she herself can be considered obese. When she goes on to argue reasonably that portion size is effective in keeping weight down and is within a chef's control, she combines logic with emotion.

An issue that involves the food industry from farm to restaurant is labor exploitation.

**STUDENT WRITING** Ashley Roosa

With the number of illegal aliens growing each day in the U.S., labor exploitation is at its all time highest. These aliens are being paid less than minimum wage for work that citizens would receive much more for. But since they are here illegally, they cannot complain due to fear of being sent back to their own country. There are two solutions for chefs today. Either don't hire illegal immigrants, or pay them honest money for honest work.

The basic argument here is ethical: all workers should receive at least minimum wage. And the choice before chefs is ethical: "Either don't hire illegal immigrants, or pay them honest money for honest work." Between these two alternatives lies an understated emotional component, as the reader is invited to enter the world of the underpaid "aliens" who "cannot complain due to fear of being sent back to their own country."

Small groups are often able to brainstorm effectively on all sides of an issue.

## EVALUATING YOUR ARGUMENTS

The writers quoted above have suggested several ways that chefs *could* address some of the food-related problems in the United States. Whether they *should* is a slightly different question.

**STUDENT WRITING** Payson S. Cushman

When one lists the responsibilities of a chef, they usually do not include addressing food-related issues or politics at all. They would, of course, include serving quality food that is safe. They would include providing an enjoyable setting to experience the food. They would include making a meal at their establishment satisfying and valuable. Thus, it depends on what both the chef and the customer define as valuable and satisfying that determines whether or not the chef has any obligations to address food-related issues.

Both the tone and the arguments here are quite reasonable: a chef's responsibilities are practical, not political. The writer of the next paragraph reaches the same conclusion and goes on to give a specific example that defines the limits of the industry's responsibility: the lawsuit brought by two teenage girls against McDonald's.

**STUDENT WRITING** Amanda Bates

Obesity is a devastating problem in this country, taking millions of lives every year through heart disease, diabetes, kidney failure, and high blood pressure. What should be seen as a call to self-action has turned into a veritable witch hunt for fast food giants, who apparently force fed some poor, unsuspecting teenagers for years until they were obese. Oh, whoops. That didn't actually happen. They just ate too much and exercised too little, the seeming ignorance of our society exemplified in one frivolous lawsuit. Just as McDonald's was not accountable for the health of those girls, chefs are not solely responsible for the health of our clientele. Of course, in the long run, we want them to live long, full lives (preferably buying dinner from us multiple times), but if they want to eat foie gras and french fries every day, so be it. All that can be expected of us is to offer healthy options for the more discerning client.

The writer's lively rebuke of her opponents' emotional arguments is quite successful, while her statement that "all that can be expected of us is to offer healthy options" seems quite fair and reasonable.

The next writer, however, has a different opinion concerning a chef's responsibilities.

**STUDENT WRITING** Justine A. Frantz

Every occupation entails some sort of civil duties. Some of these duties are quite obvious, such as a soldier's duty to defend his or her own country; however, other jobs include dealing with less obvious societal issues. Ordinary people may not think that chefs have much to do with society other than feeding the general public, but there are more responsibilities in being a chef than meet the eye. Food-related problems are rapidly increasing world-wide, and who better to help conquer these issues, such as obesity, eating disorders, and world hunger, than people who have a strong passion for and understanding of food.

In this paragraph, the responsibilities of a chef are linked with those of a soldier defending his or her country. Chefs are called to help in areas in which they have special expertise, she argues reasonably. Further, there is an emotional component: the chefs are also called to help because of their "strong passion" for food.

As with all aspects of your writing, your decision on how to approach the topic will depend on your understanding of the audience and on your individual style. Does your particular audience like "just the facts"? Are they especially sensitive to issues of right and wrong? Or can you sway them by stirring up powerful emotions?

**EXERCISE 14.4** **Evaluating Your Arguments**

Look back at your brainstorming from Exercise 14.3. Which arguments seem to be the strongest on each side? Why? Write two paragraphs, one on each side of the issue. In each case, imagine that your reader is undecided and present your strongest case.

## ORGANIZING A PERSUASIVE ESSAY

A formal persuasive essay written for an American college course typically offers "evidence" for the position, that is, it favors the appeal to reason. Great emphasis is placed on supporting your opinion with facts and statistics that you have found in well-respected sources and that you have documented appropriately (see Chapters 16 and 17). This type of persuasive writing—the essay that takes a side and uses evidence to support it—is often organized in the following way:

1. Introduction—explain what the issue is and which side you favor
2. Persuasive arguments—present the arguments on your side
3. Refutation—outline some of the arguments *on the opposite side* and explain why these arguments are unreasonable, unethical, or otherwise less persuasive than your arguments.
4. Conclusion—restate your point

This organizational plan draws on some of the methodology of a compare and contrast essay. You are essentially contrasting the arguments on different sides. Thus it is often helpful to list the arguments for and against a position, as we listed differences between the appearance, texture, and flavor of apples in Chapter 7.

Once you have assembled the evidence and decided on your position, you will need to organize your arguments in some type of logical sequence. You will often wish to emphasize the most important argument by placing it either first or last. The same is true when you are outlining the arguments on the other side. As in all your writing, you must be the judge of what sequence best suits your purpose, your audience, and your point of view on the topic.

Let's look at an example. Another controversy involving the food industry is the government ban on smoking in restaurants and bars. Figure 14.2 lays out arguments on both sides of this issue, including appeals to reason, emotion, and ethics.

The writer of the following full-length essay has chosen to support the smoking ban, and her next task is to organize her appeals in an effective way. Her first appeal is to emotion: people can't enjoy their meals in a smoke-filled restaurant. She then moves to reason: even secondhand smoke is dangerous. And she ends with an ethical appeal: we should avoid hurting others. As you will see, it is the ethical argument that weighs most strongly with this writer.

**FIGURE 14-2 Brainstorming about the Smoking Ban.**

| | For the Smoking Ban | Against the Smoking Ban |
|---|---|---|
| Reason | Secondhand smoke is dangerous. | Smoke is not the only pollutant. |
| Emotion | People can't enjoy their meal in a smoky environment. | Smoking relieves stress. |
| Ethics | The safety of all patrons is more important than the pleasure of some. | All customers should be entitled to enjoy their favorite mealtime activities, whether smoking a cigarette or drinking a coffee. |

**Annotated Student Essay.**

*Should the Government Ban Smoking in Restaurants and Bars?*

The issue

Should the government ban smoking in restaurants and bars? There absolutely will be controversial opinions. One is in favor of the smoking ban; the other is against the ban. We should carefully think about two different aspects without any discrimination. Nowadays people care so much about their health, more than before. The rate of smokers is getting down by much, and people who have smoked try to quit eventually. Why do some people still keep smoking? Does it mean they don't care about their health at all? How about other people's? Everybody knows about the risk from secondhand smoke. I strongly support that the government should ban smoking in restaurants and bars.

statement of position

persuasive arguments

First, let's look at the reasons why the government should ban smoking in restaurants and bars. The restaurants and bars are public places which are used by many people together. All the people who come to the places have a purpose to spend their time pleasantly without any discomfort. They pay money for their valuable time in the restaurants and bars. If the place is full of smoke, how can people enjoy their meals and drinks? I know that someone who smokes even doesn't like smoke from other's cigarette at all. In addition to the issue about the atmosphere, there is a huge issue about health for people who would share the smoke in the same places. It's about second/hand smoke. In general, it's already

1. appeal to emotion

(continues)

**Annotated Student Essay** *(continued)*

an issue well known to the mass media. A lot of research proved the harm of secondhand smoke. Secondhand smoke even can result in death. For example, I have read that children who grow up in the homes of smokers are much more likely to have frequent colds and respiratory diseases. Besides, it is reported that some women are suffering from lung cancer, though they have never smoked before. What a sad story is that! Can you imagine that you could get a disease then finally die due to other people's disregarding acts? I think it is not an overstatement at all that the smoker who smokes in public places is extremely selfish. Smoking is just a habit of someone who is looking for enjoyment. It shouldn't be admitted since it has a possibility to give other people damage.

*2. appeal to reason*

*3. appeal to ethics*

However, there could be an opposite opinion about banning smoking in restaurants and bars. Heavy smokers might argue that smoking a cigar or a cigarette is just one of their favorite mealtime activities, such as drinking coffee or eating chocolate bars. Like non-smokers, they also have the right to enjoy their meal. They too pay money in restaurants and bars. Moreover, they can think illness can be caused by any other factors also. Polluted air from cars and factories already exists all over the world. Smokers don't understand why people only consider smoke by cigarettes. In addition, they believe smoking can release their stress, so it probably helps to prevent their mental disease as well.

*refutation*

*1. appeal to emotion*

*2. appeal to ethics*

*3. appeal to reason*

*4. second appeal to emotion*

As I mentioned above, I agree that the government should ban smoking in restaurants and bars. I tried to see both sides equally because there would be definite reasons why people have different aspects. Although smokers might possess free will, they must be careful when they smoke in public places. There even can be children coming with their parents who would not have any means to avoid the unpleasant situation. Before insisting on their right to smoke, smokers should regard others' right to live in a safe environment first. Furthermore, smokers can choose better methods to release their stress than smoking. They can do some productive and non-toxic activities such as working out, talking with close friends, counseling from therapist, shopping, and so on. The habit of smoking is an addiction, so they might not notice why they are smoking at that time. It is a kind of unconscious behavior. Therefore the government should do the arbitral role for the society to make it great to live for everyone.

*response to opposing arguments*

*1. right to live trumps right to smoke*

*2. counters emotion with logic-better methods for stress relief*

*Conclusion 1. shows compassion*

*2. restates position and major ethical appeal*

—Hyun Ah Seong

## EXERCISE 14.5 Organizing a Persuasive Essay

Using your work from the previous exercises, write an essay on the controversial topic. Your instructor may wish you to choose one side, or, if you are undecided, you may want to weigh both sides impartially. It may be helpful to begin with a rough outline of the arguments you will use.

## RECIPE FOR REVIEW Persuasive Writing

### Informative versus Persuasive Writing

1. Informative writing intends to communicate information.
2. Persuasive writing seeks to compel a response from the audience, whether it is sympathy, agreement, or action.
3. Propaganda also seeks to compel a response but may use manipulation rather than the appeals to reason, ethics, and emotion used by persuasive writing.

### Types of Persuasive Arguments

1. The appeal to logic uses reasoning, supported by facts and examples, to draw logical conclusions. The reader is invited to follow the logic and accept the conclusion.
2. The appeal to ethics shows the reader that your proposal meets the needs of something they value.
3. The appeal to emotion seeks to invoke pity or courage or hope, which in turn will move the reader to agreement and possibly action.

### Organizing a Persuasive Essay

A formal persuasive essay written for an American college course typically favors the appeal to reason and is often organized in the following way:

1. Introduction—explain what the issue is and which side you favor
2. Persuasive arguments—present the arguments on your side

3. Refutation—outline some of the arguments *on the opposite side* and explain why these arguments are unreasonable, unethical, or otherwise less persuasive than your arguments

4. Conclusion—restate your point

## A Taste for Reading

1. In the excerpt from *Fast Food Nation* in Appendix VII, does Eric Schlosser try to influence the reader's opinion in any way? If so, how? Can you identify any appeals to emotion, logic, or ethics?

2. Go to the website *American Rhetoric* **www.american.rhetoric.com** and read three or four of the speeches quoted there. Then choose one speech and answer the following questions: What types of arguments are used? Which do you think are the most effective? Why?

## Ideas for Writing

1. Watch the documentary *Super Size Me* by Morgan Spurlock, and identify examples of appeals to reason, emotion, and ethics. Are his methods effective? Explain.

2. The Food and Drug Administration has various rules and regulations that govern advertisements for food. Find some examples of these regulations, and do some research on how some advertisements try to bend them. For example, see Bonnie Liebman's article "Designed to Sell" in the *Nutrition Action Healthletter*.[25]

3. Research a bill that is about to come before the state or federal legislature. Brainstorm the arguments for and against the bill, and choose a side. Then write a letter to your senator or representative in which you persuade him or her to vote your way.

4. How does propaganda differ from the kind of persuasive writing we've been looking at in this chapter?

chapter 15

# Writing about Literature and Film

Dinner and a movie—the classic first date. We want everything to go well, to be the perfect backdrop to this special moment. Some restaurants are better than others for that vulnerable early date, while some movies are potentially disastrous! To ensure a pleasant experience, therefore, we may wish to confirm that both the meal and the film have received good reviews. Fortunately, it's not difficult to find such recommendations in the newspaper or on the Internet. But what if we have to write the review ourselves?

Writing about literature and film doesn't always sound appealing, even though we might have enjoyed reading the book or watching the movie. Perhaps we have been pressured in the past to analyze the meaning and the symbolism of the piece before we have had a chance to respond to the story or even to understand what happened. If you were sent to review a restaurant, would you begin by analyzing the recipe of every dish? Or, would you look at each item carefully and then taste it, reacting to its unique combination of aromas, flavors, and textures? Probably you would do the second one. Once we are grounded in these specifics, we can begin to understand why we responded as we did.

Similarly, when we are reading a poem or watching a movie that we're going to write about, we don't need to leap immediately into what the poem "means" or what certain images "symbolize." Just as you would in a restaurant, simply *watch* the movie and *taste* each character and each scene. Ask yourself what happened in the story. Then take note of your response. What do you like about the poem or film? What don't you like? Let your thinking unfold gradually as you explore your response to the details of character, event, setting, word choice—and in the film, the music and images.

## UNDERSTANDING THE STORY

Our first step in writing about literature and film is to figure out what that story is, what *happens*. In putting the events in order, we will also need to sort through the characters and try to understand their relationships and motives. The story is so central to our response to the film that *re-telling* it often takes up the major portion of a typical film review. However, while a brief outline of the movie's characters, plot, and setting is very useful in preparing the readers to understand our main point, it is not the *focus* of our essay, and it should not monopolize the space (see Figure 15.1). A summary might be contained entirely within the introductory paragraph, or in a second paragraph that immediately follows. Even when we recount a series of events from the film, we try to link it to our main idea. In an academic essay or good movie review, our focus is not so much on *what* happened as on *why* it happened.

Look at the example below, in which a student briefly outlines the story of *Mystic River.*

**STUDENT WRITING** James Wenzel

In the beginning of the movie, Sean, Dave, and Jimmy are all friends playing hockey in the road when Dave is taken away by two men. At the young age of 11, he is molested and escapes days after. Twenty-five years later, the death of Jimmy's daughter Katie brings these old friends back together. The last time Katie is seen alive, she is spotted by Dave at a bar. Dave is automatically a suspect because he comes home with blood over his clothes. Throughout the movie Dave perpetually lies whenever he is asked what happened to him after leaving the bar. Brendan, who is Katie's boyfriend, is the other prime suspect in her murder. Not only is he the boyfriend, but Jimmy hates him because years earlier Brendan's father "Just" Ray Harris "rolled" on Jimmy, sending him to jail. Dave, Jimmy, and Brendan all seek justice through violent acts of revenge.

**FIGURE 15-1 Writing about Literature and Film.**

Introductory Paragraph
lead into the title
and/or theme
of the film

Setting the Scene
explain or summarize just enough
of the story of the film
so that the reader can understand
your main idea

Exploring the Main Idea
use vivid details of the characters, events,
music, and images in the film
to explore, explain, illustrate, and support
your main idea
(one or more paragraphs)

Concluding Paragraph
bring us back to the opening image
or scene of the paper and make the reader
see it in a new way:
explain the motivation of a character
or the meaning of events:
answer a question, solve a problem



The writer uses this synopsis of the story partly to set the scene for a reader who may not have seen the film and partly to ground himself in the details. It is through the details that we are led to questions, discussion, and finally conclusions. In this particular essay, the writer draws a couple of conclusions based on the details. For example, "If only Jimmy had let the police do their jobs, Dave might still be alive.

But because Jimmy took the law into his own hands, an innocent man is dead." This understanding of cause and effect within the story is very satisfying to both writer and reader, and it might lead outside the story as well, perhaps to another film directed by Clint Eastwood and a comparison of Jimmy's tragically mistaken vigilantism with Dirty Harry's own activities outside the law. The details of Jimmy's story might also lead us to more general questions and conclusions about the whole concept of "taking the law into your own hands."

Another essay summarizes the story of the film *Chocolat*. "The power of chocolate leads me to a small town of France," the essay begins. Vianne and her daughter Anouk enter the village, followed by a mysterious wind.

**STUDENT WRITING** Jina Chun

After a few days, Vianne opens a small chocolate shop to settle down in the new town. Unfortunately, the highly conservative and traditional townspeople are unwelcoming about a new chocolate shop, especially during the time of Lent. But Vianne doesn't seem to care about people's lack of interest. Soon, Vianne and her chocolate shop become a center of gossip to the townspeople.

The mayor of the town doesn't like Vianne because he thinks she shakes people's minds. Since most important decisions are made by the mayor, he has more authority than any others in the town. He even writes the church's sermon by himself. He persuades people not to go to the chocolate shop. In that way, he thinks he can send Vianne away from the town.

However, the mayor soon needs to confront another problem. The new character Roux enters, a gypsy who is played by Johnny Depp. Vianne gets closer with him very fast since both of them are treated as outsiders. Roux and Vianne work together at the chocolate shop and deliver happiness to people. In the beginning, people consider Vianne as a witch and think she might give them bad influences. But they end up loving her chocolate, and some are addicted to it.

An analysis of this film leads the writer to conclude that "I think we should not have a prejudice against people because they are a little different from us." She doesn't *jump* to that conclusion, however; it evolves naturally through paying attention to the details of the story.

Like Vianne, these students enjoy preparing special chocolates.

4

> **EXERCISE 15.1** **Understanding the Story**
>
> Choose a short story, poem, or movie, and summarize the story in two to three paragraphs. Then list at least three questions you have about the story or the characters.

## RESPONDING TO THE STORY

At the same time that we're trying to understand the story, we are reacting or *responding* to it. In preparing to write about a film, it is helpful to talk about or jot down our specific reactions, perhaps in a journal (see sample student journal entry in Appendix VII). Ask yourself what you liked about the movie and why. Ask yourself what you didn't like. Sometimes this personal response to a poem or film becomes part of the final essay, as in the following example.

**STUDENT WRITING** Trevor Brunet

"About A Boy" is a movie I thought would be completely dreadful to watch. I thought it was going to be more of a movie that my girlfriend would like, some kind of love story or something gushy that would just bore me to extreme insanity, resulting in me begging the teacher every five minutes if I could use the rest room.

Thankfully, the movie proved me wrong. It was a movie that was meant for both females and males to enjoy together. To attract the attention of the male viewers, and entertain their minds, there was a movie bachelor who was somewhat of a player. The character of "Will" has lots of money, a nice house, a nice car, and a new girl every week. For the female viewers' own interest, the player is finally caught up in his lies, and women really start getting angry with him. This results in a conscience building up inside him.

The writer goes on to describe a scene that he likes, the scene in which Will tries to pick up "single mums" at a support group and consequently meets 12-year-old Marcus. Marcus makes our author remember his own childhood, while Will offers him an idyllic image of future possibilities. Yet as the writer *reacts* to the characters and events, he draws the following conclusion.

**STUDENT WRITING**

You can clearly see that the lifestyle of a player may not be the best lifestyle after all. Although it may be a lot of fun, in the end it can come back to haunt you. Chances are you will grow old and lonely with many enemies. And for people who do suffer with this problem, you should start to pray every night for a kid named Marcus to come into your life.

The writer's response to the details of characters and events lead him in a surprising direction. Far from suffering boredom, as he initially feared, he has become so involved in and appreciative of the story that he would like to extend its benefits to his readers: "You should start to pray every night for a kid named Marcus to come into your life."

In this next example, the writer had a strong reaction to the very *title* of the movie *Good Will Hunting*.

**STUDENT WRITING** Euijin Kim

Many years past, I saw this movie. However, whenever I saw it, I felt different impressions. It was about two years ago when I thought seriously about the theme that this movie wants to tell the audience. Throughout this time, my mind was full of deep emotion.

The writer explains that she is moved by Will's attempts to step outside his "fearful mind" and form real friendships.

**STUDENT WRITING** Euijin Kim

In my opinion, the last scene was the greatest scene in this movie. Will rides his car toward the future, and the road—which looks immeasurably long—gave me a deep impression. Toward his future, toward his girlfriend, it seems that everything will be on his side. I did enjoy freedom with Will.

The details of the story continue to touch the writer, and she concludes:

**STUDENT WRITING** Euijin Kim

Every time after seeing this movie, I always repeat these words in my mind, "I have a possibility. Forget the past things. It's not your fault!" I always try to encourage myself, and these words lightened my heart in many ways . . . I am haunted by the movie *Good Will Hunting*.

**EXERCISE 15.2** **Responding to the Story**

After watching the story, poem, or movie you chose in Exercise 15.1, make a list of two or three things you liked about it—a character, a scene, the music—and briefly explain why. Then list two or three things you didn't like, and explain why.

## STICKING TO THE DETAILS

When we're writing about food, we make use of the concrete details of flavor, aroma, and texture. Remember the peaches in Chapter 5?

**STUDENT WRITING** Elizabeth Best

The aroma of the peach is very complex, with many notes intertwining in the nostrils. You can pick up a floral note, almost like a honeysuckle. Next you may pick up a sweet, almost musky scent. Perhaps the strongest aroma is honey. It smells like a very rich, wild honey, like that of the wild clover.

When we're writing about books or movies, we use the concrete details of word and image. Think of the movie *Crash,* for example. The word *crash* is one of the first things the audience knows about the film, and it is among the first words spoken by the actors. The first scene illustrates its use not only literally—the scene is that of a minor car accident—but figuratively, as the word *crash* comes to represent the painful conflict of culture and race.

In another film, *Mystic River,* the recurring metaphor of vampires and werewolves leads us to a fuller understanding of the story. The "mystic creatures" that obsess Dave Boyle reveal the extent of his wounds and create both fear and pity in the audience. Having been preyed upon as a child, Dave now fears that he himself may become a predator. In the passage that follows, the writer begins his analysis of these creatures by placing us directly inside the horrifying transformation from human to werewolf.

STUDENT WRITING | Idan Bitton

### Mystic Creatures

It was a full moon night. As the clock hit midnight, he could feel it trying to come out of him again. It was uncontrollable. He did not have a chance resisting it. It was trying to make its way out, as if it were a little creature trying to crawl outside of his guts. As sickening as the feeling was, he had no ability to fight it. It was stronger than he. Pointy claws that can cut through metal came right out of his skin; they were at least four times longer than a human being's fingernails. Hair started covering his body from head to toe. In a heartbeat, even his palms were covered with thick gray fur. He could feel the sharp-edged teeth growing so quickly in his mouth that he could not close his jaw. His big furry ears were standing like they could hear for miles. His eyes turned red like human blood.

4

At the end of the essay, the writer summarizes the film's use of the werewolf metaphor, from the pedophiles who tormented the 12-year-old Dave to the dread that afflicts him as an adult.

STUDENT WRITING | Idan Bitton

During his life, Dave falls into the hands of evil and vicious people who build the unbalanced man he grew up to be later on. All of these metaphors throughout the story make us understand better what Dave has been through as a kid and what he is going through as a grown man. Without seeing one clear shot of what happened in that basement, these images make us picture it better than anything else. In Dave's mind, the 400-year-old legend is alive and kicking. They were living inside him. He named them werewolves.

Like the recurring themes of *Crash* and *Mystic River*, words and images in *The Shipping News* set the scene and focus our attention on important ideas. In the essay that follows, the writer immediately grounds us in specifics: "The film *The Shipping News* is about a man named Quoyle who is relocated to Newfoundland." He quotes the main character's early summation of his life—"My failure to dog paddle was only the first of my many failures"—and offers his own: "At this point [Quoyle] is a 'broken' man."

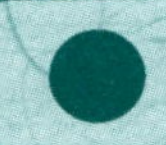

**STUDENT WRITING** David Wilson

## The Shipping News

The film *The Shipping News* is about a man named Quoyle who is relocated to Newfoundland. He goes through a number of poor choices to get into his current situation. His marriage to a girl that just jumped into his car was untrustworthy. The sudden death of his parents was surely unfortunate. Quoyle says in the beginning of the film: "My failure to dog paddle was only the first of my many failures."[26] At this point he is a "broken" man.

Quoyle begins his brokenness from birth. During his youth he was tossed off the dock to learn how to swim. This shows how ruthless his father was. His father always told him that he was never going to be good at anything. Quoyle, of course, believed what his father told him. He works a series of dead-end jobs as a dishwasher, a cashier at a movie theater, and an inksetter, a job where he did more napping than working. Upon moving to Newfoundland, he finds a job at the local newspaper, which begins his healing process.

At the *Gammy Bird* he has a mentor, Billy Pretty, who shows him how to be a good reporter. Billy tells Quoyle that he needs to find the beating heart of a story. This happens with his article about a boat previously owned by Hitler. With this story he finally realizes how to get ahead and get out of this slump of bad jobs.

His relationship with his aunt has also been an aid to his healing. She just shows up at his doorstep and tells him he should move to Newfoundland so he can get over the death of his wife and the suicidal death of his parents. Agnis can also help Quoyle raise his daughter, Bunny. She can look after Bunny while Quoyle is out doing a story.

In meeting a new woman named Wavey, Quoyle also continues to heal. She has a lot more going for her than his first wife, Petal. From this point on, Quoyle begins doing things for himself rather than having to rely on others. For example, he makes an effort to talk to Wavey.

Quoyle went from being a broken man to a brand new healed person. This process took him from a bunch of bad jobs, the death of his parents and wife, to the new job and the start of a relationship with Wavey and his aunt to finish the renewal process. He says at the end of the film, "If a piece of knotted string can unleash the wind [. . .], then I believe a broken man can heal."

The essay uses the specific details of Quoyle's "dead-end jobs," his "mentor, Billy Pretty," and his relationships with his aunt and his new love, Wavey, to picture for the reader how this broken man is able to heal. Notice how the direct quotes and the characters' names add flavor and texture as well as simply feeding us the information. It's all in the details.

**EXERCISE 15.3** **Sticking to the Details**

Based on your work in Exercises 15.1 and 15.2, write an essay in which you explore a character, event, quote, or image through its details. Include the details of your reaction to the story, poem or film.

## WRITING AS DISCOVERY

As with all kinds of writing, the process itself leads us to think more carefully and often more deeply. Many writers find that the intertwining stories of *Crash* draw them into new and uncomfortable areas of reflection. Erin Lee, for example, studies the two police officers, Ryan and Hanson. "There is no one who can be all good or all bad," she concludes. "We may only know after we crash into each other." Evaluating the character of Officer Ryan, in particular, has led to a number of thoughtful conclusions, as in the example that follows.

**STUDENT WRITING** Christy Cunningham

Officer Ryan had the biggest change in the movie because he thought he knew who he was, but really, he had no idea. He was going through a tough time; he was "crashing." But with the eye-opener he got a new look at life and gained respect for himself and others. Officer Ryan realized that what he had said to Officer Hanson was true for himself: He did not yet know himself.

Notice how the writer uses concrete details like the word *crashing* to explore the character of Officer Ryan. Student essays may also use quotes from the film or book, quotes that in this type of textbook require written permission from the owner of the copyright. Such permission was obtained, for example, for the quotes from *Finding Forrester* and *The Shipping News*.

For another student, *Crash* was a painful reflection of her experience. As you read this essay, notice how each of the first three paragraphs centers around the concrete details of a particular quote or scene.

**STUDENT WRITING** HyeWon Shin

## Crash

When I participated in orientation at the C.I.A., I noticed that I stood talking in groups of Asian people. I wanted to make some "American" friends; of course, I made some good friends, but it was hard to get involved in their community. Some students asked me, "Do Asian people eat cat?" or "Does your country have toilets?" I was speechless with those kinds of questions. I understand that they were just curious about Asian cultures or traditions, but sometimes it makes me feel sad. Just at the right time, I watched the movie called *Crash.* It was a movie that brought out many emotions and thoughts in me because I felt anger at the way stories developed. I felt sad and depressed after watching the movie.

*Crash* circles around the different stories of the most diverse groups of people living in Los Angeles. Separate stories all come together, and every scene and every character is connected to the other stories. I enjoy these ways of story lines. However, I don't like the scenes in which Asians do a vicious thing. In *Crash*, they called the man "Chinaman" who was selling illegal immigrants for money. Actually, he was speaking to his wife in Korean, and he told her to cash the check immediately. Koreans, Chinese, or any different groups from Asia, these people are usually depicted as tricky or dishonest in some movies or dramas. Why are Asian people described like that? Do American people really know about Asian people? Or do they truly care about Asian people? The more I watched this movie, the more my feelings were depressed.

This movie's major issue is racial discrimination and how people deal with it. I loved the character of the key repairman who is a hard-working man and has a warm heart to his family. Also, it would be very interesting to see how this story develops, especially the scene in which his daughter was going to be killed by the store owner. I couldn't keep my eyes off the scene. On the other hand, I didn't like the last scene of the movie in which Shaniqua was rude to some people because they were not Americans. I can't say she is such a racist, but she makes an enemy of people who can't speak English well or come from other countries. I felt sorry because she might not be the only person who thinks like that; some Americans might have the same feeling as she does when they face a foreigner.

Therefore, even though this movie helped me to understand about issues that related to the colors of our skin, the countries that we are from, and different lifestyles that are based on economic status, this movie was emotionally painful. It is not because I am an international student and moving from a different country, but because *Crash* portrays a real human condition that still exists in our society.



The first quote—sadly—is from the writer's own experience. A Korean student attending school in the United States, she was asked "Do Asian people eat cat?" The second paragraph follows up on this painful stereotyping of Asians, and it records the writer's deepening discomfort during the film. Then the third paragraph takes the discussion of racial discrimination from the film and moves it back to the writer's world: "I felt sorry because she might not be the only person who thinks like that; some Americans might have the same feeling as she does when they face a foreigner." In the essay's powerful conclusion, the writer travels from her feelings about her personal situation to her concern for the "real human condition." It's a great film that can take the audience this far, and a courageous writer who holds on for the ride.

### EXERCISE 15.4 Writing as Discovery

What is a film or book that made *you* think about the "real human condition"? Why?

## WRITING ABOUT FOOD

Writing about a restaurant has some similarities to writing about a movie. We don't jump to conclusions; instead, we smell and taste each dish. We ask ourselves, What is the *story* of the meal? What happened? Who was there? What details did we notice about the food? What did we like and dislike? As you read the next essay, notice how the writer answers these questions.

**STUDENT WRITING** | Robert A. Hannon

### Salt Cod, Alaska

I strode into Alumni Hall, the main dining room at the Culinary Institute of America, eagerly anticipating our "stage meal," which was served to us in the haute cuisine style. We were served this meal daily during the first three weeks of our curriculum. Haute cuisine is a French concept based upon culinary principles and methods that the school (as well as the French!) considers the fundamental building blocks of fine cooking and dining.

I glanced at the small mauve menu, which described the three courses we were to be served. I was slightly anxious, because the meal would be consumed with the knowledge that I'd be writing about it for a Gastronomy assignment. The appetizer was Brandade de Morue, which sounded exotic and intriguing in French. Then I read the translation, and the words "salt cod" splashed through me like the frigid Alaskan ocean in the state where I'd first tasted it. Instantly, in this haute cuisine setting (fine French food served by skilled cooks to discerning gourmands on tables bedecked with crystal, linen, and perfectly placed silverware), I was overwhelmed by memories of my friend Bebe and my college summer working in a salmon processing plant in Kodiak, Alaska. I felt a tinge of irritation as well. I didn't like salt cod, based on my summer experience with Bebe, and yet here I was staring at salt cod, when there were so many more desirable, less mundane items to cook and serve.

Bebe was my roommate in the salmon plant, where we worked up to twenty hours a day gutting salmon on a slime line. We became friends, enjoying fishing for Coho (Silver) salmon in Kodiak Sound, hiking the emerald grassy hills, and exploring what little night life existed in Kodiak. Only one thing about Bebe was bothersome: his Philippine Bacalao, the salt cod that he cooked in a rice cooker in our room every morning. He'd always offer it to me, and I'd always decline. It smelled like the rotting Chum salmon that we'd encounter on our fishing excursions. I did promise Bebe I'd try it, though as the summer progressed and he ate his accursed Bacalao every morning, I found myself fervently hoping he'd forget my vow.

Unfortunately, Bebe had an excellent memory, and a few days before my departure back to Seattle, he appeared with a foul bowl of Bacalao. "Come on, Bobby, it is very good for you," he said. Tentative, I lifted a tiny spoonful to my mouth and tasted it. A potent saltiness filled my mouth, followed by a rank fishiness. The flavor was more concentrated and far more disagreeable than the smell. I spit the spoonful into the bowl, and, as Bebe howled with amusement, I ran down the narrow wooden hallway of the dormitory, grabbed my toothbrush, and scrubbed frantically. Despite my efforts, the taste persisted through the day. I brushed my teeth, and washed my face and lips continuously, like an innocent Lady Macbeth. Finally the taste subsided, and I made a solemn vow never to taste salt cod again.

Yet here was salt cod on the menu, and I really had no choice but to sample it: we were on stage, and I needed to write about it for an assignment. When the dish arrived, I sniffed nervously, but could detect no fishy odor. The dish itself was visually appealing, a colorful arrangement of complementary colors and designs. A single delicate cod cake, golden brown and perfectly round, rested atop a salad of micro greens. The light and dark greens were commingled with cubes of red and yellow peppers. The dish was encircled with an almost fluorescent lime-green parsley dressing. The cake crackled between my teeth as I bit into it. The flavor was clean and sweet, tinted with garlic, and was only slightly salty. The micro greens added an appealing bitterness and were a piquant counterbalance to the cod. The sour tang of the dressing cleansed my palate.

Despite my previous experience with salt cod, I really enjoyed the dish, and finished the entire plate. I'm sure Bebe would never believe that I could enjoy salt cod, but then he's probably never eaten Brandade de Morue. I doubt he'd like it, though; I'm quite sure he'd find it insipid and egregiously lacking in flavor.

---

This essay is a delight with its sharp details, elegant word choice, and abundant energy. It begins vigorously: "I strode into Alumni Hall." He didn't *walk* or *stroll,* notice; he *strode.* He's excited, eagerly anticipating the meal to come. Like many restaurant reviews, the essay tells the story of the dining experience, and it has characters, a setting, and a conflict. Here the conflict revolves around one of the menu items.

> The appetizer was Brandade de Morue, which sounded exotic and intriguing in French. Then I read the translation, and the words "salt cod" splashed through me like the frigid Alaskan ocean in the state where I'd first tasted it.

The writer has an aversion to salt cod and is presented with that very dish in a situation that will make it impossible for him to decline it!

**EXERCISE 15.5**

### Describing an Experience with Food

Do you have any food aversions, that is, is there any food that you strongly dislike? Write a paragraph in which you describe the food in vivid detail. Make the reader dislike it, too. Now write another paragraph in which you imagine that you've had a new experience with the food and have come to enjoy it. Make the reader enjoy it, too.

In that moment where he sits staring at the salt cod, he shares the story of Bebe, his roommate in a fish factory one summer in Alaska. The writer names names: Brandade de Morue, Bebe, Kodiak, Bacalao. The sensory details are vivid: the emerald grassy hills, the rotting Chum salmon, a potent saltiness. When he returns to the present, he describes the dish in even greater detail. After all, this is the dish he's reviewing. And it ends well; the cod is good, though perhaps a little insipid for Bebe's taste.

The essay has all the elements of a good story—interesting characters, exotic settings, vivid details, and a problem to solve. As it moves easily back and forth from the meal to the memory, it leads us to consider the basis of aversion and the possibility of change. The best kind of food writing is simply good writing that happens to be about food.

## Organizing "Salt Cod, Alaska."

"Setting" the Scene

at the Culinary Institute of America
eagerly anticipating an experience
of haute cuisine

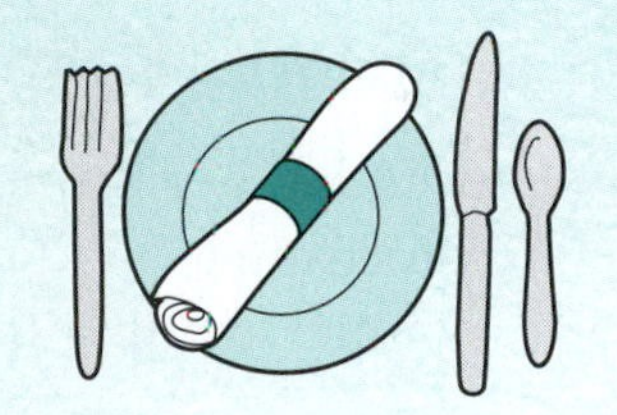

Outlining the Problem

small mauve menu reveals a dish to which
the writer has a strong aversion:
salt cod
but he has to write an essay about it
for a Gastronomy course

Memory

the source of the problem
Bebe and his Bacalaw
it smelled like rotting fish

I lifted a tiny spoonful
a potent saltiness filled my mouth
Bebe howled with amusement
I vowed never to taste salt cod again

Climax

a single delicate cod cake

Problem Solved

despite my previous experience
I really enjoyed the salt cod
Bebe would never believe it

4

## RECIPE FOR REVIEW Writing about Literature and Film

### Understanding the Story

1. As you begin to think and write about a story, poem or film, focus first on identifying the characters and setting and on understanding the sequence of events.
2. Jot down any ideas you have about "meanings" or "symbols," but don't go looking for them just yet.

### Responding to the Story

1. Ask yourself what you liked about the story, poem or film, whether it's as small as a single word or the dress on the little girl next door, or as large as "the suspense" or "the music." Now ask yourself *why* you liked it.
2. What did you *dislike* about the story? Why?
3. What questions do you have about the story, poem or film?

### Sticking to the Details

The words, events, and images of a story, poem or film are like the textures, flavors, and aromas of a meal. As you write about the film, use these concrete details to guide and illustrate your discussion.

### Writing as Discovery

Your paper most likely will try to answer a question or a solve a problem about the story, poem or film. Don't recycle some old plastic container of minestrone soup. Find the raw ingredients and cook something fresh.

### Writing about Food

1. Taste the food; tell the story.
2. Understand and evaluate the elements of the experience. Describe the location and decor of the restaurant, its purpose and theme. Evaluate the menu's appearance and contents. Describe the food items. Assess the service. Was the experience worth the price?

## A Taste for Reading

1. Read one of the poems in Appendix VII, either Lucille Clifton's "Cutting Greens" or Galway Kinnell's "Blackberry Eating." Use the punctuation to focus on one sentence at a time. Try to think about what the poem says first without worrying about what it might *mean*. What did you like and/or dislike about the poem? Why? What questions do you have? (See Alicia Lacey's journal entry on "Cutting Greens," printed after the poem.)
2. Sometimes it's helpful to have information about the author and historical background of a story. Read the introduction to "Borderland" in Appendix VII; then read the story itself. What "outside" information helped you understand and appreciate the story? Explain.

## Ideas for Writing

1. What movie (or story or poem) kept you thinking about it days later? Why? Was it because of a particular character or moment in the film? Was it because of a particular idea or response? Explain.
2. What movie (or story or poem) left you with many questions? Summarize the movie for an audience that hasn't seen it. What were the questions? Can you answer them now? Explain.
3. Describe a restaurant experience that taught you something new.
4. Write an essay based on your responses to question 1 or 2 in A Taste for Reading above.

# Nutrition

# Supporting Your Ideas with Research

5 UNIT



5

chapter 16

# An Introduction to Research I: Finding and Evaluating Your Sources

One of the most popular television series of the last few years has been *CSI,* which is an abbreviation for Crime Scene Investigation. We love to watch Grissom pursue a theory of the crime with one of his gruesome experiments—the popsicle made of freshly ground beef, for example. We marvel at the increasing effectiveness of DNA testing. We applaud the team's attention to detail—mapping out the blood spatter, shining their little flashlights into every corner and under every sheet. Based on the high ratings of this show and its spinoffs, we should be a nation passionately devoted to research. Because that's what research is—investigation. Research is as cool as the CSIs—and most of it can be done without crawling across the floor with a pair of tweezers!

**Research** simply means "look carefully" or "search again." You've probably already done quite a bit of research in the kitchen as you experimented with different ingredients and cooking methods. Does this cookie taste better with white chocolate or dark? What happens if I cook this sauce another three minutes? You have probably also looked at many, many recipes, sometimes using them as written, sometimes altering them. You've most likely explored the quality and pricing of various food items at different supermarkets. When you write a research paper, you are doing this same type of activity.

Perhaps also you've had to make a medical decision that involved research. The doctor presents you with a choice of treatments with various success rates and side effects, and you have to choose between them. Ideally, you will do some research at that point. What do other doctors recommend? What do other patients say about the treatments? Perhaps your librarian directs you to a medical database, where you search professional journals for information on your condition. Your goal is to make an "informed" choice in an area without money-back guarantees.

In so many parts of our lives, research is cool, fun, and important. Yet when it comes to writing research papers, many students see not the bright lights and glamour of investigating a murder in Las Vegas but only a bleak landscape of note cards and impossibly long URLs. Two aspects of the research paper are particularly challenging. The first is finding good information. The second, which is explored in Chapter 17, is adding that information to what's already in your mind, mixing well, and cooking up a tasty and nutritious essay.

**EXERCISE 16.1** **Identifying Research**

List three investigations you have conducted—whether in a kitchen or classroom, a business setting, or your home or library. Then write a paragraph in which you describe one of these investigations in more detail.

## FINDING SOURCES OF INFORMATION

Students sometimes believe that research involves typing a few words into a popular search engine and copying the first Web page that comes up. While a search engine can get you started with some ideas and terminology, it is not necessarily the best source of accurate information; be prepared to evaluate all Web sites carefully. And there are more sophisticated ways to find high quality sources, for example, through subscription databases that are available through your school or public library. Further, *most* sources of accurate information—the kind that is useful in serious research—are only available in *print*. Finally, your most valuable initial source of information and guidance is a good reference librarian.

Librarians are experts at finding useful resources. Just as you would turn to a dairy farmer to get fresh milk, or to a chef to get a made-from-scratch hollandaise sauce, it is wise to ask a librarian to help you find information. Libraries are vast orchards of books and articles and

**FIGURE 16-1** Libraries are vast orchards of books and articles and databases, ripe with facts and ideas.



databases, ripe with facts and ideas (Figure 16.1). Librarians can show you how to harvest this data. To produce top-notch research, get acquainted with the staff at your school or public library and learn how to find your way around.

Let's look at research in another way. Information sources are like ingredients you've assembled for a dish. You can select prepackaged items, which may or may not be of good quality, or you can choose to shop carefully for the freshest raw ingredients, knowing they will make your final product especially good. Similarly, the choices you make about your information will affect the quality of your research paper.

Once you've chosen a topic, begin your search for information with your textbook or with reference books in the library. You know already that these books are likely to contain accurate information since they were chosen by your instructors or by the librarians. Although they may not contain as much detail as you will eventually need for your paper, they will provide some background information and give you an idea of the vocabulary used in a study of the topic. Your textbook is also likely to include a list of useful books, articles, and *reliable* Web sites.

Suppose you are just beginning a research project on a food-related topic. An excellent place to start is with the reference section

**FIGURE 16-2 General Books for Food Research.**

Compiled by Christine Crawford-Oppenheimer at the Culinary Institute of America


Davidson, Alan. *The Oxford Companion to Food*. Illustrated by Soun Vannithone. Oxford: Oxford University Press, 1999.

*Encyclopedia of Food and Culture*. Solomon H. Katz, editor in chief. New York: Charles Scribner's Sons, 2003.

Herbst, Sharon Tyler. *The New Food Lover's Companion: Comprehensive Definitions of nearly 6,000 Food, Drink, and Culinary Terms*. 3rd ed. Hauppauge, NY: Barron's Cooking Guide, 2001.

Herbst, Sharon Tyler. *The New Food Lover's Tiptionary: More than 6,000 Food and Drink Tips, Secrets, Shortcuts, and Other Things Cookbooks Never Tell You*. New York: W. Morrow, 2002.

McGee, Harold. *On Food and Cooking: The Science and Lore of the Kitchen*. New York: Scribner, 2004.

Montagné, Prosper. *Larousse Gastronomique*. New English edition. New York: Clarkson Potter, 2001.

*The Oxford Encyclopedia of Food and Drink in America*. Andrew F. Smith, editor in chief. Oxford; New York: Oxford University Press, 2004.

*Webster's New World Dictionary of Culinary Arts*. Compiled by Steven Labensky, Gaye G. Ingram, and Sarah R. Labensky. Illustrated by William E. Ingram. 2nd ed. Upper Saddle River, NJ: Prentice Hall, 2001.

*Williams-Sonoma Kitchen Companion: The A to Z of Everyday Cooking, Equipment & Ingredients*. Chuck Williams, general editor; text by Mary Goodbody, Carolyn Miller, and Thy Tran. Illus. by Alice Harth. Alexandria, VA: Time-Life Books, 2000.


of your library, in which you might find general books about food like those listed in Figure 16.2.

Glancing into one of these books—perhaps *On Food and Cooking*—we find 14 pages of books and articles that author Harold McGee used in preparing his own book. McGee lists sources he used throughout the book, sources written both for general and for specialized audiences. He also lists sources for each chapter. For example, for Chapter 6, "A Survey of Common Vegetables," McGee lists eight books for the

**FIGURE 16-3 More Specialized Books for Food Research.**

Compiled by Christine Crawford-Oppenheimer at the Culinary Institute of America

Bladholm, Linda. *The Asian Grocery Store Demystified*. Los Angeles: Renaissance Books, 1999.

Claiborne, Craig. *New York Times Food Encyclopedia*. Compiled by Joan Whitman. New York: Times Books, 1985.

Mariani, John F. *The Dictionary of Italian Food and Drink: An A to Z Guide with 2,300 Authentic Definitions and 50 Classic Recipes*. New York: Broadway Books, 1998.

Mariani, John F. *The Encyclopedia of American Food and Drink*. New York: Lebhar-Friedman Books, 1999.

McClane, A.J. (Albert Jules) *The Encyclopedia of Fish Cookery*. Photography by Arie deZanger; designed by Albert Squillace. New York: Holt, Rinehart and Winston, 1977.

Passmore, Jacki. *The Encyclopedia of Asian Food and Cooking*. U.S. ed. New York: Hearst Books, 1991.

Snodgrass, Mary Ellen. *Encyclopedia of Kitchen History*. New York: Fitzroy Dearborn, 2004.

Solomon, Charmaine. *Charmaine Solomon's Encyclopedia of Asian Food*. With Nina Solomon. Boston: Periplus Editions; North Clarendon, VT: distributed by C.E. Tuttle, 1998.

Zibart, Eve. *The Ethnic Food Lover's Companion*. Birmingham, AL: Menasha Ridge Press; Old Saybrook, CT: distributed by The Globe Pequot Press, 2001.

general reader, such as *The Random House Book of Vegetables*, and thirty-five more technical books and articles, such as "Effects of salts and pH on heating-related softening of snap beans" in the *Journal of Food Science*. Each information source can lead you to ten more!

As you continue your research, check your library's online catalog for more *specialized* books about your topic. Note the narrower focus of the books listed in Figure 16.3.

Like *On Food and Cooking*, these books will lead you to others. Browse through the table of contents and the list of references. Skim a few chapters for evidence that this book might be useful to your research. Don't stop with the first quotable paragraph—there might be riper fruit on the next page.

5

**FIGURE 16-4** **Selected Online Databases.**

**Encyclopedia Britannica Online**

**EBSCOhost Web** includes the following indexes:

- EBSCO Animals
- EBSCOhost Español
- Funk & Wagnalls New World Encyclopedia
- General Science Collection
- Hospitality & Tourism Index Full Text
- MasterFile Select: general interest articles and reference works
- Primary Search via Searchasaurus
- TOPIC Search

**InfoTrac** includes several databases:

- Biography Resource Center
- Business and Company ASAP
- Business and Company Resource Center
- Custom Newspapers
- Expanded Academic ASAP
- Health and Wellness Resource Center and Alternative Health Module
- Health Reference Center Academic
- Informe
- Junior Edition, K-12
- National Newspaper Index
- Opposing Viewpoints Resource Center
- The Twayne Authors Series

**ProQuest** is an index to thousands of current periodicals and newspapers, including many full-text articles.

Your library or school also offers access to certain subscription databases, such as EBSCO's *Hospitality & Tourism Index Full Text* or *Encyclopedia Britannica Online* (see Figure 16.4). An advanced search of these databases can quickly produce reliable information. It's like shopping at a convenient yet quite specialized farm co-op. The ingredients are fresh—but you have to belong to the group to buy them. In the case of subscription databases, your school or library does belong, and you are able to access the information through the institution's account.

Although it's easy to find information on the Internet, you may not always get the quality results you need. Subscription databases can help, as can your librarian. See the next section on evaluating your sources.

**EXERCISE 16.2** **What's in Your Library?**

Pay a visit to a library—perhaps the one at your school, or a public library. What subscription databases does this library offer? What periodicals are available in print? Choose a few titles from Figure 16.2 or 16.3, and look them up in the library's online catalog. Which ones does the library own? Where are they shelved? What other services does the library provide, such as interlibrary loan or photocopying?



## EVALUATING YOUR SOURCES: BOOKS AND ARTICLES

Not all sources are equally reliable. Just think about the *people* you know. Whether it's the date of a mutual friend's birthday or the best wine to serve with oysters, some of your friends are probably more likely than others to know the real facts. In doing research for a paper, you must also seek out the most reliable information sources. In the preceding section, we looked at where these can be found: in your textbook, library reference books, and sources listed in those works. Of course, you will use other sources of information, but be prepared to evaluate them carefully. It's the difference between choosing a wine recommended by the sommelier and grabbing the nearest bottle in a budget liquor store. It's the difference between dating someone who's been an acquaintance for years and someone whom you've just met at the supermarket in front of the doughnuts. It might work out just fine in the latter case, but it might be a disaster. You're clearly taking a risk.

In evaluating a source of information, ask yourself, ***Who is the author? What are his credentials for writing on this topic?*** Information is more likely to be accurate if the author has an advanced degree or significant experience in the field, has published other books and articles on this topic, is cited by other experts in the field, or is affiliated with a reputable institution. Who is more likely to have accurate information on healthy eating—the winner of an all-you-can-eat contest at the state fair or a registered dietitian? Whose opinion on a new restaurant seems more valuable—an established food writer who publishes regularly in *The New York Times* or a beginning reporter who occasionally has a restaurant review accepted by a local paper? If you were looking for medical advice, would you seek out a professor at Yale Medical School or a student pursuing a certificate in medical transcription?

Now, of course, there are going to be exceptions. There might be unhelpful doctors at a prestigious medical school and brilliant diagnosticians elsewhere. But when your knowledge of the author is limited, it is wise to take degrees, affiliations, and other publications into account. You can often find information about the author inside the book or on the jacket, or in the article or magazine. You can also search the catalog and databases for additional books and articles published by the author.

## EXERCISE 16.3 Evaluating the Author

Evaluate each person below as a source of information on foodborne illness. Which one is least likely to be reliable? Why?

a) a Ph.D. in microbiology from Cornell University

b) the author of a food safety textbook

c) the author of three books on food photography

d) an official at the Food and Drug Administration who worked on the 2001 Model Food Code

***Where does the information come from?*** Consider, for example, the array of magazines lined up by the cashier. Tabloids offer such improbable headlines as "Three-headed Calf Born in Department Store Restroom." Yet would you really look in this type of publication for information about calves, department stores, or even restrooms? No. The tabloids are about entertainment, not information.

One step up, perhaps, but also unreliable, are gossip magazines. Packed with photos of celebrities, these magazines are sources of entertainment, though they may also record popular trends in food and dining. For many in the hospitality industry, this type of information might be useful. For example, if celebrity chef Wolfgang Puck has success with a new dish or a new restaurant, many others around the country might wish to follow his lead. The information is only going to be useful, however, if it is *true*. You would be far more likely to find accurate information on current restaurant trends in a periodical dedicated to the industry, such as *Nation's Restaurant News* or the *Journal of Restaurant and Foodservice Marketing*.

A level above the gossip magazine is the news magazine. While this type of publication is more than just entertainment, its information may be less complete and less accurate than that of a serious, specialized journal. A news magazine can raise our awareness of certain medical issues, for example, but you would be wise to consult your physician or a medical journal before beginning or ending a course of treatment. The most reliable journals are published by universities or professional organizations and contain articles by specialists in the field. In general, they do not contain advertisements, and they have titles like the *Journal of the American Medical Association, International Journal of Food Science Technology*, and *Annals of Botany*.

***What is reported, and how is it presented?*** Good information tends to be reported in some depth, supported by facts and citations, and presented in coherent sentences without glaring grammatical errors. A three-sentence teaser in a popular magazine's column on health and fitness will probably not contain the amount of detail you'd need for a research paper on trans fats.

***When was it published?*** Not only the source but the *date* of the publication is important in determining the accuracy or relevance of the information. If you need to know about the change in the 2001 FDA Model Food Code regarding the holding temperature of hot food, an article published in 2000 could not help you. If you were looking for current trends in restaurant menus, an article published even last year might be outdated. For many questions, current information is critically important. For other questions, the date does not matter. For example, if you were looking for information on Carême's vertical designs for the banquet table, a book published in 1995 could be just as useful as one published in 2005.

### EXERCISE 16.4 Choosing the Best Source

Which of the following would contain the most accurate and current information on the innovative cooking techniques of Thomas Keller, executive chef of the French Laundry?

a) A book about promising young chefs published in 1978

b) An article in *Food and Wine* published in 2006

c) An article about the French Laundry's menu design published in 2006

d) A review of his book *The French Laundry* published in 1997

Explain your answer.

***Why was it written?*** A critical question in many situations is "Why is he telling me that?" People's motives for telling us something range from an altruistic desire to help or a passion for the subject itself to a need to show off or to deceive and manipulate us. We must try to understand how the author's purpose may affect the quality of his information. Think back to that headline about the calf in the department store. What is the purpose of such a story? Clearly, it is not to inform, or to persuade. It is purely entertainment. A news magazine might also want to entertain—as well as inform—in order to boost sales. It is in competition with other similar magazines and must ask itself "What will sell?" as frequently as "What is accurate and important information?" A scholarly journal, however, is not expecting to make a great deal of money. Rather, it hopes to build its reputation through the publication of important, accurate information. Thus these journals can be excellent tools for research.

### EXERCISE 16.5 Evaluating Books and Articles

Find an article about trans fats. Now evaluate its author (whether he has appropriate credentials, what institutions he may be affiliated with), its source (what magazine published the article), coverage (what is reported and how it is presented), its currency (when it was published), and its objectivity (why it was written). Would you use this article for a research paper on trans fats? Explain.

## EVALUATING YOUR SOURCES: THE INTERNET

Evaluating the accuracy of information on the Internet is much more difficult than evaluating the accuracy of print sources. The Internet is like a giant, disorganized store in which all kinds of products of all levels of quality are strewn carelessly about the aisles. There definitely is some gold there, but there's a whole lot of garbage as well. Even worse, there's a lot of garbage pretending to be gold. And since no one really owns the store, there's no hope of formally separating the gold from the garbage. You must sift through on your own and assess each item.

***Who is the author? What are his credentials?*** Evaluating the author of an article posted on the Internet can be especially difficult; in fact, it may be impossible. The Web site might be careless in documenting its authorship, or it might even wish to conceal it. If you cannot easily find the name of the person who wrote the article, or the name of the organization that sponsors the Web site, the information may not be legitimate, and you might be wise to avoid it. If you do find the author, try to evaluate his credentials in terms of degrees, work experience, other publications, and affiliations. For example, an author who is affiliated with the National Institutes of Health appears to be more objective about trans fats than an author who works with a company that sells them!

***Where does the information come from?*** Producing a book costs money, and reputable publishers are very careful to spend money only on solid, useful information. The key word, of course, is "reputable." Just as many magazines are produced only for entertainment, so are many books. As you're researching your topic, look for well-known publishers such as Random House or Houghton Mifflin. A university press, such as Oxford University Press, is another good source of reliable books because their authors are "screened" for reputation and reliability.

On the Internet, however, there is no screening. Anyone can post information. It's we who must decide which is reliable. One way of evaluating the accuracy or "authority" of a Web site is to look at its domain, the three letters that follow the "dot" and provide some information about the sponsor of the site. Authoritative Web sites generally have one of the following domains:

.edu (sponsored by a college, university, or school)
.gov (sponsored by a nonmilitary government agency)
.org (sponsored by a nonprofit organization)

5

**FIGURE 16-5 Evaluating an Internet Domain.**

| Domain | Sponsor | Example |
|---|---|---|
| .com | $$$ | www.triswimcoach.com |
| .edu |  | www.umm.edu |
| .gov |  | www.fda.gov |
| .net | ? | www.freedomfly.net |
| .org | 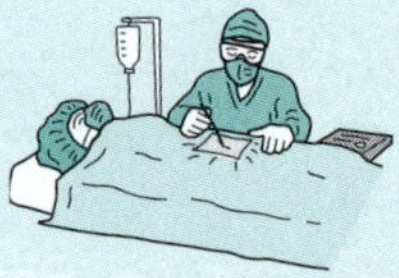 | www.americanheart.org |

Other domains exist, such as the popular *.com* (commercial site—think $$$) or the more neutral *.net*. Since anyone can use these domains, however, we need to be careful. Further, a *dot com* may be more about profitability than accuracy. In deciding which of two medications to take, would you take the advice of a drug company or of your physician? While the drug company wants to make a sale, your physician wants to keep you healthy. It makes sense to evaluate a *dot com* with particular care for evidence of bias or inaccuracy.

## EXERCISE 16.6 Evaluating Internet Domains

Look up the articles on trans fat in the sample Web sites listed in Figure 16.5. Who is the sponsor of each site? What does that suggest about its accuracy?

***What is reported, and how is it presented?*** A good source covers the topic with some breadth and depth. Three short lines on the FDA regulations regarding trans fat on nutrition labels probably do not contain information relevant to a study of the effect of the substance on human beings. Look for sources that are comprehensive enough to be useful. In addition, look for sources that cite the sources of *their* information. Reliable sources on the Internet, just as those in print do, will cite factual information in footnotes or lists of references. Finally, ask yourself whether the text is well written and the information well documented. Most likely, the care that is taken in presenting the information reflects the value of the information itself.

***When was it written? And when was it last updated?*** Again, this information may be difficult to find on a Web site. If that's the case, consider finding another site. A legitimate sponsor will want the public to know both the source and currency of its information. Check also that links to other Web pages are still working.

***Why was it written?*** It is always important to understand *why* someone is giving you certain information. A salesperson wants you to buy the product and perhaps will slant the facts to achieve that purpose. A consumer organization wants you to make a safe, informed choice and will probably be more objective. Be careful, though, that there's no secret commercial sponsorship behind an ostensibly consumer-oriented site. Even an otherwise believable author might be suspect if linked to a commercial sponsor.

## EXERCISE 16.7 Evaluating Internet Sources

Find an article about trans fat on the Internet. Then evaluate its author (whether he has appropriate credentials), its domain (who sponsors the site), coverage (what is reported and how it is presented), currency (when it was written, when last updated), and objectivity (why it was written). Would you use this article for a research paper on trans fats? Explain.

Your dish is only as good as its ingredients, and your research paper is only as good as its sources of information. Evaluate them carefully.

Your dish is only as good as its ingredients. In the same way, your research is only as good as its information. Take the time to find accurate, objective sources. Dare to visit the library, even the librarian. And most important—don't believe everything you read just because it's in print!

## RECIPE FOR REVIEW Research I

### Finding Sources of Information

1. Once you've chosen a topic, begin your search for information with your textbook or with reference books in the library. Familiarize yourself with the background and vocabulary of the topic.
2. Check your library for books on the topic. These books will also have valuable lists of other references.
3. Talk to the reference librarian about where to find specific information.
4. Search the subscription databases available through your library.
5. Remember that while it's easy to "google" a topic, steps one through four above are more likely to yield accurate information!

### Evaluating Your Sources

1. Who is the author? What are his or her credentials for writing on this topic? What are his or her connections to the product, if any?
2. Where does the information come from? Check out an Internet source with particular care.
3. What is reported, and how is it presented?

4. When was the information published? Is the date relevant to your topic?

5. Why was it written? If to entertain or persuade, is the information and/or author reliable?

## A Taste for Reading

1. The excerpt from Eric Schlosser's *Fast Food Nation* reprinted in Appendix VII incorporates data from such sources as the National Restaurant Association, the U.S. Commerce Department, the U.S. Census Bureau, the McDonald's Corporation, *Forbes* magazine, and Jim Hightower's *Eat Your Heart Out.* Explain why you think each of these sources is or isn't reliable.

2. Evaluate the information sources on the Works Cited page of Travis Becket's research paper in Appendix VI.

## Ideas for Writing

1. What do you think of when you hear the term "research paper"? Why? What have you learned about research papers in previous experiences?

2. What was the topic of your last research paper? How did you find the information? In light of the discussion in this chapter, do you think you found the *best* information? Explain.

3. When you ask a salesperson for advice on which brand or variety to choose, how do you judge the accuracy and objectivity of his response? Include specific examples in your answer.

4. Pick a topic and "google" it. Print the first page of hits. Next, log on to an appropriate subscription database and search for recently published articles in reputable journals. Print the first page of citations. How were the results of each search similar? How were they different?

chapter 17

# An Introduction to Research II: Using and Citing Your Sources

Once the investigators have processed a crime scene, they begin to analyze and reflect on the evidence they've gathered. They will draw conclusions based on this evidence and write them up in a report. In some cases they will appear in the courtroom to testify as to the accuracy of their information. Similarly, once you've assembled your sources, you must read and *think* about the information they provide. Finally, you will write up a report, your research paper, which should reflect your own interpretation of the information and should draw its basic focus and organization from you—not your sources! Like the crime scene investigators, you will make an independent assessment of the facts. To keep your own thinking central to the paper, follow the steps outlined below.

## USING YOUR SOURCES

First, be sure to leave enough time to do the research, that is, to find useful material and to read it carefully. Many problems with plagiarism are caused by procrastination. Students wait until the last minute, then cut and paste paragraphs from hastily retrieved Internet sources. That is the culinary equivalent of opening three or four cans of processed food, arranging them on a plate, and leading the customer to believe you spent the day laboring over a hot stove!

Second, as you're doing the research, take notes on the main ideas and interesting details. You can certainly start with a yellow highlighter. However, once you've identified useful passages in your material, it's helpful to write down the main points on a separate sheet of paper or on individual index cards. If you copy the words exactly, enclose them in quotation marks. This is a **direct quote**. If you use your own words to restate or **paraphrase** the information, don't use quotation marks. Be certain that your own words are sufficiently different from the original; it is not acceptable to change a word or two and pass the sentence off as your own. At the same time, the paraphrase should offer the same information

**Quote, Paraphrase, and Summary.**

**Direct Quote:**
"Whan that April with his showres soote
The droughte of March hath perced to the roote,
And bathed every veine in swich licour,
Of which vertu engendred is the flowr. . .
Thanne longen folk to goon on pilgrimages."
(Chaucer, "Prologue," lines 1–4, 12)

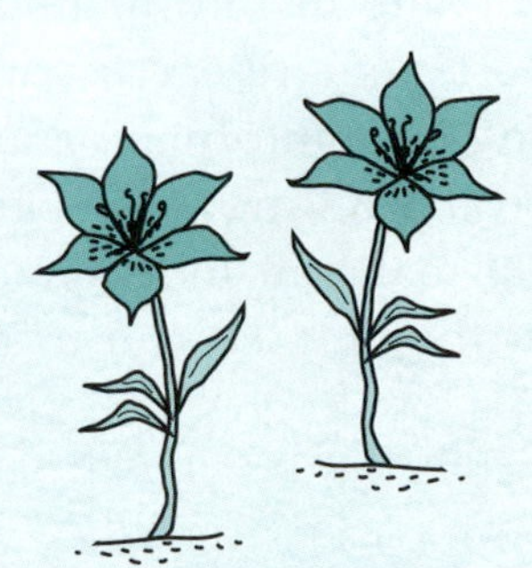

**Paraphrase:** When April rains drench the parched earth and wash each root with pure water, from whose strength the flowers grow . . . people feel like taking trips.

**Summary:** April showers bring May flowers. . . and spring fever.

**Work Cited**

Chauer, Geoffrey. "The General Prologue to the Canterbury Tales". *Chaucer's Poetry: An Anthology for the Modern Reader*. Selected and edited by E. T. Donaldson. New York: Ronald Press, 1975.

and point of view as the original. The third option for your notes is a **summary** of useful information, that is, a condensed statement—again, entirely in your own words—of the main idea and key supporting points.

When you take notes, be sure to write the name of the book or article and the name of the author at the top of each page or note card. Include all publishing information as well: the publishing company, city, and date for books; the periodical title, volume, number, and date for articles; and the URL, site sponsor, and dates posted and accessed for Internet sites. Another option is to keep a separate card (or list) that includes the title and publishing information for each source. Whether you quote, paraphrase, or summarize, be sure to write the page number of the original text in your notes.

Next, after you finish reading each chapter or article, it is especially helpful to stop and ask yourself, What did I learn? Without looking back at your notes or the text itself, write down what you discovered. In this way you begin to integrate the new information into your own thinking. Once you've written a brief summary, refer back to the text to check its accuracy and make any necessary corrections. If you add lines of text directly to your notes, however, be certain to enclose them in quotation marks. Any sloppiness in this area could result in a charge of plagiarism.

5

## Taking Notes

As we read, it is important to take notes on the main ideas and interesting details presented in each source. Whether you use notebook paper or note cards, you must be sure to write down the author, title, publishing information, and page number. You must also be absolutely certain to distinguish between sentences that you have copied exactly from the source and those that you have paraphrased or summarized. Let's try an example. Suppose, like student Travis Becket, whose research paper is printed in Appendix VI, we are interested in environmentally-friendly agriculture. Our research might lead to *The Real Green Revolution: Organic and Agroecological Farming in the South*[27] by Nicholas Parrott and Terry Marsden, which contains the following passage:

> Zaï (or tassa) is a traditional agricultural method used in Burkina Faso to restore arid and crusted areas of fields. The technique involves making seed holes 20–30 cm wide and deep and using the earth to make a raised 'demi-lune' barrier on the downslope side. Compost and/or natural phosphate is placed in

> each hole and sorghum or millet seeds is planted when it rains. This technique improves the organic structure of the soil, helps retain moisture and, through promoting termite activity, increases water filtration into the soil. The crops are planted relatively densely to increase ground cover and prevent water loss through evapotranspiration. Stones removed from the field while digging the holes are often used to make contour bunds to further stabilize the soil and reduce run-off and erosion. (39)

Note that we must copy the source's punctuation as well as its words; thus we use the single quotation marks around *demi-lune.* The number in parentheses refers to the page on which the passage is found.

Once we read the paragraph, we stop and think about what it means and about how to record the main ideas and relevant details. The first step is to copy the necessary information about the source. We'll need the authors' names, the title, and the publishing information, which for a book is city, publisher, and year.

| | |
|---|---|
| **Authors** | Nicholas Parrott and Terry Marsden |
| **Title** | *The Real Green Revolution: Organic and Agroecological Farming in the South* |
| **Publishing information** | London: Greenpeace Environmental Trust, 2002 |

Since this book was viewed on the Internet, we will need that information as well.

| | |
|---|---|
| **Date accessed** | 05 Oct. 2005 |
| **URL** | http://www.greenpeace.org.uk/MultimediaFiles/Live/FullReport/4526.pdf |

Next, we think about the passage and decide what information seems important enough to put on our note card. We must also decide whether to copy the text exactly, or whether to write a paraphrase or summary of the material.

## Quoting the Original Text

Words that are copied directly from the text (direct quote) should be enclosed in quotation marks. You *must* be able to distinguish which words are yours and which belong to the original source. For example, the second sentence looks difficult to paraphrase because of such

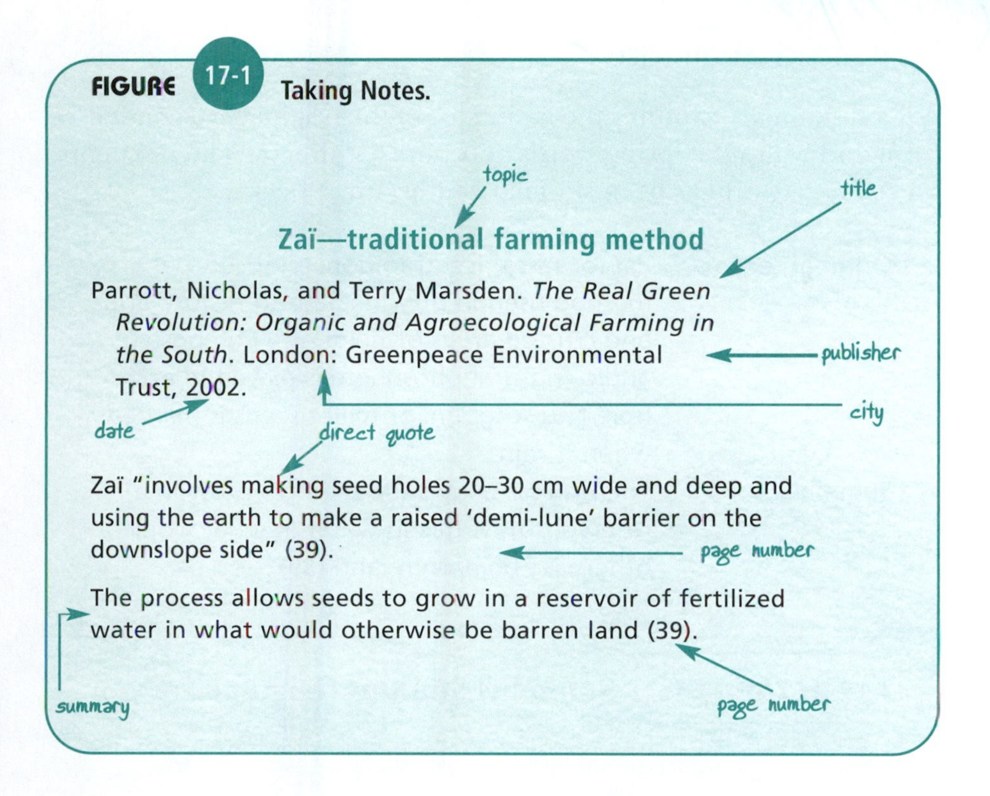


technical terms as *demi-lune* and *downslope*. Let's copy the sentence exactly onto our note card (Figure 17.1) and enclose it in quotation marks.

On the note card, the phrases copied directly from the original passage are in quotation marks.

| | |
|---|---|
| Original text: | [Zaï] involves making seed holes 20–30 cm wide and deep and using the earth to make a raised 'demi-lune' barrier on the downslope side. |
| Direct quote: | Zaï "involves making seed holes 20–30 cm wide and deep and using the earth to make a raised 'demi-lune' barrier on the downslope side"(39). |

Whatever the length of the direct quote, be sure that you have copied words and punctuation exactly and noted the page number(s). The brackets around *Zaï* indicate that the word does not appear in the original sentence. Double check the correctness of each quotation before you submit the paper. Any carelessness in the use of your sources may look like plagiarism.

## Writing a Summary

A good summary captures the main idea of the original in a condensed form and without copying words and phrases directly. The next entry on the note card has been summarized by Travis Becket.

| | |
|---|---|
| Original text: | Zaï (or tassa) is a traditional agricultural method used in Burkina Faso to restore arid and crusted areas of fields. . . . Compost and/or natural phosphate is placed in each hole and sorghum or millet seeds is planted when it rains. |
| Summary: | The process allows the seeds to grow in a reservoir of fertilized water in what would otherwise be barren land (39). |

### EXERCISE 17.1 Summarizing the Original Text

Write a one sentence summary of the following passage, also from *The Real Green Revolution*.[28]

> Egypt has what is probably the most developed organic sector in North Africa. Initial interest in organic production was triggered as a reaction to increasing health problems experienced by farmers and rural dwellers from pesticide poisoning, and cotton yields remaining constant or declining despite increased use of pesticides. Aerial spraying of cotton is now banned in Egypt and much pest control now done through the use of pheromones, even though systems are not wholly organic.

## Writing a Paraphrase

Sometimes neither the summary nor the direct quote is appropriate. We may need more detail than the summary provides, yet we cannot write a coherent paper by simply adding a bunch of quotes together. The solution may be a paraphrase, which means changing the structure and wording of the original text to fit in with our own style and to support our own points more effectively. Remember that a paraphrase translates the exact meaning of the original while using your own vocabulary and sentence structure. Look at the following example:

| | |
|---|---|
| Original text: | Thus this simple and traditional technique is proving of multiple benefit in increasing yields, restoring degraded land and, by generating employment opportunities, helping slow the process of rural/urban migration. |
| Draft paraphrase: | The "simple and traditional" process of Zaï has been proven to restore degraded land, increase crop yields and simultaneously slow the process of urban migration (39). |

Although the phrase "simple and traditional" is correctly enclosed within quotation marks, there are too many other identical or similar words, for example, "is proving" versus "has been proven," "in increasing yields" versus "increase crop yields," and "helping slow the process of rural/urban migration" versus "slow the process of urban migration." Note how the revised paraphrase below changes both the structure of the original sentence and the words themselves.

| | |
|---|---|
| Revised paraphrase: | As the soil continues to improve through this ancient, environmentally-friendly method, both crop production and jobs have expanded while the movement of the population away from the countryside and toward the cities has decelerated (39). |

In addition to the structural changes, the wording has been appropriately altered. Where the original reads "this simple and traditional technique," the revised paraphrase reads "this ancient, environmentally-friendly method." Where the original says "restoring degraded land," the revision says "as the soil continues to improve." Finally, where the original reads "helping slow the process of rural/urban migration," the revision reads "the movement of the population away from the countryside and toward the cities has decelerated." Let's look at another example:

| | |
|---|---|
| Original: | This technique improves the organic structure of the soil, helps retain moisture and, through promoting termite activity, increases water filtration into the soil. |

Paraphrase: As a result of using the Zaï method, more water seeps into and is retained in the earth because of its enhanced composition and proliferating insect population (39).

Again, notice that both the sentence structure of the paraphrase and the words themselves are different from the original text. The original begins with the main subject, the paraphrase with a pair of phrases. Instead of the original "helps retain moisture," the paraphrase reads "more water seeps into and is retained in the earth."

Study your paraphrases carefully, and remember that it is *imperative* that you can distinguish in your notes between direct quote on the one hand and paraphrase or summary on the other. As a final check before submitting your essay, be sure to compare your quotes, paraphrases, and summaries with the original text.

### EXERCISE 17.2 Paraphrasing the Original Text

Write a paraphrase of each of the following sentences.[29] Use your own words and your own word order or sentence structure, but include the important details. Try to translate the quote's meaning exactly.

1. "The system [Zaï] is labour intensive and best suited to farms with a labour surplus."
2. "Initial interest in organic production was triggered as a reaction to increasing health problems experienced by farmers and rural dwellers from pesticide poisoning."

## Processing the Information in Your Sources

At each stage of your research, try to process new information and connect it to what you already know about your topic. Think about the focus of your paper. Jot down questions, ideas, outlines. Remember step one of the writing process, brainstorming? When you ask yourself what a passage from an information source means and how it might relate to what you already know and to what you want to write about, you are using a similar process. This kind of creative reflection helps to make the essay a product of your own thinking, supported but not dominated by other people's ideas.

As your research progresses, do some freewriting or diagramming of the scope of the paper. Perhaps even do a draft that incorporates the major bits of information, however loosely. But remember—that information has already been processed, paraphrased, or summarized. Do NOT begin a draft by copying, although you may begin your notes that way. A draft should come straight from you. Then, as you rewrite, you can add quoted or paraphrased material to back up your points and give credit to the authors from whom you've drawn information.

Once you have a rough draft of the research paper, think about whether it says what you want it to say, just as you would with any other piece of writing. The major difference between an essay and a research paper is the role of outside sources in forming and answering your research question. While your paper is your own, you are responsible for showing the reader where and how you have used other people's words and ideas. Again, try to avoid writing the draft by pasting together chunks of information from outside sources. Mix this information well with your own thoughts and ideas. If you don't mix the batter well enough, you'll end up serving a cake with gobs of flour in it!

In his research paper on environmentally-friendly agriculture, Travis Becket begins with a story of his childhood, riding with his mother every Saturday to the farmers' market to sell their fresh herbs and eating up what didn't sell. "I remember wondering if everybody else in America ate as well as I did," he writes in the introduction.

5

**STUDENT WRITING** Travis P. Becket

I wondered this same idea at the University of California Santa Cruz while eating a lunch that, once again, consisted of foods that had been submerged in dirt just a few hours prior. The apprentices who worked the fields agreed with me about the superior quality of the organic food they had grown. As I saw the way that the apprentices relied on organic foods and how they supported organic farming and eating, a new question arose. "If this is the way that food should be produced in America, is it possible to produce enough for everybody?" For the rest of my time at UCSC, this was the thought in my mind and the question on my lips. If organics is best for the soil and the body, can it be utilized at a level that will feed everybody?

Travis begins his research with a question that has its roots in his childhood. Like the agriculture it explores, the question itself is organic.

While the paper goes on to use many different sources and to explore farming techniques on the other side of the globe, its origin in the "small arid plot of land just outside of town where [his mother] would slave on her hands and knees bringing up delicate flowers and herbs from the dusty soil" remains with the reader. We hear the writer's own voice throughout the paper, supported but not dominated by his outside experts.

## INCORPORATING SOURCES INTO YOUR PAPER

As we mentioned above, you can use outside sources in your writing in three ways: direct quote, paraphrase, and summary. Most information from outside sources will be incorporated in the form of paraphrases or summaries. Your research paper should not be packed full of direct quotes; they are a strong seasoning and should be used sparingly. In general, save direct quotes for material that is especially important to your topic or that is explained particularly well in the original text. There is one exception, however. If you are writing about a literary text—where the text is the subject—you will tend to use many more direct quotes.

Much of the information you use from outside sources will be paraphrased, that is, you will translate the ideas into your own words, sentence structure, and style. In this way your paper will read more smoothly as a product of your thinking, rather than as an assortment of other people's ideas. It will be a casserole made of fresh ingredients rather than a collection of prepackaged foods arranged on a plate.

### Writing Tag Lines

In order to integrate quotes and paraphrases smoothly into your paper, it is extremely important to introduce them with an appropriate **tag line**. A good server doesn't plop the plate down in front of the customer and stalk away with a toss of the head. Instead, the plate is introduced with a small remark, such as "Here we are," or "Enjoy your dinner," and perhaps a smile. The tag line often names the **author of the original text**, as in the following example:

> *Nicholas Parrott and Terry Marsden* note that the traditional farming method of Zaï "involves making seed holes 20–30 cm wide and deep and using the earth to make a raised 'demi-lune' barrier on the downslope side" (39).

The tag line may also contain the **title of the original text**, particularly the first time it is cited or if you are citing more than one work by the same author.

> In *The Real Green Revolution: Organic and Agroecological Farming in the South*, Nicholas Parrott and Terry Marsden note that the traditional farming method of Zaï "involves making seed holes 20–30 cm wide and deep and using the earth to make a raised 'demi-lune' barrier on the downslope side" (39).

The tag line may **emphasize the author's credentials**.

> Nicholas Parrott and Terry Marsden, *both at Cardiff University*, note that the traditional farming method of Zaï "involves making seed holes 20-30 cm wide and deep and using the earth to make a raise 'demi-lune' barrier on the downslope side" (39).

In the next example, the tag line **suggests the tone** of the quote.

**STUDENT WRITING** | Travis P. Becket

> *It was Kate Posey, the wistful tour guide at Santa Cruz, who said,* "We see things so neatly arranged in grocery stores, sometimes we forget what the plant looks like."

In addition to variations in content, the tag line may vary in placement. As in the example above, the tag line may precede the quoted or paraphrased material. It may also interrupt it.

> "We see things so neatly arranged in grocery stores," *said Kate Posey, the wistful tour guide at Santa Cruz,* "sometimes we forget what the plant looks like."

Or the tag line may follow the quoted or paraphrased material.

> "We see things so neatly arranged in grocery stores, sometimes we forget what the plant looks like," *said Kate Posey, the wistful tour guide at Santa Cruz.*

It is also important to incorporate borrowed ideas smoothly into the flow of your *own* thinking. For example, Travis Becket cites Posey within the following passage in a paper on environmentally-friendly agriculture:

**STUDENT WRITING** Travis P. Becket

It is often said that "United we stand, divided we fall." I believe that this old adage holds true for mankind's relationship with the environment. It was Kate Posey, the wistful tour guide at Santa Cruz, who said, "We see things so neatly arranged in grocery stores, sometimes we forget what the plant looks like." She was speaking of the disconnect between mankind and its food, how we as a society often forget that food comes not from a grocery store, but from being submerged in very unappetizing dirt.

Travis uses the quote to illustrate and emphasize the point he has already made. Quotes must not *substitute* for your own statements and interpretations; instead, they are *added* as support, illustration, or proof.

This culinary student is in no danger of forgetting fennel's connection with the earth.

**EXERCISE 17.3** **Writing Tag Lines**

Write three different tag lines for the following quote from Dr. Christos Vasilikiotis' paper "Can Organic Farming 'Feed the World'?"[30] written when he was at the University of California, Berkeley. Include the following quote: "Organic farming systems have proven that they can prevent crop loss to pests without any synthetic pesticides."

## CITING YOUR SOURCES

You **cite** your source—that is, you tell the reader where you used outside information and where you found it originally—in two places: first, within the text itself and second, at the end of your paper. The **format** of these citations is very precise. In this textbook, examples will be given in **MLA style**; however, your instructor or publisher may prefer that you use a different style, such as **APA** (American Psychological Association) or **Chicago** (*Chicago Manual of Style*). Documentation can be quite a complex process, and this introductory chapter will cover only the basics. For further information, visit the Web sites listed at the end of this chapter or consult one of the several excellent books devoted exclusively to this topic.

### Using Parenthetical Citations

The purpose of a parenthetical citation is to let the reader know that you've used another source for this information and to direct her to the complete citation on the **Works Cited page**. Thus in MLA format only two bits of information are generally included in the parenthetical citation: **the author's last name and the page number**.

> "I learned to cook by sight as the colors of the spices turned and then by smell—sweet, earthy, heady, sharp if they are roasting correctly, or the unforgiving acrid smells if they burn" (Bhide 103).[31]

Note that the parentheses lie outside the quotation mark but inside the period. However, if the quote is long enough to be set in a separate, indented paragraph, the parentheses begin one space *after* the period. See the passage quoted under Taking Notes earlier in this chapter.

If you've used **the author's name in a tag line**, put just the page number in parentheses.

> Monica Bhide explains, "I learned to cook by sight as the colors of the spices turned and then by smell—sweet, earthy, heady, sharp if they are roasting correctly, or the unforgiving acrid smells if they burn" (103).

If you use **more than one source by the same author**, write both the author's name and the title, as well as the page number, in parentheses.

> "I learned to cook by sight as the colors of the spices turned and then by smell—sweet, earthy, heady, sharp if they are roasting correctly, or the unforgiving acrid smells if they burn" (Bhide, "A Question of Taste" 103).

If the source has **two or three authors**, put all the names in the parenthetical citation or in the tag line.

> (Kirszner and Mandell 3)

If the source has **four or more authors**, you may list them all or you may use only the first author's name followed by *et al.*, which means "and others."

> (Alasalvar *et al.* 1411)

If the *source* **has no author**, use just the title and page number:

> (*Encyclopedia of Food and Culture* 193)

If you're quoting **text that is found in another source**, use the names of both.

Travis P. Becket

> According to Monsanto CEO Robert Shapiro, the end result of these stressors is "loss of topsoil, of salinity of soil as a result of irrigation, and ultimate reliance on petrochemicals" (qtd. in Vasilikiotis).

If the source **has no page numbers**, as is the case with many Internet sites, you may leave them out or put *n.pag.* for "no pagination." For example, since the article by Christos Vasilikiotis quoted above is from an Internet site and has no page numbers, none appear in the parenthetical citation. If paragraphs or sections are clearly numbered, you may cite these in parentheses. Consult your instructor, publisher, or librarian for further information.

While parenthetical citations are most common in research papers, a book like this one may use a note format, that is, a small number raised slightly above the line of type that refers to a list of citations in the back of the book. Follow the preference of your instructor or publisher.

### EXERCISE 17.4 Parenthetical Citations

Write a parenthetical citation for the following items:

1. A quote from page 96 of Anthony Bourdain's *Kitchen Confidential*
2. A quote from page 96 of Anthony Bourdain's *Kitchen Confidential* in a essay in which you also quote from Anthony Bourdain's *A Cook's Tour*
3. A quote from page 36 of *The Oxford Encyclopedia of Food and Drink in America*
4. A quote from page 20 of *Coffee Basics* by K. Knox and J.S. Huffaker
5. A quote from the Food and Agriculture Organization of the United Nations (FAO) website, *Organic Agriculture at FAO*

### EXERCISE 17.5 More Practice Writing Tag Lines

Now write a tag line for each of the sources in Exercise 17.4.

## What NOT to Cite

In writing your research paper, you've probably used the following types of information: your own thoughts and experiences, material from outside sources that represents the thoughts and experiences of other writers, and general knowledge, information that everyone shares. You only need to cite the middle type of information, the words and ideas of other writers.

General knowledge includes such information as scientific facts (water boils at 212°F) or historical facts (Elizabeth II is the Queen of England). Such information is general knowledge even though you yourself may not have known it before. For example, the fact that the boiling point of water is lower (203° F) at mile-high elevations is general knowledge, even though a particular individual may not know it. General knowledge also includes observations or conclusions based on common sense: Burned toast doesn't taste good. If the knowledge is more specialized, such as the precise chemical reaction that occurs when the toast burns, then you must cite the source.

However, it is not always clear what information falls into the category of general knowledge, what needs to be cited, and what does not. If you read the same information in several different sources, it is possible that it is general knowledge within that subject area. When in doubt, cite the source of your information.

### EXERCISE 17.6 To Cite or Not to Cite

Label each of the following items as "Yes" if it needs to be cited or "No" if it is general information. Be prepared to explain the reason for your answer.

_____ 1. "Nothing really matters to me" is the concluding line of Queen's "Bohemian Rhapsody."

_____ 2. The Beatles took the world by storm in the 1960s.

_____ 3. Julia Child was one of the first celebrity chefs in the United States.

_____ 4. Anthony Bourdain had a life-changing job in a restaurant on Cape Cod in 1974.

_____ 5. Anthony Bourdain is the author of *Bone in the Throat* and *Gone Bamboo*.

## Developing the Works Cited Page

The Works Cited page contains an alphabetized list of all the outside material used or **cited** in the paper, including publishing information for each one. The purpose of the Works Cited page is not only to acknowledge the outside sources you have used in your paper but also to give your readers the information they need to find the sources.

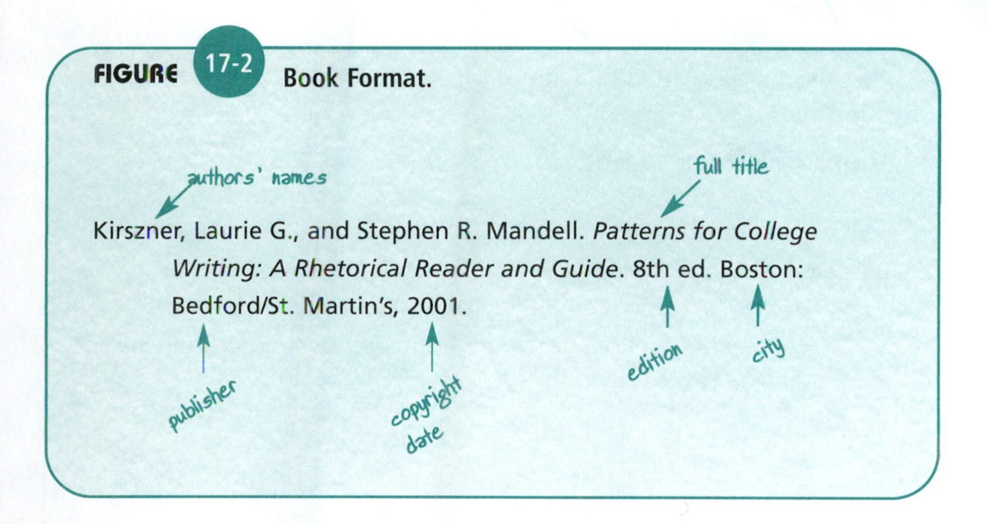

FIGURE 17-2 Book Format.

Readers see the author's name or the source's title in a parenthetical citation and can find the corresponding entry on the Works Cited page. Occasionally, as in this textbook, a writer will use a **Works Consulted page**, which includes sources that influenced her thinking whether or not they are actually cited in the text.

The Works Cited entries are like little "hanging" paragraphs, that is, the first line begins at the left hand margin, while the subsequent lines are indented. In general, the entries for print sources have three parts, each of which ends with a period.

Author. Title. Publishing information.

A **book with one author** is fairly straightforward.

McGee, Harold. *On Food and Cooking: The Science and Lore of the Kitchen.* New York: Scribner, 2004.

The same principle applies throughout the many variations described below. If the source has **more than one author**, only the name of the first author is inverted. Notice that subsequent editions of the book are added after the title (Figure 17.2).

Kirszner, Laurie G., and Stephen R. Mandell. *Patterns for College Writing: A Rhetorical Reader and Guide.* 8th ed. Boston: Bedford/St. Martin's, 2001.

If the source has an **editor instead of an author**, alphabetize the item by title, and add the editor's name after it.

*Encyclopedia of Food and Culture.* Ed. Solomon H. Katz. New York: Scribner, 2003.

5

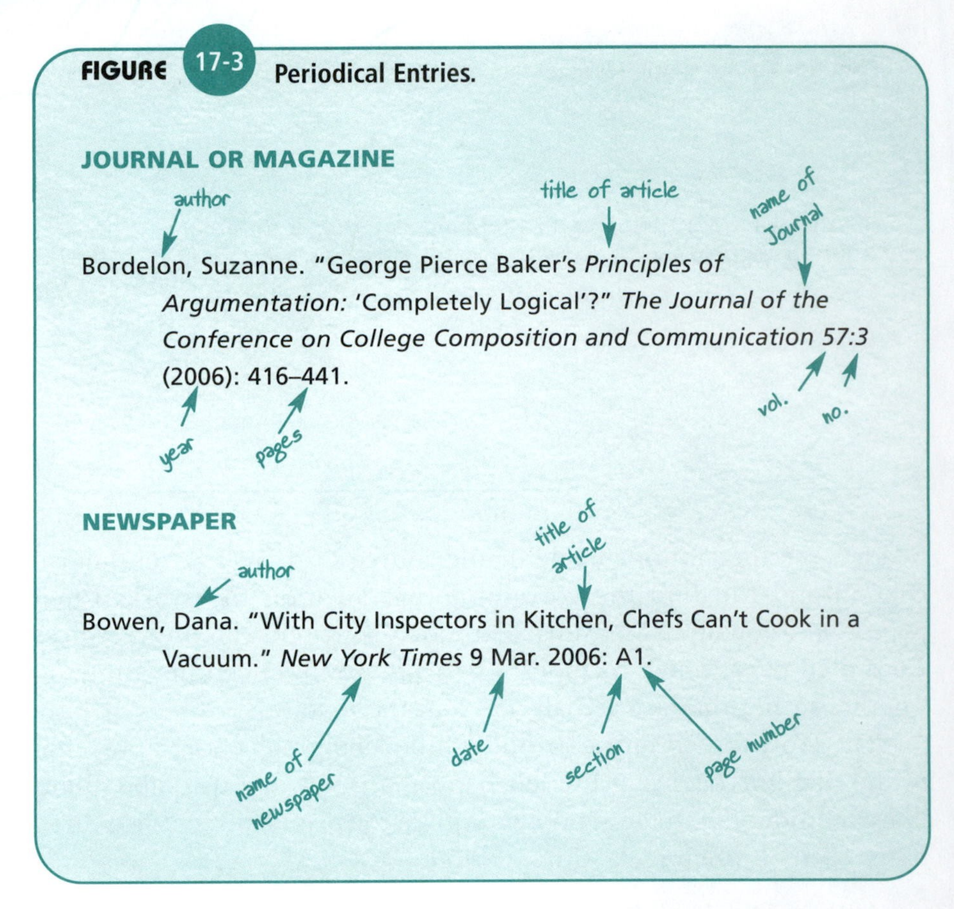
FIGURE 17-3 Periodical Entries.

JOURNAL OR MAGAZINE

Bordelon, Suzanne. "George Pierce Baker's *Principles of Argumentation:* 'Completely Logical'?" *The Journal of the Conference on College Composition and Communication 57:3* (2006): 416–441.



NEWSPAPER

Bowen, Dana. "With City Inspectors in Kitchen, Chefs Can't Cook in a Vacuum." *New York Times* 9 Mar. 2006: A1.



If the source is a **collection of works by various authors**, begin with the author and title of the work you used in your paper. Then add the information on the collection, followed by the page numbers of the particular title.

> Bhide, Monica. "A Question of Taste." *Best Food Writing 2005.* Ed. Holly Hughes. New York: Marlowe, 2005. 102–104.

For a **weekly or monthly magazine**, write the author's name, the title of the article, the title of the magazine, the date, and the page numbers of the article.

> Wemischner, Robert. "Crimson Tide." *Food Arts* Sep. 2004: 87–89.

For a **quarterly magazine**, write the volume number and issue following the title (Figure 17.3).

> Bordelon, Suzanne. "George Pierce Baker's *Principles of Argumentation:* 'Completely Logical'?" *The Journal of the Conference on College Composition and Communication* 57:3 (2006): 416–441.

For an **article from a newspaper**, include the section as well as the page number (Figure 17.3). If the newspaper's title does not name the city in which it is published, put that information in brackets.

> Bowen, Dana. "With City Inspectors in Kitchen, Chefs Can't Cook in a Vacuum." *New York Times* 9 Mar. 2006: A1.

For a **movie or video**, list the title (underlined or italicized), director, major performers, distributor, and date of release. Where applicable, add *videotape* or *DVD* before the name of the distributor.

> *Crash.* Written and directed by Paul Haggis. Perf. Don Cheadle, Matt Dillon, Terence Howard, Thandie Newton, Ryan Phillippe. Lion's Gate, 2005.

**Electronic sources** present some unique issues. Often it is not clear who the author is, or who has sponsored the site, or even when it was written (Figure 17.4). There may be no page numbers. Further, the site itself may change over time so that it is important to include the date

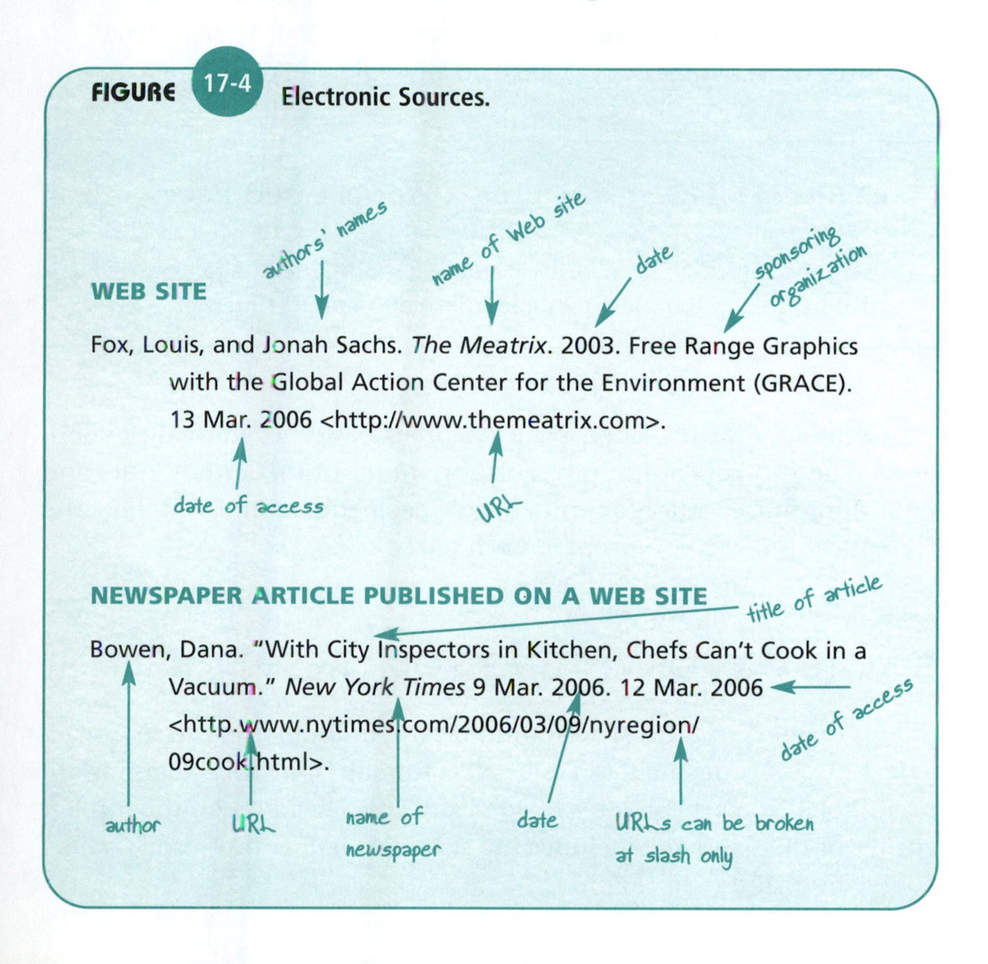


FIGURE 17-4 Electronic Sources.

on which you last viewed the material, called the **date of access**, in your notes. Consult an MLA reference work, MLA's Web site (http://www.mla.org), your instructor, or a librarian if you get stuck.

For a **Web site**, list the name of the author, if known, then the title of the site, italicized or underlined, date of publication or update, sponsoring organization, date of access, and **URL**.

> Fox, Louis, and Jonah Sachs. *The Meatrix.* 2003. Free Range Graphics with the Global Action Center for the Environment (GRACE). 13 Mar. 2006 <http://www.themeatrix.com>.

For an **online magazine**, list as much of the same information as you can.

> *StarChefs.com: The Magazine for Culinary Insiders.* StarChefs. 13 Mar. 2006 <http://www.starchefs.com>.

For an **article published or republished online**, add information on the Web site.

> Bowen, Dana. "With City Inspectors in Kitchen, Chefs Can't Cook in a Vacuum." *New York Times* 9 Mar. 2006. 12 Mar. 2006 <http.www.nytimes.com/2006/03/09/nyregion/ 09cook.html>.

### EXERCISE 17.7 Creating a Works Cited Page

Using the sources in Exercise 17.4, create a Works Cited page. You will have to look up the full publishing information on the Internet.

Creating a Works Cited page becomes easier as you gain experience. The two keys are to take accurate notes of the author, title, and publishing information as you consult each source and to follow the format carefully as you prepare each entry.

## A NOTE ON FORMATTING YOUR RESEARCH PAPER

Student papers in MLA style should be typed double spaced in a readable 12 point font, such as Times New Roman. One inch margins are preferable. No cover sheet is needed since the heading on the top left corner of the first page includes the student's name, professor's name, course title, and date (Figure 17.5).

FIGURE 17-5 **First Page of an MLA Research Paper.**

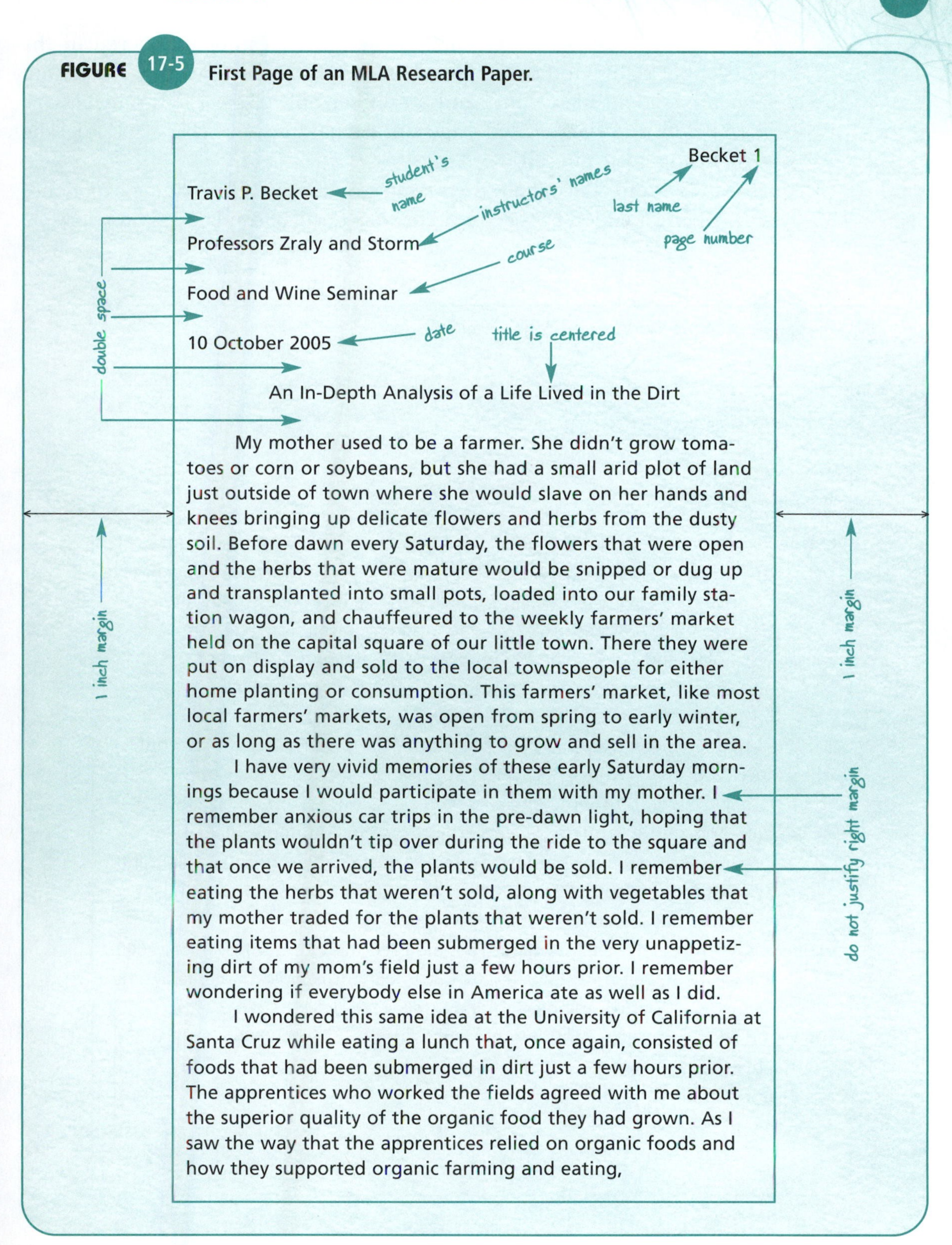

The title of the paper is centered. Pages are numbered in the upper right hand corner, one half inch from the top of the paper, with the student's last name and a number only. *Page or p.* is unnecessary. Begin the Works Cited page on a separate sheet (Figure 17.6). The heading is centered.

Proofread carefully. A research paper is like a fine dining experience where attention has been paid to every detail of content and presentation.

**FIGURE 17-6 Sample Works Cited Page.**

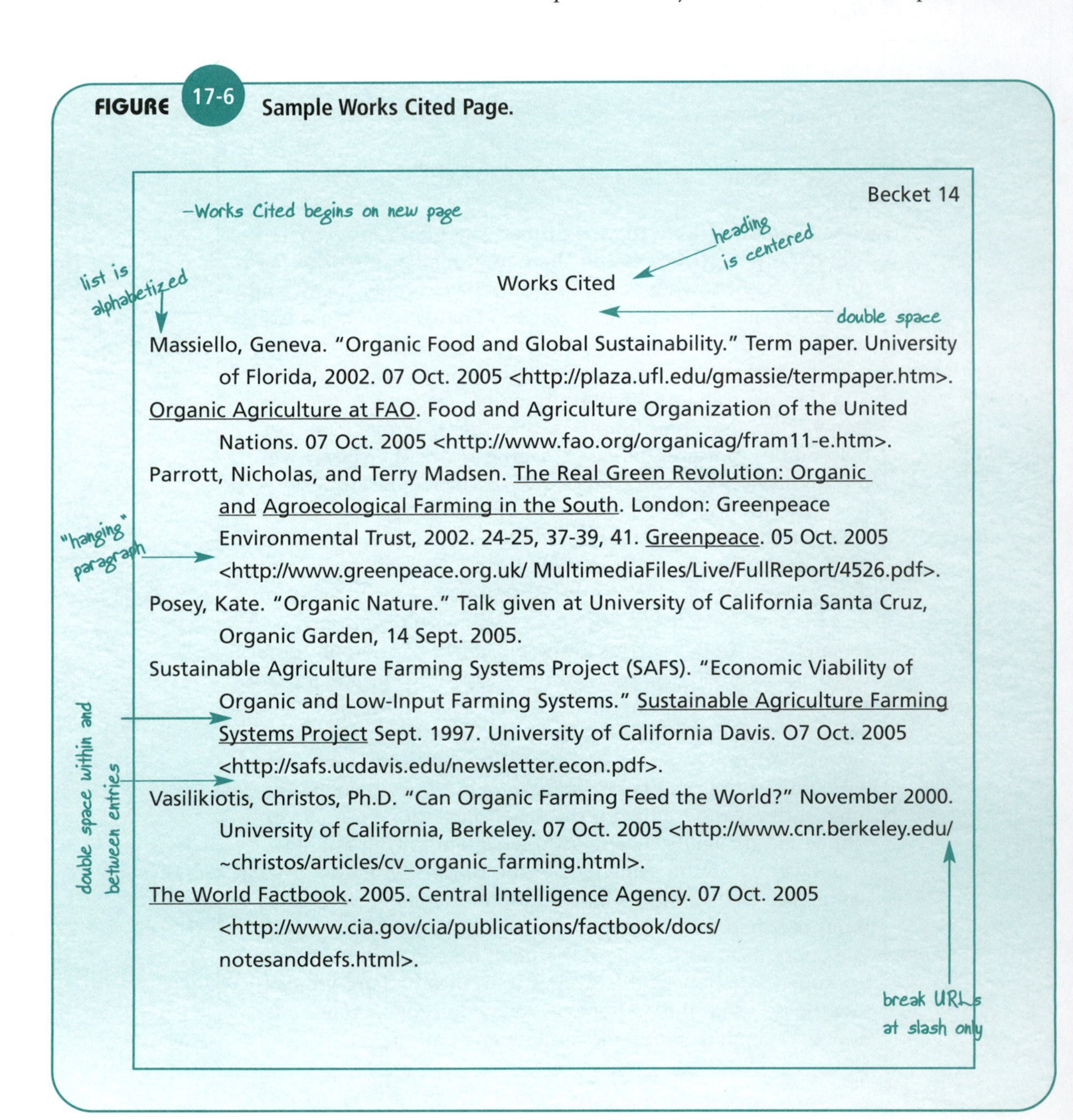

Becket 14

Works Cited

Massiello, Geneva. "Organic Food and Global Sustainability." Term paper. University of Florida, 2002. 07 Oct. 2005 <http://plaza.ufl.edu/gmassie/termpaper.htm>.

Organic Agriculture at FAO. Food and Agriculture Organization of the United Nations. 07 Oct. 2005 <http://www.fao.org/organicag/fram11-e.htm>.

Parrott, Nicholas, and Terry Madsen. The Real Green Revolution: Organic and Agroecological Farming in the South. London: Greenpeace Environmental Trust, 2002. 24-25, 37-39, 41. Greenpeace. 05 Oct. 2005 <http://www.greenpeace.org.uk/ MultimediaFiles/Live/FullReport/4526.pdf>.

Posey, Kate. "Organic Nature." Talk given at University of California Santa Cruz, Organic Garden, 14 Sept. 2005.

Sustainable Agriculture Farming Systems Project (SAFS). "Economic Viability of Organic and Low-Input Farming Systems." Sustainable Agriculture Farming Systems Project Sept. 1997. University of California Davis. O7 Oct. 2005 <http://safs.ucdavis.edu/newsletter.econ.pdf>.

Vasilikiotis, Christos, Ph.D. "Can Organic Farming Feed the World?" November 2000. University of California, Berkeley. 07 Oct. 2005 <http://www.cnr.berkeley.edu/ ~christos/articles/cv_organic_farming.html>.

The World Factbook. 2005. Central Intelligence Agency. 07 Oct. 2005 <http://www.cia.gov/cia/publications/factbook/docs/ notesanddefs.html>.

## RECIPE FOR REVIEW Research II: Using and Citing Your Sources

### Using Your Sources

1. Read and *think* about the information in your sources.
2. Take careful notes. Distinguish quoted material from paraphrased. Your notes must always include the name and author of the source, as well as the page number and publishing information.
3. Write the paper from scratch; do not cut and paste quotes from your sources.
4. Add **quotes, paraphrases, and summaries** of outside sources to support or illustrate your points.

### Citing Your Sources

1. **Cite** material from outside sources in parentheses in the text.
2. You *must* cite any information or wording that you borrowed from another source, unless it is general knowledge.
3. List all the outside sources you cited in your paper on the **Works Cited page**.
4. Use the **format** required by your instructor or publisher.

### Helpful Web Sites

1. For additional help with **MLA style**, visit http://www.mla.org.style.
2. For **APA style**, see http://www.apastyle.org.
3. For **Chicago style**, see http://www.chicagomanualofstyle.org or *The Chicago Manual of Style: The Essential Guide for Writers, Editors, and Publishers.* 15th ed. Chicago: U of Chicago P, 2003.

5

## A Taste for Reading

1. Eric Schlosser's *Fast Food Nation* is a model of good research and good writing. As you read the excerpt from his Introduction in Appendix VII, note how smoothly he incorporates information from sources as various as the National Restaurant Association, the U.S. Commerce Department, the U.S. Census Bureau, the McDonald's Corporation, *Forbes* magazine, and Jim Hightower's *Eat Your Heart Out.*
2. Read Travis Becket's research paper in Appendix VI. What principles of research discussed in this chapter does he apply in his writing?

## Ideas for Writing

1. Explore the Web site *StarChefs.com* (http://www.starchefs.com), and find an article of interest. Summarize the main points of the article in a paragraph. Then write a one-page summary in which you use two direct quotes from the article. Be sure to cite the information parenthetically, as well as in a complete Works Cited entry at the end.
2. Watch "The Meatrix" at http://www.themeatrix.com. Then pick one of the subtopics, such as Antibiotics, Buy Local, or Genetic Engineering. Take notes on the information, and develop a summary of the main points.
3. Do further research on one of the topics above or on another of your choice. Consult three reputable sources, using the evaluation techniques discussed in Chapter 16. Take notes on the information, and summarize the main points of each article. Then write a page or two about the topic, incorporating the information you learned as quotes or paraphrases. Use parenthetical citations, and create a Works Cited page.
4. What food-related jobs involve research? Consult at least three information sources as you investigate this question, and incorporate at least two in your answer.

# Plating I

## Editing for Grammar

6

UNIT

chapter 18

# Reviewing the Parts of Speech

In the earlier sections of this book, we explored the first four steps of the writing process: brainstorming, organizing, drafting, and revising. We worked on developing and organizing our ideas and on writing transitional expressions that would guide our readers smoothly through these ideas. In Units 6 and 7, we will focus on the next step, editing, which requires some understanding of the ingredients and structure of a sentence.

If we didn't have onions, celery and carrots, it would be difficult to make *mire poix*. Without tomatoes, lettuce, and bread, we couldn't build the sandwiches in Figure 18.1. Similarly, sentences are constructed from certain ingredients, the parts of speech. Each of these eight parts of speech represents a *role* or *job* to be performed in the sentence.

## THE NOUN

Perhaps the first or most basic part of speech is the noun. As children, we begin to speak by learning the names of people and things that are important in our lives: mama, kitty, juice. A **noun** is a word that names something—a person, place, thing, or idea. Some nouns refer to categories of persons, places, or things. They are called **common nouns**, and they are not capitalized unless they are used at the beginning of a sentence. *Chef, toque, tomato,* and *preparation* are examples of common nouns. **Proper nouns**, on the other hand, refer to specific persons,

**FIGURE 18-1** Like these sandwiches, a sentence is constructed from certain ingredients.

places, or things, and they are always capitalized. *Julia Child, New York,* and *Cheerios* are examples of proper nouns.

In writing class, we are interested in nouns because most often one of them—or a pronoun—will be the **subject** of a sentence, that is, it will identify who or what is doing something in the sentence, or who or what is being described by the rest of the sentence. It is important to be able to identify the subject of the sentence in order to make sure that the verb agrees with it (Chapter 22). It is also important to identify the subject so that we can avoid writing sentence fragments (Chapter 20), run-on sentences, and comma splices (Chapter 21).

A noun can also be used as the **object of a preposition**. It is especially useful to recognize prepositional phrases because neither the subject nor the verb of the sentence can be found inside them. We will look more closely at prepositional phrases later in the chapter.

## EXERCISE 18.1 Identifying Nouns

Identify the nouns in each of the following sentences.

1. In *The Lord of the Rings,* J. R. R. Tolkien creates a world inhabited by such creatures as hobbits, elves, wizards, trolls, and orcs, as well as by men.
2. Many of the creatures have unusual, evocative names, for example, Frodo or Galadriel or Gollum.
3. Over the course of the trilogy, the characters cross mountains and rivers, dead marshes and strangely alive forests.
4. Proud heroes brandish glittering swords and magic rings, while the evil Sauron wields a paralyzing fear.
5. It is a story of courage, strength, and friendship tested by the forces of greed, betrayal, and despair.

**EXERCISE 18.2** **Using Nouns**

On a separate sheet of paper, write a sentence for each group of nouns.

1. Uncle Bob, snowshoes, coffee
2. lizard, fly, tongue

## THE PRONOUN

**Pronouns** are words that can be used in the place of nouns, usually to avoid repetition. For example, in the sentence below, the proper noun *Ivan* is repeated so often that it sounds bad.

As *Ivan* walked into *Ivan's* kitchen, *Ivan* noticed that *Ivan* had left *Ivan's* books on the table.

The sentence is so bogged down with *Ivan*s that we can't even focus on what Ivan has been doing. By substituting appropriate pronouns for *Ivan,* we can produce a sentence that sounds better and makes its point more clearly.

As Ivan walked into *his* kitchen, *he* noticed that *he* had left *his* books on the table.

There are several varieties of pronouns (see also Chapter 24). *He* is one of the **personal pronouns**, which include *I, we, you, he, she, it,* and *they.* These pronouns also have **possessive** forms, such as *my, our, your, his, her, its,* and *their: his* kitchen, *his* books.

**Relative pronouns** are words that "relate" or connect one clause to another. A **clause** is a group of words that contains a subject and verb. Relative pronouns include *who, whom, whose, which, what*, and *that.* For example:

Ivan is a student <u>*who*</u> *prefers to study late at night.*

Here the relative pronoun *who* refers to the noun *student* and begins the relative clause *who prefers to study late at night.* Like other subordinate or dependent clauses, a relative clause must be attached to another sentence; it cannot stand alone. Another example:

> We saw the baker <u>*whose*</u> *recipe for crème brûlée had won first prize*.

Here the relative pronoun *whose* refers to *baker* and begins the relative clause *whose recipe for crème brûlée had won first prize.*

**Interrogative pronouns,** such as *who, what,* and *which,* are used in questions, such as <u>*Who*</u> *is planning to enter the chili competition?* The **demonstrative pronouns** *this, these, that,* and *those* direct the reader's attention to particular nouns. For example, <u>*That*</u> *is the best recipe for ratatouille.*

**Indefinite pronouns**, the largest group, do not refer to a specific person, place, or thing. Often, they suggest an amount. The indefinite pronouns include *all, many, most, some, few, none, one, each, anyone, everybody,* and *nothing.*

> *Many* of the most delicious dishes on the table were made with chocolate.

> The lecturer saw that *nobody* was listening.

**Table of Pronouns.**

| PRONOUN TYPE | EXAMPLES |
|---|---|
| Personal | *I, you, he, she, it*<br>*we, you, they* |
| Possessive | *my, mine, your, yours, his, her, hers, its*<br>*our, ours, your, yours, theirs* |
| Relative | *who, whom, whose*<br>*which, that* |
| Interrogative | *what, which, who, whom, whose* |
| Demonstrative | *this, these, that, those* |
| Indefinite | *each, one, anyone, something, nobody*<br>*everyone, nothing,*<br>*somebody*<br>*either, neither*<br>*both, few, several, many*<br>*all, any, most, some* |

**EXERCISE 18.3** **Identifying Pronouns**

Identify the pronouns in each sentence below.

1. One of the story lines in the movie *Crash* follows a police officer named Ryan and his partner.
2. Officer Ryan, who is stressed by his father's illness, has a hostile encounter with a woman named Christine and her husband.
3. The scene in which Ryan searches her is one of the most uncomfortable in the film.
4. Both of the officers are affected by it, and they dissolve their partnership.
5. Each has a subsequent scene that reverses the audience's assessment of his character.

## THE VERB

Like the noun or pronoun that forms the subject, the **verb** is also an essential ingredient in the sentence. The verb tells what the subject of the sentence is doing, or connects the subject with some more information later in the sentence. Verbs that tell what the subject is *doing* are called **action verbs.** A strong, precise action verb can pull the reader right into the sentence.

The chef *whisked* the eggs for the Greek omelet.

**EXERCISE 18.4** **Identifying Action Verbs**

Identify the action verbs in the sentences below.

1. The professional chef lightly seared the veal shoulder in the skillet.
2. Cordelia slices onions more carefully after she cut her hand last week.

3. Before he put the baking sheet in the oven, Ivan dusted the cookie dough with chopped nuts and powdered sugar.
4. The student read a chapter in the math textbook and wrote the answers to the problems on a sheet of notebook paper.
5. For each of his classes, Javier makes a set of flashcards.

Another kind of verb is the **linking verb**, which connects the subject of a sentence to some additional information later in the sentence. Look at the following example:

The eggs *are* ready to be served.

Here *the eggs* are connected to the information that they are *ready to be served* by the linking verb *are*. Common linking verbs include *is, are, was, were, appear, feel*, and *seem*.

## EXERCISE 18.5 Identifying Linking Verbs

Identify the linking verbs in the sentences below.

1. This chef is quite knowledgeable about veal.
2. Cordelia was the first student to visit the new restaurant.
3. Fresh out of the oven, the cookies seemed perfect.
4. The students were excited about their next unit in Product Knowledge.
5. Javier's brightly colored flashcards appear to be very useful to him.

A third kind of verb is the **helping verb**, a word that is added to the **main verb** to form a **verb phrase.** For example, in the sentence "Fernanda has decided to major in baking and pastry arts," the word *has* is a helping verb, *decided* is the main verb, and the verb phrase is *has decided*. Helping verbs are often used to show **tense**, that is, the time at which the action took place.

The chef is *whisking* the eggs. [present progressive/*is*]

Yesterday the chef *was whisking* the eggs for an omelet. [past progressive/*was*]

The chef *had whisked* eggs many times before. [past perfect/*had*]

Perhaps the chef *will whisk* eggs again tomorrow. [future/*will*]

Helping verbs are also used to form the **passive voice**, in which the subject of the sentence *receives* the action of the verb rather than performs it (see also Chapter 23).

Active: The baker *dusted* the cookie dough with chopped nuts and powdered sugar. [The subject is doing the dusting.]

Passive: The cookie dough *was dusted* with chopped nuts and powdered sugar. [The subject is *receiving* the dusting; *was* is the helping verb.]

Helping verbs always come before the main verb in the sentence, though they may sometimes be separated by other words, as in the question "*Did* the baker *dust* the cookie dough with powdered sugar?" Here the verb phrase *did dust* is split by the subject *the baker. Did* is the helping verb.

**Common Helping Verbs**

| | | | | | |
|---|---|---|---|---|---|
| am | are | be | been | being | can |
| could | did | do | does | had | has |
| have | is | may | might | must | shall |
| should | was | were | will | would | |

Most of the common helping verbs are used *only* as verbs (with the exception of *being, can, will*, and *might,* which may also be used as nouns) and should be memorized as such. Sometimes a verb phrase will contain more than one helping verb, as in the following examples:

*should have* studied

*will be* driving

*must have been* grilled

Verb phrases are very common. Once you find one verb in the sentence, be sure to look both to the left and the right to find any helping verbs. For example, if the verb you find might be a helping verb, look to the right to see if there is a main verb following it, and look to the left to see if there are other helping verbs preceding it (see Figure 18.2).

6

**FIGURE 18-2 Finding a Verb Phrase**

Suppose you see the word *have* first.

After the quiz, Fernanda realized that she should have studied more carefully.

Next look to the left and you'll find the helping verb should.

After the quiz, Fernanda realized that she should have studied more carefully.

Then look to the right and you'll see the main verb studied.

After the quiz, Fernanda realized that she should have studied more carefully.

Now you have the full verb phrase: should have studied.

After the quiz, Fernanda realized that she should have studied more carefully.

## EXERCISE 18.6 Identifying Helping Verbs

Identify the helping verbs in the sentences below.

1. The terrier has stolen one of Harry's shoes again.
2. He had eaten the other one the day before.

3. Harry should have put his shoes away in the closet.
4. That dog will be stealing shoes every day unless Harry learns to put them away.
5. Harry's shoes must have been almost completely destroyed by now.

Notice the word *almost* in sentence 5 above. *Almost* is an **adverb**, a word that describes or "modifies" a verb, adjective, or another adverb. Technically, adverbs are not part of the verb. One possible exception is the adverb *not,* which is so closely related to the meaning of the verb that in some languages it becomes a part of the verb itself. While students may include *not* as part of the verb phrase, they should remember that no other adverbs can be treated in this way.

### EXERCISE 18.7 Identifying Verb Phrases

Identify the verb phrases in the sentences below.

1. Geraldo has been invited to play on a special post season baseball team.
2. His family members could not contain their excitement.
3. Geraldo was chosen because of his excellent fielding skills.
4. He had caught many difficult ground balls during the regular season.
5. The team will be practicing almost every day.

## SPOTTING THE TERMINATOR

We've now covered two of the essential ingredients of the sentence, the subject (noun or pronoun) and verb. While nouns and action verbs tend to be recognized fairly easily, linking and helping verbs seem to fade into the background. They are often small words like *is* that don't appear to carry a particular meaning, certainly not the clear and dramatic meaning of *cut* or *burn.* It is therefore important to *memorize* these common verbs and watch out for them as you read a sentence. Otherwise, you may not be able to distinguish between a correct sentence and a fragment.

This inability to distinguish between two similar structures is the type of danger faced by the protagonists in *Terminator 2.* In this film, Arnold Schwarzenegger is a highly functioning robot, the model T100, and he is trying to protect John Connor from another robot, the newer model T1000, called a "terminator" because its sole purpose is to terminate or kill. Now the T1000 is a more sophisticated mechanism and can camouflage itself by becoming part of an inanimate object, or hoodwink its adversaries by assuming a specific human identity. Remember the scene at the mental hospital where John Connor's mother is imprisoned. A mild-looking security guard locks the outside doors and walks back down the corridor. The floor of this corridor is tiled in large black and white squares, a floor we've seen in many buildings. Black and white. Simple, ordinary. These tiles are like words, in fact—black type on a white page. Simple, ordinary. But not every word in a sentence is the same.

Think back to the film. The security guard continues his stroll down the corridor, and the camera scrolls down his uniformed leg to the floor. And suddenly—with a *frisson* of strings—the outline of a face swells up from the tiles. It is the face of the T1000—still covered in black and white squares—but swiftly and silently rising up, forming a body, and striding after the guard. In fact, the T1000 becomes an exact replica of the guard, who has been getting himself a cup of coffee from a vending machine, completely oblivious of the danger behind him. Suddenly he wheels around and confronts, open-mouthed, his own self. There is a pause; then the T1000 slowly extends its arm toward the face of its human twin. As the arm begins to lengthen, it turns into a sword and finally skewers the hapless guard through the eyeball!

In a sentence, there are some words that are different, some words like verbs that are so ordinary yet so central to the structure of the sentence that they should stand out from the other words just like the face of the T1000 emerges from the tiles. In particular, the words *is, are, has, have, was, were, does,* and *do* are common, ordinary, and crucial (see Figure 18.3). Memorize them.

## THE PREPOSITION

**FIGURE 18.3** The "face" of the T1000.

Every sentence contains a subject, which will be a noun or a pronoun (or a construction used like a noun, such as the infinitive phrase *to err* in the sentence "To err is human") and a verb. Other parts of speech may be almost as common and may add a great deal to the meaning of the sentence but are not required for its structural completeness. **Prepositions** are words that show how other words relate to each other—the "position" of one word with regard to another—and are usually found in **prepositional phrases**.

As Ivan walked *into his kitchen*, he dropped his books *on the table*.

Prepositional phrases begin with a preposition and end with a noun or pronoun, which is called the **object** of the preposition. They may also contain adjectives and adverbs, for example, *on the really messy table*. Prepositional phrases may have two or more objects, as in *between Samuel and Veronica*—and the prepositions themselves may consist of more than one word, such as *according to* or *in addition to*.

Prepositional phrases very often answer the question *Where?* Where is the knife? *In the drawer.* Where are you going? *To the walk-in.* They may also answer the question *When?* When did you do your homework? *After lunch.* A third group of prepositions *with, of, for, to* occurs frequently and might be said to answer the question *WOFT,* a made-up word or **acronym** constructed from the first letters of those four prepositions: *with, of, for,* and *to.*

In the paragraph below, the prepositional phrases are underlined.

Jane went to see *Terminator 3* with her two sisters. The main character of the movie was John Connor, a young man who was told he would be very important to the world's safety when he was older. For this reason, various robots from the future were sent to kill or protect him. To his surprise, one of those robots was Arnold Schwarzenegger, who had protected him ten years earlier in *Terminator 2*.

6

It is important for us to recognize prepositional phrases as we study sentence structure because **the subject and verb are never found within a prepositional phrase**. Thus the many nouns and pronouns in prepositional phrases can be eliminated from our search for the subject of the sentence. Familiarize yourself with the list of prepositions below. Note that some of them can be used as other parts of speech, for example, *for* (coordinating conjunction) and *until* (subordinating conjunction).

**Common Prepositions**

| | | | | | |
|---|---|---|---|---|---|
| about | above | across | after | against | along |
| among | around | as | at | before | behind |
| below | beneath | beside | between | beyond | but (except) |
| by | despite | down | during | except | for |
| from | in | inside | into | like | near |
| of | off | on | onto | out | outside |
| over | past | since | than | through | throughout |
| to | toward | under | underneath | unlike | until |
| up | upon | with | within | without | |

### EXERCISE 18.8 Recognizing Prepositional Phrases

Copy the sentences below on a separate sheet of paper and place brackets [ ] around the prepositional phrases.

1. Over the years, three films have been made about the "terminator."
2. The best of them may be *Terminator 2*.
3. In that movie John Connor and his mother are initially confused about the intentions of Arnold Schwarzenegger.
4. One of the biggest appeals of that film is the combination of Schwarzenegger's stony face and dry humor.
5. During the course of the film, he rescues the mother and son from a "terminator" sent from the future and becomes an important father figure to the boy.

6

## THE REMAINING PARTS OF SPEECH

Although we noted at the beginning of this chapter that there are eight parts of speech, we have only covered four here: noun, pronoun, verb, and preposition. **Conjunctions** will be discussed at length in Chapter 21, and **adjectives** and **adverbs** in Chapter 25.

**Parts of Speech in the Kitchen.**

### NOUNS

Pots, knife, colander

Onions, potatoes, stock, chicken

Chef, cook, waiter

### ADJECTIVES

*Iron* pots, *ceramic* knife, *plastic* colander

*Yellow* onions, *new* potatoes, *veal* shoulder, *barbecued* chicken

*Professional* chef, *line* cook, *experienced* waiter

### VERBS

Chop, stir, boil, sear

Set, pour, clear

### ADVERBS

Chop *finely,* boil *rapidly, lightly* sear

Set *correctly,* pour *carefully,* clear *immediately*

### PREPOSITIONAL PHRASES

*Into* the glass, *on* the stove, *from* the walk-in, *over* the sink, *in* the skillet, *in* the weeds

The professional chef lightly seared the veal shoulder in the skillet.

The experienced waiter poured the wine carefully into the glass.

The line cook in the weeds chopped the yellow onions finely.

The eighth part of speech is the **interjection**, something that is thrown into the middle. Interjections add emotion or emphasis and are often followed by an exclamation point. For example:

Wow! That is a great chardonnay.

Hey, watch what you're doing!

Ouch, I didn't realize the pan was hot.

Uh oh!

Interjections are not part of the structure of a sentence or connected to any other words. We tend to avoid them in academic writing, unless they're part of a direct quote.

The customer exclaimed over the superb chardonnay.

"Wow, what a great chardonnay!" exclaimed the customer.

## RECIPE FOR REVIEW Reviewing the Parts of Speech

1. A **noun** is the name of a person, place, thing, or idea. A **common noun** indicates a general category, such as *restaurants,* while a **proper noun** names a specific item, such as *The French Laundry.* Nouns can be used as subjects of a sentence and as objects of a preposition.

2. Pronouns are words that can be used in the place of nouns, usually to avoid repetition. The **personal pronouns** include *I, we, you, he, she, it,* and *they.* The **possessive** forms include *my, our, your, his, hers, its,* and *their.* **Relative pronouns** include *who, what, that,* and *which.* **Interrogative pronouns**, such as *who, what,* and *which,* are used in questions, while the **demonstrative pronouns** include *this, these, that,* and *those.* **Indefinite pronouns** include *all, many, some, none, one, each, somebody, nothing.*

3. Every sentence must contain a **verb**. **Action verbs** tell what the subject of the sentence is doing, for example, *slice, jump, ask.* **Linking verbs** connect the subject of a sentence to some additional information and include *is, are, was,* and *were.* **Helping verbs** are added to the main verb to form a **verb phrase**. Helping verbs include *is, am, are, was, were, has, have, had, can, could, did.*

6

4. **Prepositions** describe the relationship between nouns through a **prepositional phrase**. a group of words that starts with a preposition and ends with a noun or pronoun. Examples include *in the kitchen, on the table, after breakfast, with the confectioner's sugar.*

## END OF CHAPTER QUIZ Reviewing the Parts of Speech

**DIRECTIONS:** In each of the sentences below, one of the words has been underlined. For each of these words, identify the part of speech, for example, noun, pronoun, verb, or preposition.

Example: Television doctors are extremely popular and have been for decades.

Answer: *are* is a verb

______ 1. In the 1960s, two very popular television doctors were Ben Casey and Dr. Kildare.

______ 2. Young and handsome, they had a devoted female audience.

______ 3. Some years later, Dr. Marcus Welby brought a fatherly appeal to his bedside manner.

______ 4. In the 1990's, the medical drama enjoyed a renewed popularity with the smash hit *ER*.

______ 5. Unlike the medical dramas of the 60's that focused on a single doctor, *ER* featured an ensemble cast.

______ 6. Among the most popular members of *ER*'s medical staff was the smooth and confident Doug Ross, who was played by George Clooney.

______ 7. The role eventually launched Clooney into an extremely successful career on the big screen.

______ 8. Another favorite was the young medical student Carter.

______ 9. During the first episode of the series, both he and the audience were fascinated and intimidated by the fast-paced drama of the emergency room.

______ 10. Part of the fascination was with Carter's supervisor, the handsome, dedicated, and hypercritical Dr. Benton.

chapter 19

# The Structure of a Sentence

In Chapter 18 we reviewed the basic ingredients of the sentence—the parts of speech. Now we are going to look at how those ingredients work together to make a sentence. The **sentence** that you write in an essay is like the plate of food you offer to a customer. Both have certain ingredients that must be present before service. While a traditionally complete plate contains a protein, a starch, and a vegetable, a complete sentence contains a subject, a verb, and a "complete thought." We might also think of the sentence as a mathematical formula:

subject + verb + complete thought = sentence

The **subject** of the sentence contains one or more nouns or pronouns (or phrases behaving like nouns). The **verb** tells what the subject is doing (*action verb*) or links the subject (*linking verb*) to some information in the **predicate**, the part of the sentence that is not the subject. In addition to possessing a subject and verb, a sentence is also said to express a **complete thought**. We'll discuss this aspect of a sentence in more detail in Chapter 20.

Let's look at an example:

The chef in the immaculate white jacket tosses the pizza dough into the air.

The **complete subject** of this sentence is *the chef in the immaculate white jacket* and the **complete predicate** is *tosses the pizza dough into the air.*

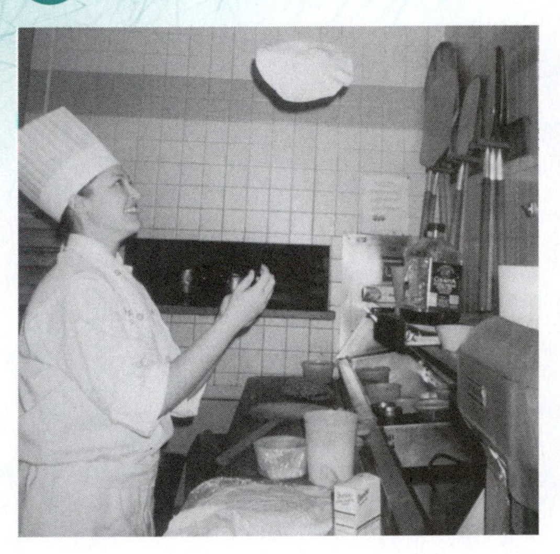

"The chef in the immaculate white jacket tosses the pizza dough into the air."

While the complete subject includes all the words that modify or describe the subject, the **simple subject** consists simply of one or more nouns or pronouns (or phrases acting like nouns—see Verbals later in the chapter). In this sentence, the simple subject is *chef*. Similarly, we want to identify the **simple predicate** of a sentence, the verb; in our example, the verb is *tosses*. Our main interest in this chapter will be finding the simple subject and simple predicate (verb) in a sentence.

There are a number of reasons that we might want to identify the subject and verb. First, since a complete sentence must contain both of these ingredients, we must be able to identify them in order to recognize and avoid incomplete or fragmented sentences (see Chapter 20). Second, identifying the subject and the verb is important in recognizing run-on sentences and comma splices, in which two complete sentences are incorrectly joined together (see Chapter 21). Third, we need to be able to find the subject and the verb in order to check that they agree (see Chapter 22).

## FINDING THE SUBJECT AND VERB

In looking for the subject and the verb, first remember that they are generally restricted to certain parts of speech. The verb must be a verb—not a verbal like *to slice* or *baking* (see the section on Verbals later in the chapter)—while the subject must be a noun or pronoun (or a word or phrase acting like a noun). If we look at a word in a sentence and it is an adverb, adjective, preposition, or conjunction, then it cannot be the subject or the verb. Second, finding the subject and verb is easier when we realize that each word in the sentence can have only one job. If a noun is the *object* of a preposition, for example, then it cannot be the *subject* of the sentence. In our example, "The chef in the immaculate white jacket tosses the pizza dough into the air," the noun *jacket* cannot be the subject of the sentence because it is the object of the preposition *in*. Thus it makes sense to eliminate all words that cannot be the subject or the verb, such as those within prepositional phrases. *The subject and verb of a sentence will never be found within a prepositional phrase.*

Finally, since each word has a specific job, we can use job-related questions to help us. To find the verb, for example, ask "Which word or phrase refers to an action or links the subject to some additional information?" More humorously, we might ask if there are any words from the face of the T1000 in the sentence: *is, are, was, were, has, have* (see Figure 18.3). To find the subject, ask "Who or what is performing the action of the verb, or who or what is being described in the rest of the sentence?"

Some students find that they can quite easily recognize the subject and verb in any given sentence; others will have quite a bit of difficulty. You may find the following process helpful.

1. **Box out**, that is, put brackets around or cross out all the prepositional phrases in the sentence. Since the subject and verb can never be in a prepositional phrase, boxing out will narrow the possibilities and make identification easier.
2. **Find the verb**, that is, the word or phrase that refers to an action or links the subject to some additional information. Remember to check for verb phrases and to ignore all adverbs except *not.*
3. **Find the subject**, that is, ask yourself who or what is performing the action, or who or what is being connected to the additional information. Very often the subject comes near the beginning of the sentence.

Let's look again at our example:

The chef in the immaculate white jacket tosses the pizza dough into the air.

1. **Box out**: Identify the prepositional phrases *in the immaculate white jacket* and *into the air.*

   The chef [*in the immaculate white jacket*] tosses the pizza dough [*into the air*].

2. **Find the verb**: Which word or phrase refers to an action? *Tosses.*

   The chef [*in the immaculate white jacket*] <u>tosses</u> the pizza dough [*into the air*].

3. **Identify the subject**: Who or what is performing the action of the sentence, that is, who is tossing the dough? The *chef* is.

6

**FIGURE 19-1** Diagram of a sentence: "The chef in the immaculate white jacket tosses the pizza dough into the air."

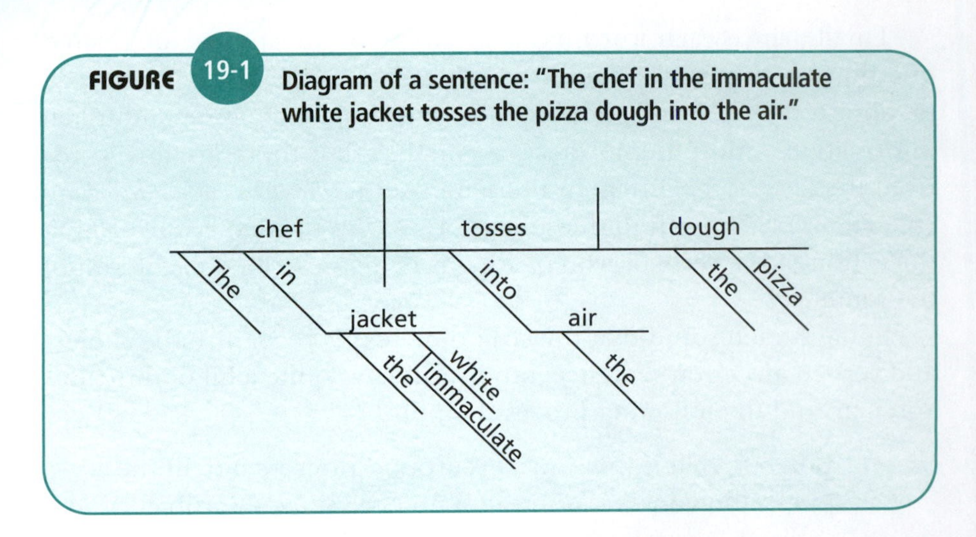


The *chef* [*in the immaculate white jacket*] tosses the pizza dough [*into the air*].

The subject and verb are like the *spine* of the sentence, or the *trunk;* other words and phrases form the limbs and branches. Look at the diagram of our sample sentence in Figure 19.1.

The spine of the sentence above is *chef tosses dough.* Although it does not contain all the information in the sentence, it does state the main point and includes the basic ingredients of simple subject (*chef*) and simple predicate (*tosses*). *Dough* is the direct object of the verb; it's what being tossed. Not every sentence has a direct object, however.

## EXERCISE 19.1 Finding the Subject and Verb

Rewrite the following sentences on a separate sheet of paper. Then box out prepositional phrases, underline the verb, and circle the subject.

1. The youthful chef in the very tall toque taught a class in Mediterranean cuisine.
2. All of the students took very careful notes on the lecture.
3. Sean asked the instructor a question about the risotto.
4. After the demonstration, the students in the class went to their stations.
5. The risotto of the day was delicious!

**Diagram of a compound verb: "Jill stirred and tasted the soup."**

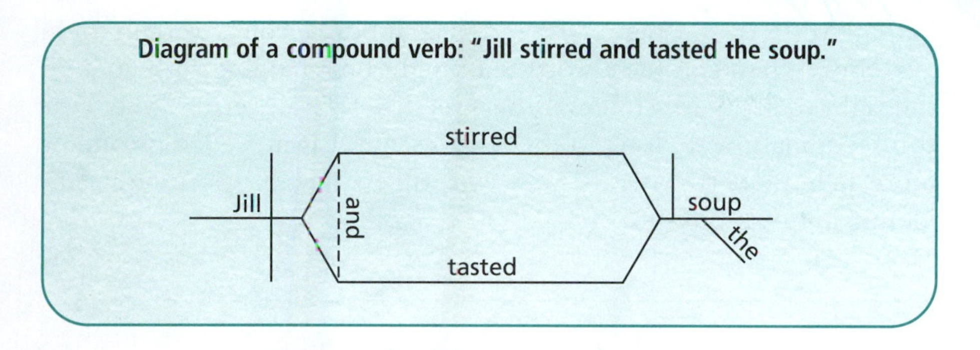


## A CLOSER LOOK AT VERBS

Sometimes a sentence has two or more verbs that refer to the same subject and are connected by conjunctions such as *and, or,* or *nor.* These are called **compound verbs**. To check a sentence for a compound verb, look carefully at all conjunctions—particularly *and* and *or*—to see what they are connecting. It might be two adjectives (for example, the black *and* white photograph), two prepositions (Did he go up *or* down the stairs?), but it might be two verbs: *Jill stirred and tasted the soup.*

### EXERCISE 19.2 Finding Compound Verbs

Identify the compound verbs in the sentences below.

1. The terrier ran to the window and barked at the mail carrier.
2. The two cats hid under the couch and hissed at the dog.
3. Three small children ran to the window and pulled the dog away.
4. The dog looked at the children and whined anxiously.
5. The mail carrier ignored the dog or perhaps did not hear him.

In that last sentence, the second verb contains an old friend from Chapter 18, the **verb phrase.** *Did* is a helping verb, and *hear* is the main verb. (*Not* is an adverb and technically not part of the verb phrase itself. For more information on adverbs, see Chapter 25.) Review the list of helping verbs from Chapter 18 and keep it beside you as you do these exercises. When you see a word that might be a helping verb, look to the right to see if it is followed by a main verb. Conversely, whenever you see a main verb, look to the left to see if it is preceded by one or more helping verbs. In this way, you will be become accustomed to identifying verb phrases.

**EXERCISE 19.3** **Identifying Verb Phrases**

Identify the verb phrases in the sentences below. Exclude any adverbs.

1. Gregory Peck has starred in a number of well-known films.
2. In *To Kill A Mockingbird*, he was cast as a Southern lawyer, the widower Atticus Finch.
3. His performance in that film is considered one of his best.
4. The actor was required to display both the tenderness of a loving parent and the courage of a social activist.
5. Contemporary audiences are still moved by Peck's courage and compassion.

## VERBALS

**Verbals** are words that look like verbs but are used in different ways. One example is the **infinitive phrase**. Look at the sentence below:

In five years Janine hopes to manage her own bakery.

Notice the phrase "to manage." Although it begins with the word *to,* it is not a prepositional phrase but an infinitive, that is, the word *to* followed by the base form of the verb (see Chapter 23). The "verb" in an infinitive phrase cannot be the verb in the sentence and might be boxed out in the same way as a prepositional phrase.

[*In five years*] Janine hopes [*to manage*] her own bakery.

In fact, infinitive phrases may—as in this instance—end with a noun or pronoun, again similarly to a prepositional phrase. The entire infinitive phrase in the sentence above is *to manage her own bakery,* and we can box out all of it.

[*In five years*] Janine hopes [*to manage her own bakery*].

Once we've boxed out prepositional and infinitive phrases, we can continue the three-step process by identifying the true verb, *hopes*, and the subject, *Janine*. Who hopes to manage her own bakery? *Janine*.

[*In five years*] **Janine** hopes [*to manage her own bakery*].

## EXERCISE 19.4 Identifying Infinitive Phrases

Box out the infinitive phrase in each sentence below. Then identify the subject and verb.

1. Eddy likes to go to the county fair every summer.
2. He tries to taste each new food, as well as to sample his old favorites.
3. His little sister, on the other hand, loves to watch Gary the Silent Clown.
4. She wants to win a balloon animal.
5. Both children like to ride on the roller coaster.

Perhaps the most famous infinitive phrase in the English language is "To be or not to be" from Shakespeare's *Hamlet*. We might modify that quote to read: "To be or not to be—that is *not* the verb."

Note, however, that although an infinitive phrase cannot be the *verb*, it might be the *subject* of a sentence. Consider the old saying "To err is human." The verb is clearly *is*. The word *human* is an adjective describing the subject of the sentence, the infinitive phrase *To err.* Another example is "To roast a turkey is traditional at Thanksgiving." The subject of this sentence is the infinitive phrase *To roast a turkey*.

Another type of verbal is the **gerund**, the base form of the verb plus *-ing*, which is used like a noun. Consider the example below:

Managing her own bakery is Janine's long-term goal.

Here the word *managing* is actually the subject of the sentence. The verb in the sentence is simply *is*. Now look at the next example.

> Janine anticipates managing her own bakery in the future.

Although *managing* looks like a verb, it is again used as a noun, in this case as the direct object of the verb *anticipates*.

Finally, **past** and **present participles** look like verbs but are often used in other ways. In fact, the participle cannot be the verb in the sentence *unless* it is part of a verb phrase and is preceded by one or more helping verbs. The past participle of regular verbs is formed by adding *-d* or *-ed* to the main verb. The present participle is formed by adding *-ing* to the main verb. Study the sentence below:

> Walking into the kitchen, Jeremy grabbed a freshly laundered side towel.

*Walking* looks like a verb, but it is not preceded by a helping verb and is instead used as an adjective modifying *Jeremy*. *Laundered* also looks like a verb, but here it is an adjective modifying *towel*. The true verb in the sentence above is *grabbed*. When helping verbs are added to the participles, however, they become verb phrases.

> Jeremy *was walking* through the kitchen, holding a side towel that *had been* freshly *laundered*.

In this sentence *walking* is preceded by the helping verb *was* and forms the verb phrase *was walking*. *Laundered* is preceded by the helping verb *had been* and forms the verb phrase *had been laundered*. (Note that *freshly* is an adverb and should not be underlined.) Look carefully at all words that look like verbs to determine how they are used in a particular sentence.

## EXERCISE 19.5 Distinguishing Verbs from Verbals

Underline the verbs in the sentences below, being careful to distinguish them from verbals.

1. *Finding Forrester* is an uplifting movie about two very different men.
2. Jamal is a very young man trying to become a better writer.

3. Finding Forrester is the key to Jamal's growth.
4. Inspired by the older man's work, Jamal writes an outstanding story.
5. Audiences still are finding Forrester a likeable and inspiring character.

## A CLOSER LOOK AT SUBJECTS

A single sentence may contain more than one subject that shares the same verb, for example, "Forrester and Jamal were both writers." In this sentence *Forrester and Jamal* is the **compound subject**. Compound subjects, like compound verbs, may also be connected with *or*, for example, "Blueberries or peaches will make an excellent filling for the cobbler." Sometimes the compound subject may contain modifiers such as adjectives or prepositional phrases.

The fresh blueberries or the peaches in the freezer will make an excellent filling for the cobbler.

For our purposes, ignore the modifiers and identify only the nouns (that is, the simple subjects), *blueberries* and *peaches*. Again, whenever you see the conjunctions *and*, *or*, or *nor*, look carefully to see what they are connecting; they might be joining the two parts of a compound subject.

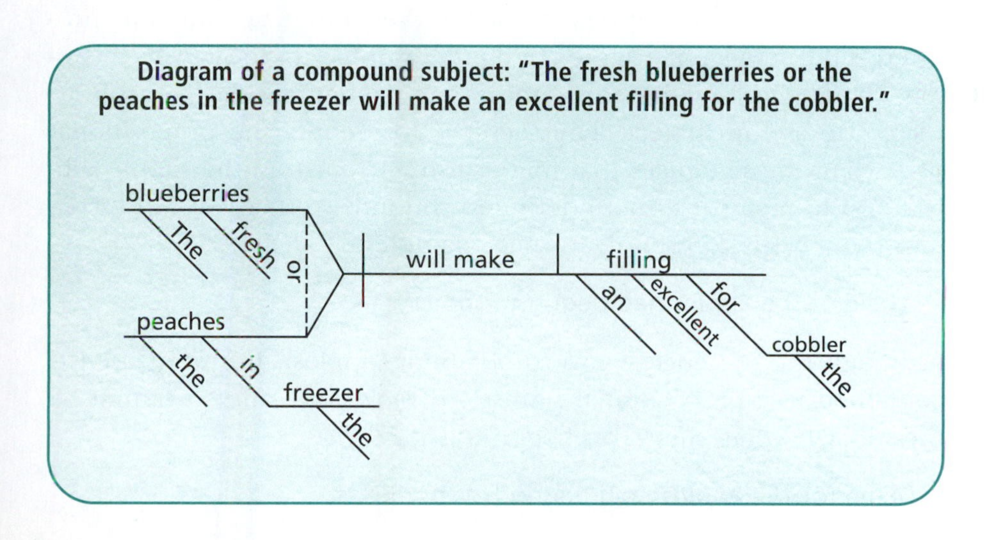


6

## EXERCISE 19.6 Identifying Compound Subjects

Identify the subject(s) in each sentence below.

1. As children, Mary and Albert liked to pick fresh blueberries in the patch behind the garden.
2. They and their little sister selected three small, tin pails from the cellar and hung them round their necks with baling twine.
3. Then the three children and their dog walked slowly through the blueberry patch, picking the ripe fruit and eating half of it.
4. Mary and Albert's mother washed the blueberries and removed their stems.
5. To the children's delight, their mother and father decided to serve the fresh blueberry pie for breakfast.

Another type of subject is the **indefinite pronoun**. We've already worked with the personal pronouns *I, we, you, he, she, it*, and *they*. While these refer to definite persons, places, or things, the indefinite pronouns describe general things or identify the amount. Indefinite pronouns include *one, each, both, all, most, none, many, everyone, something*, and *nobody* (see also Chapters 22 and 24). Very often, an indefinite pronoun is the subject of a sentence but does not seem to carry its meaning. Thus the indefinite pronoun may sometimes be difficult to recognize as the subject. The secret is to identify the prepositional phrases in the sentences that might distract you from the simple subject and to memorize the indefinite pronouns so that you will recognize them easily. Let's look at some examples:

One of the apples had begun to spoil.

Although the sentence seems to be about "apples," the word itself is contained in a prepositional phrase—*of the apples*—and so cannot be the subject. What, then, had begun to spoil? *One.*

**One** [*of the apples*] <u>had begun</u> [*to spoil*].

**EXERCISE 19.7** 

## Identifying Indefinite Pronouns as Subjects

For each sentence below, identify any prepositional phrases; then find the subject and verb.

1. Many of the holiday decorations were made by hand by the children of the family.
2. Nothing is more difficult for young children than waiting to open their presents.
3. All of the children received brightly colored sweaters from their grandmother.
4. In fact, each of the sweaters had been knit with a different color.
5. Everyone admired the group of children dressed in their new sweaters.

Did you remember to box out the infinitive phrase, *to spoil?* Here's another example:

All of the customers at the table ordered dessert.

Once you box out *of the customers* and *at the table,* ask yourself, Who ordered dessert? *All.*

**All** [of the customers] [at the table] ordered dessert.

## VARIATIONS IN WORD ORDER

In many English sentences, the subject comes before the verb and is often one of the very first words in the sentence. For example, in the sentence "The subject comes before the verb," the subject of the sentence, *subject,* is the second word in the sentence and precedes the verb, *comes.* There are also, however, types of sentences that invert this word order. One of these types begins with the words *there* or *here.*

There are several issues in finding the subject and verb in a sentence.

When a sentence begins with *there* or *here*, cross out that word (which can never be the subject) and box out the prepositional phrases in the usual way:

> There are several issues [*in finding the subject and verb*] [*in a sentence*].

Next, look for the verb to come *before* the subject. In fact, in this example the verb follows immediately after *There.*

> There are several issues [*in finding the subject and verb*] [*in a sentence*].

Finally, look for the subject to *follow* the verb.

> There are several **issues** [*in finding the subject and verb*][*in a sentence*].

The order of subject and verb may also be inverted by an initial prepositional phrase, as in the example below:

> In the pantry are several types of flour.

In following our three-step process, we would box out *in the pantry* and *of flour*, then identify the linking verb *are* and the simple subject *types.*

> [*In the pantry*] are several **types** [*of flour*].

## EXERCISE 19.8 Identifying the Subject and Verb in Inverted Sentences

Rewrite each sentence below on a separate sheet; then cross out *there* or *here*, box out prepositional phrases, underline the verb, and circle the subject.

1. Here are the main characters of Shakespeare's *Hamlet.*
2. There is a young Danish prince, sensitive and idealistic.
3. In the cast of characters also are his lustful mother and ambitious stepfather.
4. Finally, there is his father's ghost, who appears to Hamlet and tells him to kill his stepfather.
5. Here is a difficult problem for the melancholy prince.

Questions also change the order of subject and verb. Often the verb is the very first word in a question. Consider the following examples:

Are you in the kitchen?

Is breakfast ready yet?

In the first sentence, the verb *are* is followed by the subject *you*. In the second, the verb *is* followed by the subject *breakfast*. Some questions begin with a helping verb, part of a verb phrase.

Do you like to snack on fresh fruit?

Have these students made a sentence or a sandwich?

In each of these sentences, the verb phrase is split in two by the subject. *Do like* is the verb in the first sentence; *you* is the subject. *Have made* is the verb in the second sentence; *students* is the subject. A helpful way to begin analyzing a question is to rewrite it as a statement without changing or dropping any words. Thus, *Do you like to snack on fresh fruit?* becomes *You do like to snack on fresh fruit.* Notice that this brings the two parts of the verb phrase together and puts the subject first, where we most often expect to

"Have these students made a sentence or a sandwich?"

find it. Once the question is rewritten as a statement, we can box out prepositional phrases and find the subject and verb in the usual way.

**You** do like [*to snack*] [*on fresh fruit*].

**EXERCISE 19.9**

## Identifying the Subject and Verb in Questions

Rewrite each question as a statement. Then box out any prepositional phrases, underline the verb, and circle the subject.

1. Is Daniel-Day Lewis one of the greatest actors of his generation?
2. Do some people prefer the work of Kenneth Branagh, Kevin Spacey, or Denzel Washington?
3. Are there any young actors with the talent of these four men?
4. Will a child star like Haley Joel Osment have a career as an adult?
5. Have some very good actors been ignored at the Academy Awards?

Another type of question begins with a "question word," such as *why* or *how*. The order of the subject and verb is still likely to be inverted, and the question may still contain a verb phrase. Such question words may be dropped when rewriting the sentence as a statement. For example, "Why is gazpacho cold?" might be rewritten as "Gazpacho is cold [why]." Or, "How do we make gazpacho?" might be rewritten as "We do make gazpacho [how]." Finally, some types of questions begin with an interrogative pronoun, such as *who, what,* or *which*. In these cases, the subject is the pronoun, which tends to come first. Look at these examples:

Who is your favorite actor?

What is the funniest scene in that movie?

Which of the television shows are worth watching?

The subjects of these sentences are *who, what,* and *which*, respectively.

## RECIPE FOR REVIEW The Structure of a Sentence

### Definitions

Every **sentence** has a subject and a verb, and expresses a complete thought. The **simple subject** is a noun or pronoun (or a phrase acting like a noun) that is performing the action of the verb or that is linked to information in the predicate. The **complete subject** includes all the words and phrases that modify the simple subject. The **predicate** is the part of the sentence that is not the subject. It contains the **verb** (or **simple predicate**), which tells what the subject is doing or links the subject to information in the predicate.

### Finding the Subject and the Verb

1. Box out prepositional phrases, infinitive phrases, and other words that cannot be the subject or the verb.
2. Find the verb, the word that expresses an action or that appears on the face of the T1000 (*is, are, was, were, has, have, does, do*).
3. Find the subject by asking who or what is performing the action of the verb, or who or what is being described in the predicate.

### A Closer Look at Verbs

1. **Compound verbs** have two or more verbs that share the same subject.
2. Many "verbs" are actually **verb phrases**, groups of words that acts like a single word. See also Chapter 18.

**Verbals** are words that are formed from verbs but are used in different ways.

1. An **infinitive phrase** is formed with *to* plus the base form of the verb, for example, "to run."
2. A **gerund** is formed with the base form plus *–ing* and is used like a noun.
3. The **past** and **present participles** look like verbs but are often used in other ways. Unless the participle is immediately preceded by one or more helping verbs, it is not used as a verb.

## A Closer Look at Subjects

1. A **compound subject** consists of two or more subjects that share the same verb.
2. An **indefinite** pronoun can be the subject of the sentence.

## Variations in Word Order

1. When a sentence begins with *there* or *here*, the subject usually *follows* the verb.
2. In a question, the first word is most likely a verb. The subject follows this initial verb. If the verb is part of a verb phrase, the rest of the phrase follows the subject.

   Are **you** in the kitchen? (verb + **subject**)

   Do **you** like to snack on fresh fruit? (helping verb + **subject** + main verb)

## END OF CHAPTER QUIZ The Structure of a Sentence

**DIRECTIONS:** For each sentence*, identify the simple grammatical subject and the verb/verb phrase.

1. The scent of the baked goods and coffee caught my senses.
2. There are green centerpieces on every table.
3. The aroma of cloves, nutmeg, and cinnamon tickles one's nose, growing stronger with every step.
4. Was my attention distracted for a moment by the jingling of coffee cups and the sharp clamoring of silverware in the background?
5. The streusel had a very flaky texture and tasted tangy near its fruit-filled center, an equal combination of sweet and sour.

---

*Sentences adapted from various student essays.

6. The pack of students in their white coats clusters around the bright glass case filled with tantalizing, picture-perfect delights.

7. The perfect ingredients in a cup of coffee are extra sugar, extra cream, and a touch of cinnamon.

8. There was a mixture in the air of cloves, nutmeg, fresh-brewed coffee, and cinnamon.

9. The jazz music and dim lighting of the room make for a nice, warm atmosphere.

10. I picked up my cup of coffee and stepped out into the frost-bitten air.

chapter 20

# Sentence Fragments

We have said that a complete sentence requires a subject, a verb, and an independent structure or "complete thought." Thus an *incomplete* sentence, a **sentence fragment**, is missing one or more of those ingredients. It's as if you were served a plate with mashed potatoes and broccoli but no protein, or as if you tried to bake bread without the yeast. To fix the dish—or the sentence—we need to add the missing ingredients.

## IDENTIFYING SENTENCE FRAGMENTS

To identify a sentence fragment, ask yourself whether the sentence has a subject and a verb and whether it can stand alone, that is, whether it has an independent structure. Let's look at some examples.

The chef in the immaculate white jacket.

We have a subject, *chef,* but no verb. What is the chef *doing*? To fix the sentence, we need to add a verb.

The chef in the immaculate white jacket *arrived*.

Another example:

Nibbled a French pastry.

Here we have a verb, *nibbled,* but no subject. *Who* nibbled the pastry?

*The little girl in the frilly pink dress* nibbled a French pastry.

Other fragments may lack *both* a subject and a verb, like the following:

Into the frying pan.

We might add these missing ingredients as follows:

*Geraldine put the butter and onions* into the frying pan.

Finally, there is a type of fragment that contains a subject and verb but nevertheless cannot stand on its own.

Although the pork was a bit tough.

This "sentence" has a subject, *pork*, and a verb, *was*. However, it is a sentence fragment because it does not contain a complete thought. What *about* the pork?

Although the pork was a bit tough, *it had an excellent flavor*.

If the coat hanger is physically hooked over your hand, the pants are safe.

The idea of the "complete thought" is not only about missing information, however. For example, the sentence "That's right" is complete even though we don't know what "that" refers to. "Completeness" is really more about the *structure* of the sentence than about its content. When a sentence contains a subject, a verb, and a subordinating conjunction such as the word *although*—that is, when it is a subordinate clause—it is structurally incomplete unless it is joined to a main or independent clause. The word *although* is like a coat hanger, and the clause *the pork was a bit tough* hangs from it like a pair of pants. If the coat hanger is physically hooked over a rod in a closet or over your hand, the pants are safe. Similarly, if the subordinate clause is "hooked" to a main clause, such as *it had an excellent flavor*, it forms a complete sentence. But if the coat hanger tries to hover in the air on its own, it will fall to the ground, bringing the pants with it.

## EXERCISE 20.1 Identifying Sentence Fragments

Read each item below, and identify it as a complete sentence or a fragment.

_____ 1 Rachel was taking a class in identifying fruits and vegetables.

_____ 2 Learned to recognize twenty varieties of fresh herbs.

_____ (3) The differences between marjoram and oregano particularly subtle.

_____ (4) At the local farmer's market on the outskirts of town.

_____ (5) Although she studied hard for the quiz.

Of course, it's one thing to identify sentence fragments when they're listed separately, as in the exercise above. It is much more difficult to identify fragments when they are hidden in our essays.

Sentence fragments often appear in our writing because we wrongly place a period in the middle of a thought. The earlier example about Geraldine might have looked like this in our paper:

Geraldine put the butter and onions. Into the frying pan.

Or another example:

Although the pork was a bit tough. It had an excellent flavor.

Since the two parts are close together (not actually *missing* from the text), it is easy when proofreading to ignore the period and see the two parts as a whole. Therefore, it is very important to read carefully over the final draft of your paper and consider each "sentence" as a unit. Read from the capital letter to the period, and ask yourself whether the sentence contains the three necessary ingredients of a complete sentence. Sometimes it is useful to read the last sentence first, then the next to last, and so on. In that way we do not get caught up in the flow of ideas, and we are more likely to catch errors in grammar and punctuation.

6

**EXERCISE 20.2** **Identifying Sentence Fragments**

Read each item below, and identify those that contain sentence fragments.

_____ (1) Vincenzo owns a popular Italian restaurant. On New York's lower East Side.

_____ (2) After immigrating to the United States. Vincenzo's family finally gathered enough capital to open the restaurant.

_____ (3) The restaurant has been open for fifteen years. During that time it expanded into two store fronts.

_____ (4) Over time the restaurant has developed an excellent reputation. For its authentic cuisine and outstanding service.

_____ (5) Restaurant critics from *The New York Times* gave the restaurant a rave review. When they visited last year.

## MISSING SUBJECTS AND VERBS

A missing subject, somewhat rare in American English, is often rather simple to identify. Remember the plate with the missing protein at the beginning of this chapter? Clearly, if the protein is missing, the plate can be "fixed" by adding it. The same is true of the sentence below:

Does not have a pair of lamb chops.

**Fixing a "plate fragment"—add the missing protein.**

The missing subject—who or what doesn't have a pair of lamb chops?—cries out for attention, and this type of sentence fragment is less likely to be missed in editing the final draft.

Far less clear is the sentence that is missing a verb, particularly a linking verb such as *is* or *are*. Look at the example below:

> The plate with the pair of slightly rare frenched lamb chops ready to be served.

We seem to have all the information—from a description of what's on the plate to the information that it's ready to be served. What we don't have is the tiny structural essential—the verb.

> The plate with the pair of slightly rare frenched lamb chops is ready to be served.

Of course, the missing verb might well be an action verb or a verb phrase, as in the following example:

> Semi-sweet chocolate chips into the buttery cookie dough.

Add the verb:

> Semi-sweet chocolate chips were stirred into the buttery cookie dough.

Or both a subject and a verb may be added:

> *Daniel mixed* semi-sweet chocolate chips into the buttery cookie dough.

The key to identifying any fragment is to keep in mind our work from Chapter 19 on identifying the subject and verb and to remember that the verb may be one of those on the face of the T1000: *is, are, was, were, has, have, does, do.*

6

## EXERCISE 20.3 Identifying and Correcting Sentences Missing a Subject or Verb

Read each item below. Identify any sentence fragments, and write a corrected version on a separate sheet of paper.

> Example: Wilted, discolored lettuce [insert *is*] inappropriate for a salad plate.

_____ (1) Several different types of salad greens, including arugula, mache, and endive.

_____ (2) Some of the salads sprinkled with pine nuts or sugared pecans.

_____ (3) Other salads on the menu contained chunks of warm goat cheese.

_____ (4) Ordered a salad with a mix of roasted peppers and sun-dried tomatoes.

_____ (5) Mixed greens drizzled with raspberry vinaigrette.

## SUBORDINATE CLAUSE FRAGMENTS

Subordinate clause fragments—which contain a subject and a verb plus a subordinating conjunction—are extremely common. Let's look again at an earlier example.

Although the pork was a bit tough. It had an excellent flavor.

The sentence fragment *Although the pork was a bit tough* is a **subordinate** or **dependent clause**. It is a group of words that includes a subject and verb but cannot stand on its own because it lacks a complete thought or, as we said earlier, because it contains a **subordinating conjunction** or "coat hanger." The coat hanger is a structure that must be attached to a **main** or **independent clause**.

Common Subordinating Conjunctions:

| | | | | |
|---|---|---|---|---|
| after | although | as | as if | because |
| before | even though | if | in order that | once |
| since | so that | though | unless | until |
| when | whenever | where | wherever | whether |
| while | | | | |

Many sentence fragments are of this type. We remember that the formula for a complete sentence is as follows:

Subject + verb + complete thought = complete sentence
(or, independent structure)

The formula for a subordinate clause fragment looks like this:

subordinating conjunction + subject + verb = sentence fragment
*Although* + *the pork* + *was (a bit tough)* = sentence fragment

Note that subordinating conjunctions are quite different from **coordinating** conjunctions such as *for, and, nor, but, or, yet,* and *so.* (See Chapter 21 for further discussion of these words.) A sentence can start with a coordinating conjunction—yes, even *but*—and be complete. Note the formula and example below:

coordinating conjunction + subject + verb = complete sentence
*And* + *the pork* + *was* (*a bit tough*) = *complete sentence*

When evaluating a sentence that begins with a coordinating conjunction, we can essentially ignore it; the coordinating conjunction does not affect the structure of the *clause.*

Subordinate clause fragments can be corrected in a number of ways. First, they can be fixed by removing the subordinating conjunction, which in our example is the word *Although.* The sentence would become simply this:

The pork was a bit tough.

Subordinate clause fragments may also be fixed by adding an appropriate main or independent clause to complete the thought.

Although the pork was a bit tough, *the rice pilaf was delicious.*

Perhaps most often in our own writing, however, the subordinate clause fragment lies next to a main clause already, and we have simply neglected to connect the two.

Although the pork was a bit tough. It had an excellent flavor.

Here we must be careful to remove the period, replace it with a comma, and lowercase the *I* in *it.*

Although the pork was a bit tough, it had an excellent flavor.

Remember that these types of fragments are often difficult to recognize because they contain a subject and a verb and because they so often lie next to the independent clause they are meant to "depend" on. Therefore, as you're editing your papers, be especially careful of sentences that begin with one of the common subordinating conjunctions, or coat hangers. A coat hanger plus a subject plus a verb equals a sentence fragment.

## EXERCISE 20.4 Identifying and Correcting Subordinate Clause Fragments

First, identify the items that contain a sentence fragment. Then, on a separate sheet of paper, rewrite and correct these fragments by joining them to an appropriate main clause or by removing the subordinating conjunction. Adjust the capitalization as needed.

_____ 1. Some of the most daring and innovative television dramas have been police stories. One example is *Cagney and Lacey*.

_____ 2. When it premiered in the 1980s. *Cagney and Lacey* created a stir because of its female leads, who were as capable and successful as their male counterparts.

_____ 3. Ten years later, *NYPD Blue* pushed the limits on language and nudity. Although the network attempted to censor it.

_____ 4. While that show's gritty dialogue and hand-held camera technique mimicked the realism of a documentary. Today's "reality" series have taken the genre of the police drama a step farther.

_____ 5. The popular "interactive" reality show *America's Most Wanted* has helped apprehend over 850 fugitives. Because viewers respond to its compelling reenactments of real-life crimes.

## RELATIVE CLAUSE FRAGMENTS

Another type of subordinate clause, which becomes a fragment if it is not connected to a main clause, is formed with **relative pronouns**, words that *relate* the content of one clause to a word or phrase in another:

| | | | |
|---|---|---|---|
| that | which | whichever | who |
| whoever | whom | whomever | |

Relative clause fragments have a slightly different formula from the subordinate clause fragments above. Sometimes the relative pronoun acts as the subject of its own clause:

relative pronoun acting as subject + verb = sentence fragment

For example:

Who won the chili competition.

*Who* + *won the chili competition* + period = fragment

Notice that if this clause ended with a question mark, *who* would be an interrogative rather than a relative pronoun, and the sentence would be correct. However, since it ends with a period, it is a fragment. This type of relative clause fragment, in which the relative pronoun is both the subordinating conjunction and the subject of the clause, can be fixed in several different ways. First, as with all subordinate clause fragments, a main clause may be added.

*I know the students* who won the chili competition.

**Diagram of a relative clause: I know the students who won the chili competition.**

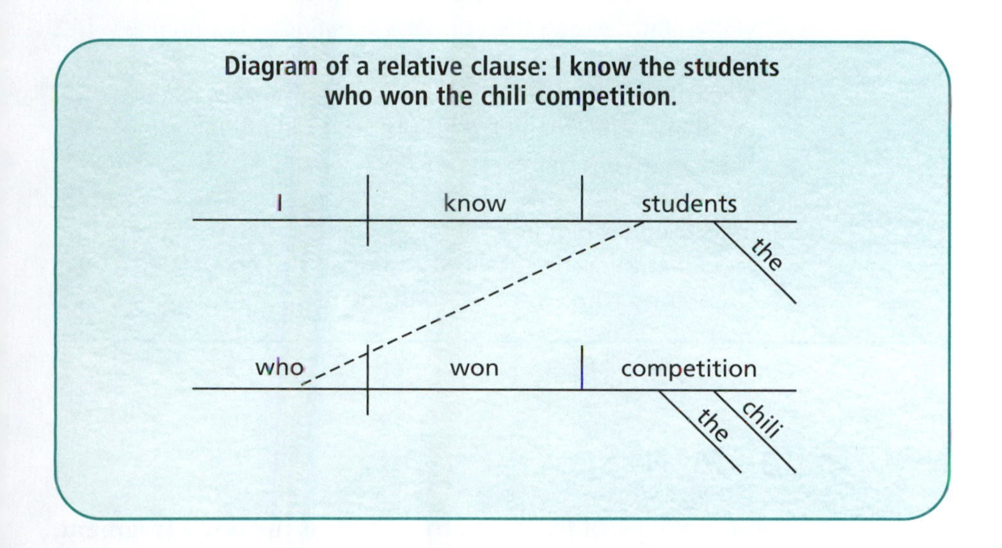


Second, the subordinate clause fragment may lie next to a main clause:

The biggest fans of hot pepper were Carlos and Cecilia. Who won the chili competition.

This fragment may be fixed by removing the period, adding a comma, and making the W lower case.

The biggest fans of hot pepper were Carlos and Cecilia, who won the chili competition.

**EXERCISE 20.5**

### More Practice with Subordinate Clause Fragments

First identify the items that contain a sentence fragment. Then, on a separate sheet of paper, rewrite and correct the fragments by adding or deleting words, or by joining the fragment to a main clause.

_____ 1 *Hill Street Blues* was another innovative police drama in the 1980s. That featured a talented ensemble cast and fast-paced, realistic action.

_____ 2 Leading the police station was Captain Frank Furillo. Who managed not only the crime-infested Hill Street district but his own alcoholism.

_____ 3 Each show began with roll call. Which reminded the viewer of the names and personalities of the officers as well as introducing the episode's storylines.

_____ 4 One of the most beloved characters was Sergeant Phil Esterhaus. Who reminded his officers each morning to "be careful out there."

_____ 5 In its complex plot and character development, *Hill Street Blues* was a forerunner of such shows as *ER* and *Law & Order*. And who hasn't heard of them?

## NOUN FRAGMENTS

A particularly difficult type of fragment to identify is the **noun fragment**. These groups of words rename or describe a noun or pronoun and may look like the following:

The story of a group of detectives in a Manhattan precinct.

Although full of information, this phrase does not have a verb and was most likely intended to describe a noun in a main clause.

One long-running police drama is *NYPD* Blue, the story of a group of detectives in a Manhattan precinct.

Sometimes a noun fragment seems to have both a subject and a verb:

An actor who has won eight or nine Emmys.

This fragment contains a dependent clause *who has won eight or nine Emmys,* in which the subject is *who* and the verb is *has won.* It also seems to contain another subject, *actor.* However, there is no verb for *actor,* and the fragment lacks a complete thought. What *about* that actor?

To fix this type of fragment, we might delete or change some of the words to make a complete sentence:

> An actor has won eight or nine Emmys.

Or we might add some words:

> He is an actor who has won eight or nine Emmys.

But most likely, in our own writing, such a fragment lies next to a main clause, and, just as we did with the subordinate clause fragments earlier, we need to join the two by removing the period and changing capital letters when necessary.

> The pivotal role of Andy Sipowicz is played by Dennis Franz. An actor who has won eight or nine Emmys.

> The pivotal role of Andy Sipowicz is played by Dennis Franz, an actor who has won eight or nine Emmys.

Here's another example.

> The young man who was kicking the football.

The verb *was kicking* belongs with the relative pronoun *who,* which is the subject of the relative clause *who was kicking the football.* But that leaves *man* without a verb and the sentence without an independent structure. By removing *who,* we can correct the problem.

> The young man was kicking the football.

Remember to read each sentence carefully, looking for the three essential ingredients. Once you have identified a sentence fragment, then decide whether to correct it by adding or deleting words, or by connecting the fragment to a main clause before or after it.

As part of a group presentation on grammar, the students act out this noun fragment: "The young man who was kicking the football."

## EXERCISE 20.6 Identifying and Correcting Noun Fragments

First, identify the items that contain a sentence fragment. Then rewrite and correct the fragments by adding or deleting words, or by joining the fragment to a main clause.

_____ 1 Another significant role in the series was that of Bobby Simone. Sipowicz's partner for a number of years.

_____ 2 When the two first met, Sipowicz found it difficult to work with a new person. He had been with his previous partner, John Kelly, for a long time.

_____ 3 The character of John Kelly left the show to follow his girlfriend. A police officer who was serving time for manslaughter.

_____ 4 Sipowicz started to warm up to Simone, however, when he heard about Simone's hobby. Raising and training homing pigeons.

_____ 5 For a man of his type, Sipowicz himself had a somewhat unusual passion. A collection of rare and beautiful tropical fish.

## RECIPE FOR REVIEW Sentence Fragments

To check that a sentence is complete, do not think only about the context, or even the content. Think about the structure. Remember that a sentence fragment is missing an independent structure rather than information. Read each sentence carefully from the capital letter to the period. Pay special attention to sentences that begin with **subordinating conjunctions** (coat hangers). To be complete, the sentence must contain a subject, a verb, and an independent structure (no coat hangers without a main clause). Use the following process to identify sentence fragments.

1. **Is there a subject?** If NO, it's a fragment. Add a subject. Then go to step 2.

   If YES, go to step 2.

2. **Is there a verb?** If NO, it's a fragment. Add a verb. Then go to step 3.

   If YES, go to step 3.

3. **Is there a coat hanger?** If NO, the sentence is most likely correct. Move to next sentence.

   If YES, it's a fragment. Correct it by dropping the coat hanger, adding a main clause, or joining the fragment to an adjacent sentence.

## ? END OF CHAPTER QUIZ Sentence Fragments

**DIRECTIONS: Part I.** *Read each item below and identify those that contain a sentence fragment. Then rewrite those items on a separate sheet of paper and correct the fragment by adding or deleting words, or by connecting the fragment to the sentence before or after it.*

_______ 1. *Law & Order* is a crime show franchise that has seen many spin-offs over the years.

_______ 2. Chris Noth plays a detective in one of these, *Criminal Intent.* Which focuses on the investigation of the crime.

_______ 3. Noth reprises his role of Detective Mike Logan, who was part of the original series in 1990.

_______ 4. His character left the series after throwing a punch at a politician. Who had made an inflammatory statement.

_______ 5. Detective Logan is one of those unusual television characters. That has evolved both on and off the screen.

_______ 6. Logan's partner Detective Carolyn Barek, played by Annabella Sciorra.

_______ 7. Detective Barek is an eccentric character, who sometimes talks to herself. And is fluent in ten languages.

_______ 8. Although Barek and Logan have very different personalities. They seem to work well as a team.

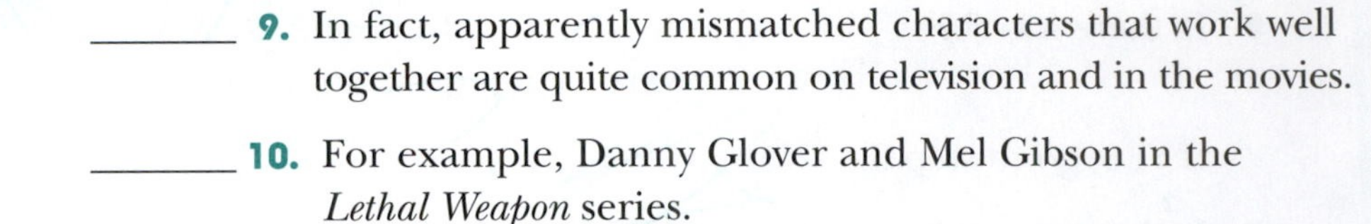

_______ **9.** In fact, apparently mismatched characters that work well together are quite common on television and in the movies.

_______ **10.** For example, Danny Glover and Mel Gibson in the *Lethal Weapon* series.

**DIRECTIONS: Part II.** *Identify the five sentence fragments in the passage below.* Then copy them on a separate sheet of paper and correct them by adding or deleting words, or by connecting the fragment to the sentence before or after it.*

When pondering what dish best describes me, at first I could think of nothing. But throughout the course of time that irresistible dessert, tiramisu, kept entering my mind. Tiramisu is my ultimate favorite dessert, and how better to describe me than with something that I love. Tiramisu best describes me. Because of its taste and intensity. Tiramisu is an open book, as am I.

Tiramisu is like tasting a piece of heaven. Each bite is evenly distributed with whipped topping, moist lady fingers, creamy marscapone cream cheese, and the enticing flavors of strong espresso and rich rum. Besides the fact that it is absolutely delicious. It has another side to it. When tasting this dessert, I find that all the flavors are strong and noticeable. Not like other desserts which are a little lighter in flavor and where some of the ingredients are slightly hidden. Everything that was used as an ingredient in this dessert I can taste and notice right away in the first bite.

Now that is kind of how I am. I have this problem. I am very opinionated and outspoken. These are great things to have as part of my personality, but sometimes they can really get me into trouble. Because I am not afraid to speak my mind. I tell things how they are, and I am not very good at covering things up or "sugar coating" situations. Which is exactly how tiramisu is, very intense with flavor as I am intense with character.

Like tiramisu, I cannot cover up who I am. In the first bite or meeting you will find out everything that you ever needed to know about the both of us.

—Gretchen L. Hardy

---

* Adapted from an essay by Gretchen L.Hardy. The errors were added to create this quiz.

chapter 21

# Run-on Sentences and Comma Splices

In Chapter 20, we looked at sentences that were missing key ingredients, such as a subject, verb or independent structure. In the case of run-on sentences and comma splices, however, the ingredients are actually doubled (or even tripled). It's as if you had two juicy 10 oz. steaks and two baked potatoes and two servings of green peas all crowded onto a single plate. Or, to use another image, it's as if we have two trains, both traveling in the same direction. But there's a problem—the one behind gets too close and actually *runs on* into the caboose of the first one (see Figure 21.1). Whether you have a bursting stomach or a derailed train, you have a problem.

## IDENTIFYING RUN-ON SENTENCES AND COMMA SPLICES

In terms of language, we have a **run-on sentence**: two independent clauses without an appropriate word or punctuation mark to join or separate them.

> One train stopped the other kept going.

*One train stopped* is a complete sentence, and *the other kept going* is another. When you read the two out loud, you will probably find that your voice drops after *stopped* as you recognize the completion of a thought, that is, the end of a sentence. On the other hand, you might have been confused

and thought that *the other* was part of the first sentence: *One train stopped the other.* But that would leave *kept going,* which doesn't make sense. To avoid such confusion, and to follow the rules of standard American English, complete sentences must be correctly separated—

One train stopped. The other kept going.

—or joined.

One train stopped, but the other kept going.

A related problem is the **comma splice**, in which two independent clauses are separated only by a comma.

One train stopped, the other kept going.

According to the rules of standard American English, a comma is not sufficient to separate (or join) two independent clauses. (Exceptions: Some writers do use commas between two very short sentences, and note that in British English, the comma *may* be used to join two independent clauses.) Let's look at the trains again. There between the two powerful trains stands Comma Man,[32] arms outstretched in a desperate attempt to prevent a collision (see Figure 21.2). He is doomed to fail, however, because no matter how many times Comma Man works out at the gym, even if he were to take an illegal steroid cocktail, he will never be strong enough to separate these two trains. Let's turn back to that plate with the two steaks. Suppose we were to separate the steaks

with a few sprigs of parsley—could we serve the plate then? No. We need to get a second plate. There's simply too much food there.

To identify these two errors, read any given "sentence" carefully from the capital letter to the period. A run-on or comma splice will have two independent clauses, two subject-verb pairs.

> Joaquin grilled two steaks Dexter added a garnish of mushrooms.

In the first clause, the subject is *Joaquin* and the verb is *grilled*. In the second clause, the subject is *Dexter* and the verb is *added*. The presence of two subject-verb pairs within one "sentence" suggests a run-on or a comma splice. Be careful, however, not to confuse these independent pairs with compound subjects or verbs. All of the following are single, correct, independent sentences:

> Joaquin *grilled* two steaks and *added* a garnish of mushrooms. [compound verb]
>
> *Joaquin and Dexter* grilled two steaks. [compound subject]
>
> *Joaquin and Dexter* grilled two steaks and added a garnish of mushrooms. [compound subject *and* compound verb]

Let's go back to the first example. After you locate two subject-verb pairs or clauses in a sentence, read each one carefully to check that each contains a complete thought. That is, look for a "coat hanger" or subordinating conjunction. (Do you remember the coat hanger from Chapter 20?) If either clause is preceded by a subordinating conjunction such as *although, because, after,* or *while,* or by a relative pronoun such as *who, which,* or *that,* it is not a complete thought. In fact, it would be a fragment if it stood on its own. But when that "fragment" is attached to another clause, the coat hanger makes for a correct complex sentence.

6

**FIGURE 21-3 Identifying run-ons and comma splices.**

If there is nothing in that box, you have a run-on sentence.

Joaquin grilled two steaks Dexter added a garnish of mushrooms.

RO

If there is just a comma in that box, you have a comma splice.

Joaquin grilled two steaks, Dexter added a garnish of mushrooms.

CS

CORRECT: *After* Joaquin grilled two steaks, Dexter added a garnish of mushrooms.

CORRECT: Joaquin grilled two steaks *before* Dexter added a garnish of mushrooms.

However, if there are no coat hangers in the sentence, look for the place where the two clauses meet. Perhaps even draw a little box there. If there is nothing in that box, you have a run-on sentence. If there is just a comma in that box, you have a comma splice (see Figure 21.3).

Let's analyze the following examples.

**A.** Jordan has an unusual job in which he killed and cleaned a ten pound octopus each day for octopus soup.

Sentence A has two subject-verb pairs: *Jordan has* and *he killed and cleaned.* At the place where they join, however, there is the relative pronoun *which* (the object of the preposition *in*). This sentence is therefore correct.

**B.** He picked it up by its large head then he dropped it into boiling water.

Sentence B also has two subject-verb pairs: *he picked* and *he dropped.* At the place where the two clauses join, we have the word *then,* an adverb rather than a conjunction, and no mark of punctuation. This sentence, therefore, is a run-on.

**C.** The octopus would make a desperate attempt to escape, it would squirt Jordan with ink and grab the sides of the pot with its tentacles.

Sentence C has two subject-verb pairs: *octopus would make* and *it would squirt.* This is a difficult but extremely common type of run-on or comma splice in which the subject of the second clause is a pronoun that refers to the subject of the first clause. At the point where the two independent clauses meet, there is a comma, and this sentence is indeed a comma splice.

**D.** Finally, Jordan cut open the head and removed the ink bladder, the ink gave the octopus soup its rich black color.

Sentence D has two subject-verb pairs: *Jordan cut and removed* and *ink gave.* Don't be fooled by the compound verb—there are still two subject-verb *pairs.* At the point where the two independent clauses meet, there is only a comma; thus, the sentence is a comma splice.

**E.** Jordan enjoyed his job, however he was very tired at the end of the day.

Sentence E has two subject-verb pairs: *Jordan enjoyed* and *he was.* Where the two independent clauses meet, we find a comma and the word *however.* It's tempting to view the sentence as correct. Remember, though, that *however* is not a conjunction; it's only an adverb and cannot connect the two clauses.

## EXERCISE 21.1 Identifying Run-ons and Comma Splices

Read each item carefully, looking for subject-verb pairs. Draw a box around the place where the sentences meet, and identify the sentences as correct, run-on, or comma splice.

______ 1 While he was growing up, Dexter's favorite dish was peanut butter and jelly.

______ 2 He was a picky eater he liked only bland, simple food.

______ 3 His mother would use plain white sandwich bread, she bought creamy peanut butter and grape jelly.

______ 4 Dexter was a picky eater as a child, however he grew to like a variety of foods as an adult.

______ 5 Now one of his favorite foods is steamed clams he also likes clam cakes.

6

There is a certain simplicity or purity about this unit. We have two closely related errors, identified in almost exactly the same way, and—almost as simply—five ways to correct them.

1. One train stopped. The other kept going.
2. One train stopped, and the other kept going.
3. One train stopped; the other kept going.
4. One train stopped; however, the other kept going.
5. One train stopped while the other kept going. OR, While one train stopped, the other kept going.

Let's practice each type of correction.

## FULL STOP: THE PERIOD

Run-ons and comma splices are two sentences that are not correctly connected. They can be fixed by separating them with a period. Unlike the comma, the period indicates the end of a complete thought, the end of a complete sentence. Where the comma is like a waving hand drawing attention to a certain group of words, the period is like the vertical palm of the traffic cop saying "Stop!" It is like a knife slicing through a beef tenderloin.

A period slices through a run-on sentence like a knife through beef tenderloin.

The correction is made at the same point where we looked for the error, the point where the two sentences meet. If there is nothing there, nothing in that box, then we add a period. If there is a comma at that point, we change it to a period. In both cases, we must then capitalize the first letter of the next sentence, unless it is already in uppercase. Note that in the second example, *Dexter* is already capitalized since it's a proper noun.

One train stopped. The other kept going.

Joaquin grilled two steaks. Dexter added a garnish of mushrooms.

With a period, the hand is no longer waving but firmly signaling a full stop.

## EXERCISE 21.2 Using the Period

Read each item carefully, and identify it as correct, run-on, or comma splice. Then correct any errors using a period and a capital letter, where needed.

_____ 1 Joaquin had worked at the restaurant every Friday night for two years he enjoyed the pressure of the line.

_____ 2 At the restaurant Joaquin especially liked the grill the heat and the danger were exciting to him.

_____ 3 Unlike his friend Dexter, Joaquin had never been a picky eater.

_____ 4 Joaquin had always enjoyed steamed clams and calamari, he liked garlic and chili peppers.

_____ 5 Although he enjoyed almost all foods, Joaquin did dislike lima beans.

## COMMA PLUS COORDINATING CONJUNCTION

Run-on sentences and comma splices often occur in our writing because there is some relationship between the two ideas they express. Thus, it is often a good idea to keep the sentences connected. One way to do so is to put a comma and one of the coordinating conjunctions at the point where the two sentences touch.

One train stopped, but the other kept going.

Joaquin grilled two steaks, and Dexter added a garnish of mushrooms.

There are seven **coordinating conjunctions**: *for, and, nor, but, or, yet, so.* Each one has a somewhat different meaning, and we'll want to use the one that best expresses the relationship between the ideas in the two sentences. *But* and *yet* suggest a contrast or contradiction between the two ideas, as in the first example above. *For* and *so* suggest a cause and effect relationship: The steaks weren't done, so Joaquin left them on the grill.

These seven coordinating conjunctions, the **FANBOYS**, are words whose job is to *join* two other words, phrases, or sentences. The words themselves are the glue. Commas are also involved, however. We place a comma before the coordinating conjunction when joining

6

two independent clauses. Like the clasped hands of Comma Man and a FANBOY, the comma signals the end of one group of words and the beginning of another (see Figure 21.4).

## EXERCISE 21.3 Using Coordinating Conjunctions

Read each item* carefully, and identify it as correct, run-on, or comma splice. Then correct any errors using a comma and an appropriate coordinating conjunction.

_____ 1. The air around me was crisp with a slight chill sweet and earthy smells came from the shrubs and flowers.

_____ 2. I could hear the bubbling of the espresso machines as they roared and steamed like the trains passing through the valley.

_____ 3. I had a good morning, I'll go with the double espresso and a flaky, buttery croissant.

_____ 4. The iced coffee looked clear, dark, and delicious, it tasted flat and stale.

_____ 5. At the back of the glass case, there were some reduced fat muffins.

*Sentences adapted from various student essays

## THE SEMICOLON

A third way to join two sentences correctly is the semicolon. The semicolon is like an equal sign; it tells the reader that on each side is an independent sentence. Yet it also suggests a relationship between the two. English teachers love it when students use the semicolon correctly. However, if students fall into the error of using the semicolon like a comma, English teachers are correspondingly outraged! By all means, deploy the semicolon appropriately. You could, though, go through your entire life without ever using one. Note that after the semicolon the next letter is not capitalized unless it is a proper noun.

One train stopped; the other kept going.

Joaquin grilled two steaks; Dexter added a garnish of mushrooms.

We know Comma Man isn't strong enough by himself to separate the two trains, but what would happen if he got into the cab of a semi?

### EXERCISE 21.4 Using a Semicolon

Read each item carefully, and identify it as correct, run-on, or comma splice. Then correct any errors using a semicolon.

______ 1 Joaquin Phoenix is an actor who has played a number of interesting characters.

______ 2 In *Gladiator* he was Commodus, he wanted to succeed his father Marcus Aurelius as Emperor of Rome.

______ 3 Joaquin played a courageous firefighter in *Ladder 49* the movie also starred John Travolta.

______ 4 More recently the actor played country singer Johnny Cash in *Walk the Line*, a film that takes its name from one of Cash's most popular songs.

______ 5 Joaquin's older brother River Phoenix was also an actor tragically he died of a drug overdose in his twenties.

6

**FIGURE 21-5 Common Conjunctive Adverbs.**

| Conjunctive Adverbs | . . . and their meaning |
|---|---|
| also, besides, furthermore, moreover | "and" |
| however, nevertheless, still | "but" or "yet" |
| accordingly, consequently, therefore, thus | "because" |
| also, likewise, similarly | "too" |
| otherwise | "or else" |
| finally, meanwhile, still, then | related to time |

## THE SEMICOLON PLUS CONJUNCTIVE ADVERB

We have said that using the semicolon to join two independent clauses suggests that the ideas are somehow related. We often choose to specify the nature of this relationship by using an appropriate **conjunctive adverb**, such as *consequently, also, besides, moreover*, or *nevertheless*. For example:

One train stopped; *however,* the other kept going.

Joaquin grilled two steaks; *next,* Dexter added a garnish of mushrooms.

Notice that in these two examples the sentences are joined by a semicolon; the adverbs describe the relationship between the sentences but do not actually join them. Figure 21.5 contains a list of commonly used conjunctive adverbs.

As with the coordinating conjunctions, we want to choose an adverb that explains the connection between the two ideas. The conjunctive adverb typically follows the semicolon and is in turn followed by a comma, as in this example:

Alicia studied diligently throughout Wines and Beverages; *consequently*, she received high marks for the course.

"Alicia studied diligently throughout Wines and Beverages; consequently, she received high marks for the course."

However, the conjunctive adverb can be placed in either clause, depending on what it modifies, as in the following example:

One of the other students in the class, *however,* did not make time to study; she had been focusing on the softball team.

## EXERCISE 21.5 Using the Semicolon with a Conjunctive Adverb

Read each item carefully, and identify it as correct, run-on, or comma splice. Then correct any errors using a semicolon and an appropriate conjunctive adverb from the list above.

_____ 1 Adam Dalgliesh, the central character in a number of books by the British mystery writer P.D. James, works as a police detective; furthermore, he writes poetry.

_____ 2 As a detective he is somewhat detached and unknowable the families of murder victims seem to like him.

_____ 3 He expresses his passionate and sometimes painful thoughts and feelings in his poems, his colleagues at New Scotland Yard are reluctant to mention these poems to him.

_____ 4 These poems seem to be an important outlet for Dalgliesh he might become overwhelmed by the grief and hopelessness that he encounters as a detective.

_____ 5 Dalgliesh himself lost his wife and infant son he listens compassionately to the victims' families.

## THE SUBORDINATING CONJUNCTION

6

The final way we might choose to connect two related sentences is by using an appropriate **subordinating conjunction** or "coat hanger," such as *because, after, although, since, when*, and *while*. These conjunctions may be placed either between the two independent clauses, that is, at the spot where we have identified the run-on or comma splice, or at the beginning of the first clause.

One train stopped *while* the other kept going.

*After* Joaquin grilled two steaks, Dexter added a garnish of mushrooms.

If the subordinating conjunction is placed between the two clauses, a comma is used only when the second clause is a contrast to the first: We ordered dessert, although we were already full. Note that in this next sentence, no comma is required after the main clause: We ordered a large appetizer because we were very hungry. Here the second clause explains rather than contrasts with the first clause, and no comma is needed. However, when the subordinating conjunction is placed at the beginning of the first clause (as in After Joaquin grilled two steaks, Dexter added a garnish of mushrooms), a comma should always be placed at the end. The waving hand signals the reader, "Look, here's the beginning of the main clause."

You will choose both the conjunction and its placement in order to best explain to the reader the relationship between the two sentences.

| | | | | |
|---|---|---|---|---|
| after | although | as | as if | because |
| before | even though | if | in order that | once |
| since | so that | though | unless | until |
| when | whenever | where | wherever | whether |
| while | | | | |

Note that the word *however* can occasionally be used as a subordinating conjunction rather than as a conjunctive adverb, as in the following sentence: *However* you look at it, English is a difficult language.

## EXERCISE 21.6 Using the Subordinating Conjunction

Read each item carefully, and identify it as correct, run-on, or comma splice. Then correct any errors using an appropriate subordinating conjunction.

_____ 1. The storeroom is an important place for students to visit it contains much of the information they will need to judge the quality of fresh ingredients.

_____ 2. Students may study pictures of fruits and vegetables in books in the storeroom they can touch and smell the produce.

_____ 3. Sometimes students are surprised by the many kinds of mushrooms, peppers, and salad greens that are available.

_____ 4. They take tests that require them to identify specific varieties, it is wise to spend time studying in the storeroom.

_____ 5. Students with previous industry experience may find this course less difficult than students without such experience.

We've practiced several different ways of joining independent clauses correctly. In your own work, of course, you will choose whatever methods fit your purpose and style. For example, shorter sentences might be connected with a coordinating or subordinating conjunction to create a more fluid and interesting rhythm. Longer sentences, on the other hand, might be more clearly separated with a period or semicolon. If the relationship between two sentences is unclear or if it needs to be emphasized, use a specific conjunctive adverb.

## RECIPE FOR REVIEW Run-ons and Comma Splices

### Identifying Run-ons and Comma Splices

To check whether a sentence is a run-on or a comma splice, read from the capital letter at the beginning to the period at the end. Find the subject and verb. Follow the steps below.

1. Are there **two separate subject-verb** pairs in the sentence?

   If NO, go on to the next sentence.

   If YES, go to step 2.

2. Is there a **subordinating conjunction** (coat hanger) attached to one of the subject-verb pairs?

   If YES, this sentence is correct. Go on to the next sentence.

   If NO, go to step 3.

3. Look at the place where the two clauses meet. Is there a **comma plus a FANBOY or a semicolon** at that point?

   If YES, the sentence is correct. Go on to the next sentence.

   If NO, go on to step 4.

4. Look again at the place where the two sentences meet. Draw a small box there.

   If there is nothing in that box, the sentence is a **run-on**.

   If there is a comma in that box, the sentence is a **comma plice**.

6

## Correcting Run-ons and Comma Splices

Correct run-ons and comma splices in one of these five ways:

1. Put a period at the place where the two independent clauses meet, and **capitalize** the first letter of the next word.

   Joaquin grilled two steaks Dexter added a garnish of mushrooms.

   Joaquin grilled two steaks. Dexter added a garnish of mushrooms.

2. Add a **comma** and a **FANBOY** at the place where the two clauses meet.

   Joaquin grilled two steaks Dexter added a garnish of mushrooms.

   Joaquin grilled two steaks, and Dexter added a garnish of mushrooms.

3. Put a **semicolon** at the place where the two independent clauses meet.

   Joaquin grilled two steaks Dexter added a garnish of mushrooms.

   Joaquin grilled two steaks; Dexter added a garnish of mushrooms.

4. Put a **semicolon**, a **conjunctive adverb**, and a **comma** at the place where the two independent clauses meet.

   Joaquin grilled two steaks Dexter added a garnish of mushrooms.

   Joaquin grilled two steaks; later, Dexter added a garnish of mushrooms.

5. Put a **subordinating conjunction** or coat hanger at the place where the two independent clauses meet, *or* put the coat hanger at the beginning of the first clause and add a comma where the two independent clauses meet.

   Joaquin grilled two steaks Dexter added a garnish of mushrooms.

   Joaquin grilled two steaks *while* Dexter added a garnish of mushrooms.

   *After* Joaquin grilled two steaks, Dexter added a garnish of mushrooms.

## END OF CHAPTER QUIZ Run-on Sentences and Comma Splices

**DIRECTIONS: Part I.** *Read each item* carefully, and identify it as correct, run-on, or comma splice. Then correct the errors in any way you choose.*

_______ **1.** An important class for culinary students is Food Safety, it offers knowledge that will keep future customers safe.

_______ **2.** Much of the information is common sense, especially for those with experience in the industry, some of the details concerning foodborne pathogens require careful study.

_______ **3.** Food Safety keeps the customers safe Mathematics keeps the proprietors in business.

_______ **4.** Introduction to Gastronomy offers an historical perspective on the industry it features lively discussion of current issues such as organic food, fast food, and slow food.

_______ **5.** The composition courses develop students' ability to express themselves in writing, students may use these skills for business proposals, restaurant reviews, and cookbooks.

_______ **6.** In one of the restaurant classes, students sit in blue bamboo chairs, the silverware is properly set on each table.

_______ **7.** They are told to hold their noses as they taste the mystery food, it has a light, smoky sweet taste.

_______ **8.** The brown rice syrup has many different flavors it tastes like earth and like butterscotch.

_______ **9.** When students tour the kitchen, they see many familiar items, such as mixers, boilers, "salamanders," and other restaurant equipment.

_______ **10.** In the background they can hear pots and pans lightly striking each other they can taste the flavors of sautéed onions and garlic on the roofs of their mouths.

**DIRECTIONS: Part II.** *Correct the five sentence errors—a mix of run-ons and comma splices–in the passage below.*

*Sentences 6–10 adapted from various student essays.

6

Basil and rosemary are aromatic herbs with different scents, flavors, and textures they can be prepared and used in different ways.

Basil is an herb that is predominately used in Italian cuisine. The sweet flavor of the basil and the acidic tomato marry well together. When basil is added to a hot pot of tomato sauce, the aroma intensifies the mixture of sweet and savory is an amazing flavor combination. Basil is a hearty plant with delicate green leaves and white flowers. The leaves are the only thing on the plant that is edible. The best way to prepare the leaves is to tear them right before adding them to a dish, the tearing helps to prevent bruising to the leaf. Another good way to cut basil is to stack several leaves on top of each other, roll them up, and run the knife through. This process is called a chiffonade, which means thin strips of ribbon.

Rosemary is more of a hearty herb it resembles a pine tree but in a smaller version. Unlike that of basil, the taste of rosemary is pungent. It has more of an earthy flavor as opposed to sweet. Potatoes and poultry work well with rosemary because of their bland taste. The rosemary adds a strong boost of flavor to anything bland. The needles are the only edible portion of rosemary, they can be removed from the stem by pulling them down. Once they are removed, I would recommend finely chopping them due to the coarse texture of the leaves. Rosemary is often added to recipes in a sachet and then removed in order to get the flavor of the herb and not the texture. Another way to use rosemary would be to make a bouquet garni, which is a bundle of herbs tied together. This is used when the flavor of the herbs is infused in stocks and sauces.

Both rosemary and basil are easy to work with. The easy tips of preparation help in keeping the herbs fresh and flavorful. Although the two are different, they're both delicious and will enhance any dish over salt any day.

—Amber Ziembiec†

†Adapted from an essay by Amber Ziembiec. The errors were added in order to create this quiz.

chapter 22

# Subject–Verb Agreement

Every sentence has a verb, as we know, and every verb has five properties called person, number, tense, voice, and mood. Verbs also have different forms that are used to coordinate with these five properties. Understanding the properties of verbs and choosing the correct form, particularly of irregular verbs, are important and often difficult parts of editing. This chapter explains the properties of person and number, the formation of the present tense, and the rules of subject-verb agreement. Chapter 23 covers the principal parts of verbs, irregular verbs, and the six basic tenses, as well as the concepts of voice and mood. An explanation of the progressive tenses and the use of helping verbs appears in Chapter 31.

## NUMBER AND PERSON

Both subjects and verbs have something called **number**, that is, the subjects refer either to one item (**singular**) or to more than one (**plural**) and the verb follows suit. For example, the subject of the sentence "The student relaxes in the gazebo" is singular: *student.* In contrast, the subject of the sentence "The students relax in the gazebo" is plural: *students.*

The students relax in the gazebo.

Look at the subjects in the list below.

| **Singular** | **Plural** |
|---|---|
| *I* relax in the gazebo. | *We* relax in the gazebo. |
| *You* relax in the gazebo. | *You* relax in the gazebo. |
| *He* relaxes in the gazebo. | *They* relax in the gazebo. |
| *The student* relaxes in the gazebo. | *The students* relax in the gazebo. |
| *Sergio* relaxes in the gazebo. | *Sergio and Jean Paul* relax in the gazebo. |

Subjects also have something called **person**. **First person** refers to the person who is speaking: the singular form *I* and the plural form *we.* **Second person** refers to the person who is spoken to directly, *you,* whether used as a singular or plural subject. **Third person** refers to the person spoken *about* and includes the singular pronouns *he, she* and *it,* as well as all singular nouns, such as *student* and *Sergio.* Third person also includes things, such as *knife, tomato,* and *pastry.* The third person plural includes the pronoun *they* and all plural nouns, such as *students, knives, tomatoes,* and *pastries,* and compound subjects, such as *Sergio and Jean Paul.* Taken together, number and person create six categories or boxes (see Figure 22.1).

**FIGURE 22-1 Singular and Plural Subjects.**

| | Singular | Plural |
|---|---|---|
| 1st person | I | we |
| 2nd person | you | you |
| 3rd person | he, she, it<br>student, Sergio<br>knife, tomato, pastry | they<br>students, Sergio and Jean Paul<br>knives, tomatoes, pastries |

You can see that while the boxes for first and second person have only one word each, subjects in the third person come in many different shapes and sizes.

## SUBJECT-VERB AGREEMENT

Now the work we've done in Chapter 19 on identifying the subject and verb really pays off, for in order to make the subject and verb "agree" we must first find them. Again, we're looking for the simple grammatical subject. Next, we must decide whether the subject is singular or plural and whether it is in the first, second, or third person. That is, we must put the subject into one of the six boxes. Finally, the verb must be made to agree with the subject by using the correct form in relation to number and person. In other words, a verb must be in the same box as its subject (see Figure 22.2).

**FIGURE 22-2 Subject/Verb Agreement.**

| | Singular | Plural |
|---|---|---|
| 1st person | I *cook* | we *cook* |
| 2nd person | you *cook* | you *cook* |
| 3rd person | he, she, it *cooks*<br>student, chef *cooks*<br>Sergio *cooks*<br>tomato cooks | they *cook*<br>students, chefs cook<br>Sergio and Jean Paul *cook*<br>tomatoes *cook* |

6

You can see that the *form of the verb* changes in only one of the boxes, the third person singular. The verbs in every other box use the base form *cook*, but the verbs in the third person singular box add an *–s, cooks.* Note the possibility for confusion here. While most *nouns* form the *plural* by adding *–s, verbs* form the third person *singular* by adding *–s* or *–es.* Don't make the mistake of thinking that verbs must end in *–s* to be plural; in fact, the opposite is true. No plural verbs end *–s.* Thus in most subject-verb pairs, only one of them can have the *s.* (There are exceptions, but we'll deal with them in a moment.) Look again at this pair of sentences:

The student relaxes in the gazebo.

The students relax in the gazebo.

Only one gets the *–s,* the subject or the verb. To remember which is which, keep in mind that the verb must agree with the subject, not the other way around. So if the subject is singular, the verb gets the *–s.*

However, not all nouns form the plural by adding *–s,* for example, *children.* In the sentence "The children relax in the gazebo," neither the subject *children* nor the verb *relax* ends in *–s.* Further, some singular nouns end in *–s,* and not all of them even have a plural form, for example, *news* (see Nouns Ending in *–s* below). In the sentence "The news about the teaching assistantship is good," both the subject *news* and the verb *is* end in *–s.*

Subject–verb agreement is a topic filled with such issues and exceptions. As with all formal writing, one potentially confusing aspect is that in our academic writing we are going to follow the rules of standard *written* English. Many dialects of *spoken* English use different rules, and that's perfectly okay for *speaking.* Writing, however, is another matter and requires the more formal usage. Another issue is that English contains both regular and irregular verbs. While the forms of regular verbs can be predicted, those of irregular verbs must be memorized individually.

## VERB FORMS

All verbs have a **base form** or **stem** that follows the word *to* in an infinitive phrase. For example, *cook* is the base form of *to cook* and *relax* is the base form of *to relax.* To put a verb in the **present tense**, that is, to make the verb indicate that its action is occurring right now, in the present, we use either the base form (*cook, relax*) or the base form plus *–s* or *–es* (*cooks, relaxes*). For regular verbs, like *cook* and *relax,* adding the suffix is straightforward. For *irregular* verbs, however, additional and unpredictable spelling changes occur.

**FIGURE 22-3** The Irregular Verb *to be.*

| | Singular | Plural |
|---|---|---|
| 1st person | I *am* | we *are* |
| 2nd person | you *are* | you *are* |
| 3rd person | he, she, it is<br>student, chef is<br>Sergio is<br>knife, tomato,pastry is | they *are*<br>students, chefs *are*<br>Sergio and Jean Paul *are*<br>knives, tomatoes, pastries *are* |

The verbs *have* and *do* both use irregular forms in the third person singular. The infinitive *to have* has the base form *have* and uses this base form for the present tense: *I have, we have, you have, they have.* When it adds the *–s* for the third person singular, though, it happens also to drop the *ve* and becomes *has.* The irregular verb *to do* is similar: *I do, we do, you do, they do.* However, it forming the third person singular, it adds the letter *e* along with the suffix –s and becomes *does.*

Most irregular is the verb *to be* (see Figure 22.3), which does not even use its base form in the present tense but rather the forms *am, are,* and *is.* Like regular verbs, irregular verbs have an ending with *–s* and an ending without *–s,* but we will need to memorize the other changes that occur within the word.

6

## EXERCISE 22.1 Choosing the Correct Verb

Identify the form of the verb that agrees with the subject in the left hand column.

1. kitchen is/are
2. onions has/have
3. Julia Child does/do
4. children is/are
5. it has/have

The verb *to be* also changes form in the past tense.

| | Singular | Plural |
| --- | --- | --- |
| 1st person | I *was* | we *were* |
| 2nd person | you *were* | you *were* |
| 3rd person | he, she, it *was* | they *were* |
| | chef *was* | chefs *were* |
| | Sergio *was* | Sergio and Jean Paul *were* |

## EXERCISE 22.2 The Past Tense of *to be*

Read each sentence, find the subject, and identify the form of the verb that agrees with it.

1. *10 Things I Hate About You* (was/were) loosely based on Shakespeare's *The Taming of the Shrew*.
2. Cat and Bianca (was/were) two sisters with very different temperaments.
3. Everyone agreed that Bianca, the younger sister, (was/were) pretty and easygoing.
4. In contrast, most of the boys at school (was/were) convinced that Cat was bossy and bad-tempered.
5. At the end of the movie, though, each of the sisters (was/were) happy.

## VERB PHRASES

As we saw in Chapter 18, many sentences contain verb phrases, that is, two or more words acting as one. For example, Samuel *is reading* his textbook, or Veronica *has planned* to visit bakeries in France. In such cases, the first helping verb in the phrase is the one that must agree with the subject. The other parts of the phrase do not change form.

Samuel *is reading* his textbook.

Samuel and Veronica *are reading* their textbooks.

The choice is between the singular *is* and the plural *are*. The participle *reading* remains unchanged.

Note that only some of the helping verbs change form: *is/are, was/were, has/have, does/do*. Other helping verbs do not: *can, could, did, had, may, might, must, should, will, would.*

| | |
|---|---|
| Samuel could read the French menu. | Samuel and Veronica *could read* the menu. |
| The waiter *must take* the drink order. | The waiters *must take* the drink orders. |

**EXERCISE 22.3** **Making Verb Phrases Agree**

Read each sentence carefully, and classify the subject as singular or plural. Then identify the correct form of the helping verb.

1. kitchen — has/have been redecorated
2. onions — is/are cooked
3. Julia Child — was/were cooking
4. children — does/do enjoy
5. it — has/have been simmering

6

## FINDING THE SUBJECT

All this time we've been talking about simply "finding the subject" as if that were the easy part of the job. Yet we remember from Chapter 19 that finding the subject of the sentence can be tricky. Sometimes prepositional phrases or other groups of words can come between the subject and verb and cause confusion. For example, in the sentence—

> The host in the grey suit and shiny black shoes took the reservations.

—there are three nouns before the verb: *man, suit,* and *shoes*. In looking carefully, we see that *suit* and *shoes* are both objects of the preposition *in* and therefore cannot be the subject. Nor can the word *reservations* later in the sentence. Thus when we ask "Who took the reservations?" the word *host* is the only possible answer and is, in fact, the subject of the sentence. Similarly, the subject and verb may be separated by an entire clause, as in the following sentence:

> The host, who was wearing a grey suit and shiny black shoes, took the reservations.

Here the relative clause *who was wearing a grey suit and shiny black shoes* may be boxed out—it has its own subject *who* and its own verb *was wearing*—and *host* can be correctly identified as the subject of the main clause and the verb *took*.

The reason such phrases and clauses can cause confusion is that we instinctively want the verb to agree with the noun nearest to it. We must be aware, though, that the nearest *noun* is not necessarily the simple grammatical *subject* of the sentence.

### EXERCISE 22.4 Finding the Subject and Making the Verb Agree

Read each sentence carefully, identify the simple grammatical subject, and choose the verb that agrees with it.

1. Veronica, who is a student in baking and pastry arts, (plans/plan) to make a chocolate mousse cake for her project.
2. Her friend Samuel, on the other hand, (prefers/prefer) to try a soufflé.
3. The difficulty with soufflés (is/are) the possibility that they will collapse.
4. The purpose of these class projects (was/were) to experiment with a number of different recipes.
5. The students in Chef Vaughn's class (has/have) chosen popular desserts for their class projects.

## COMPOUND SUBJECTS

In Chapter 21 we also talked about **compound subjects**, two or more nouns or pronouns that are connected by *and, or,* or *nor* and share the same verb. When compound subjects are connected by *and*, they are treated as plural.

Samuel and Veronica like to make apple strudel. [no *–s* on *like*]

It's almost like a mathematical equation.

Samuel and Veronica like
1 + 1 = more than one, or plural

One *tiny* exception is two nouns that are so often used together that they are thought of as a unit, for example, macaroni and cheese or rock and roll. You wouldn't say "Macaroni and cheese *are* my favorite lunch." No, instead you would say:

Macaroni and cheese *is* my favorite lunch.

These exceptions are rare, however.

## EXERCISE 22.5 Compound Subjects with *and*

Read each sentence below, identify the subject, and choose the verb that agrees with it.

1. Samuel and Veronica (eats/eat) at the restaurant around the corner every Friday night.
2. Rock and roll (plays/play) softly in the background.
3. The salads and steaks (is/are) their favorite items on the menu.
4. The freshness of the greens and the originality of the house vinaigrette (makes/make) ordering the house salad a no-brainer.
5. Meanwhile, the steak and its caramelized onions (is/are) a special favorite of Veronica's.

Compound subjects may also be connected by *or* or *nor*, and here the rule is different. Let's look at the equation.

Samuel or Veronica likes
1 or 1 = 1

But,

Samuel or his friends likes or like?
1 or 2 = ?

To clear up this confusion, a simple rule exists. When a compound subject is connected by *or* or *nor*, the verb agrees with the subject nearest

6

to it. This, remember, is our instinctive preference, to have the verb agree with the nearest noun, or in this case, the nearest *subject*. In the first example above, *Veronica* is nearest to the verb, which takes the singular form *likes*. In the second example, *friends* is nearest to the verb, which therefore takes the plural form *like*.

Samuel or his friends like to make apple strudel.

If the subjects were reversed, however, it would be a different matter.

His friends or Samuel likes to make apple strudel.

Since *Samuel* is closer to the verb, it takes the singular form *likes* (see Figure 22.4). However, in most cases where one subject is singular and

**FIGURE 22-4 Compound Subjects.**

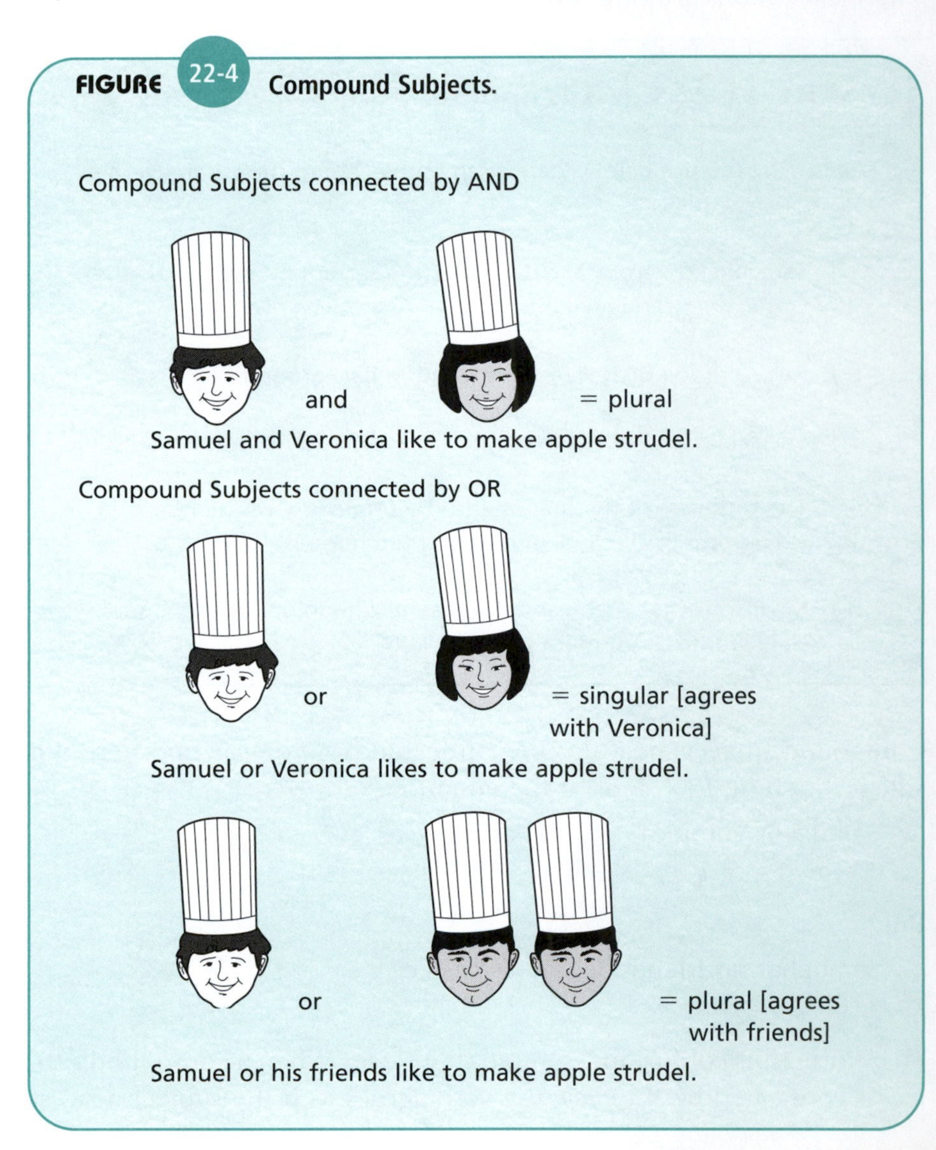

the other plural, many writers choose to put the plural subject closer to the verb.

**EXERCISE 22.6** **Compound Subjects with *Or***

Read each sentence below and identify the subject. Then circle the verb that agrees with it.

1. Chocolate or vanilla (is/are) a possible frosting for this yellow cake.
2. Rainbow sprinkles or chocolate shavings also (looks/look) good on this dessert.
3. One large cake or individual cupcakes (has/have) been successful birthday treats.
4. Ice cream or whipped cream (makes/make) a good addition to any cake.
5. The guests or the birthday child (is/are) likely to complain if no cake at all is served.

6

## INDEFINITE PRONOUNS AS SUBJECTS

We remember **indefinite pronouns** from earlier chapters. These words refer to general rather than specific persons, places, or things and include *one, each, both, none, anything, somebody*, and so on. They are sometimes difficult to identify as the subject simply because they are general words and do not seem to carry the meaning of the sentence in the way that *the baker* would. But remember—the simple grammatical subject has to do with the *structure* of the sentence rather than with its meaning.

In terms of number, indefinite pronouns can be divided into three categories. The largest group is **singular** and contains such words as *each, one, anything, anyone, anybody, everything, everyone, everybody, someone, something, somebody, nothing, nobody, either*, and *neither.* The form itself of these words seems to refer to one "thing"—not "things."

*Each* of the leading actors *is* effective.

*One* of the supporting actors *has* an especially difficult role.

*Everything was* ready for the shooting of the last scene.

*Nobody was* prepared for the film's success.

*Neither* of the screenwriters *expects* to win an Oscar.

The second group of indefinite pronouns is always **plural**: *both, few, several, many*. These words clearly refer to more than one. *Both* is two, a *few* is perhaps three, *several* is three to five, and *many* is probably more than five.

*Both* of the movies *were* nominated for an Academy Award.

*Few* of the experts *predict* the comedy to win.

*Several like* the suspenseful and well-acted *Apollo 13.*

*Many prefer* the moving, fact-based *Philadelphia.*

Indefinite Pronouns, Singular and Plural.

**One** of the actors **was** in *Apollo 13.*

**Both** of the actors **were** in *Apollo 13..*

The final group of indefinite pronouns can be *either singular or plural* depending on their use in a given sentence: *all, any, more, most,* and *some.* And here, having carefully encouraged you to box out prepositional phrases and ignore their contents, we must now ask you to look around for clues as to whether the pronoun is being used in the sense of singular or plural. Consider the following pair of sentences:

> Most *of the book* looks interesting.
>
> Most *of the book* look interesting.

The grammatical subject of both examples is *most.* Yet in the first sentence *most* refers to a single *book* and is treated as singular, while in the second *most* refers to many *books* and is treated as plural.

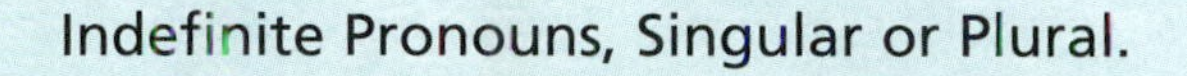

Indefinite Pronouns, Singular or Plural.

*all, any, more, most, none, some*

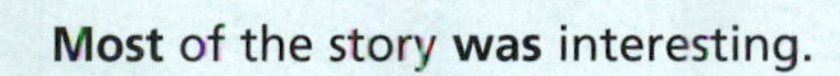

**Most** of the story **was** interesting.

**Most** of the stories **were** interesting.

**EXERCISE 22.7**

### Making Verbs Agree with Indefinite Pronouns

Read each sentence below and identify the subject. Then choose the verb that agrees with it.

1. Most of the fans of *The Lord of the Rings,* a trio of novels by J. R. R. Tolkien, (was/were) looking forward to the movie.
2. Some of the tickets to the popular film series (was/were) sold in advance.
3. In fact, some of the movie version (does/do) not follow the story in the books.
4. Despite these changes, most of the story (is/are) satisfying to many Tolkien fans.
5. Not all of the fans, however, (appreciates/appreciate) Peter Jackson's interpretation of the well-loved books.

## COLLECTIVE NOUNS, TITLES, AND AMOUNTS AS SUBJECTS

**Collective nouns** are words that name a group with several members, such as *audience, class, flock, jury,* and *team.* In American usage, collective nouns are most often treated as singular, that is, the group is thought of as a unit. Even though more than one person is in the audience, for example, we would write The audience is applauding enthusiastically. Study the following examples:

The basketball *team was* ranked number one in the poll.

The *board* of directors *has* decided to approve the budget.

The *flock* of sheep *is* to be sold at the end of the summer.

On the rare occasion when we wish to highlight the individual members of the group, we may use a plural verb.

In other dialects of English, such as British English, collective nouns are typically plural: The family *are* sitting down to breakfast. However, in American English collective nouns are most often treated as singular: The family *is* sitting down to breakfast.

Collective Nouns.

The **flock** of sheep **was** sold at the end of the summer.

Similarly, expressions that refer to an *amount*, whether of money, weight, time, distance, or fractions, are considered singular when the amount is considered as a unit. For example,

> *Three dollars was* too much to pay for that cup of coffee.
>
> *Five pounds* of sugar *is* needed for this recipe.
>
> *Six years seems* like a long engagement.
>
> *Four miles is* a good length for the dog's walk.
>
> *Two thirds* of the money *was* lost in a bad investment.

However, as with collective nouns, there are rare occasions when these concepts are considered plural. For example, in the sentence "Three dollars *are* lying on the table," we're thinking of three separate dollar bills.

Finally, titles of organizations, nations, books, or films always take a singular verb, even when they seem to be plural. For example,

> *Simon & Schuster is* a well-known publishing company.
>
> *The United States was* fortunate in its first President.
>
> *The Grapes of Wrath has* been made into a film.

What collective nouns, amounts, and titles share is an apparently plural form or meaning that nevertheless takes a singular verb.

6

### EXERCISE 22.8 Collective Nouns, Titles, and Amounts

Read each sentence below, and identify the subject and the verb. If the verb does not agree, write the correct form on a separate sheet.

Example: ___was___ The pack of dogs were running wild outside of town.

______ 1 The men's basketball team at Duke University is often ranked among the top five teams in the nation.

______ 2 Three fourths of the birthday money were put into the bank.

______ 3 The Netherlands are famous for growing magnificent tulips.

______ 4 The jury were sequestered in a nearby hotel for the duration of the trial.

______ 5 The class has many homework assignments over the long weekend.

______ 6 *War and Peace* is perhaps Tolstoy's best novel as well as his most famous.

______ 7 Three hours were far too long for that movie.

______ 8 Ten dollars were too much to pay for that sandwich.

______ 9 The crowd of shoppers is moving slowly through the mall.

______ 10 In spite of its complicated plot, *The Hours* have been quite a popular film.

## NOUNS ENDING IN –S

Some nouns that end in *–s* are actually singular and take a singular verb, for example, *economics, mathematics, measles, mumps,* and *news.* You don't get sick with a "mump." You don't turn on a "new" at six o'clock.

Yet, although they end with *–s, mumps* and *news* both take a singular verb.

> Mathematics *was* my favorite subject in high school.
>
> The mumps *is* a potentially serious disease for adolescent males.
>
> The news concerning universal health care *remains* rather discouraging.

Conversely, some other nouns that always end in *–s* and take a plural verb actually refer to single entities. Examples are *scissors, shears, tweezers, pants,* and *trousers.*

> These scissors *seem* dangerously sharp.
>
> The young chef's pants *were* too long.

## INVERTED WORD ORDER

Finally, finding the subject may be difficult when the most common word order of subject followed by verb is inverted, as in questions or in sentences that begin with *there* or *here.* Questions are frequently formed with a verb phrase that is split in two by the subject, making both harder to find.

> Does that young man wish to order his dinner now?

One way to tackle such a question is to change it into a statement without dropping or changing any words: That young man does wish to order his dinner now. This change brings the two parts of the verb together–*does wish*—and puts the subject first, which is where we are more accustomed to finding it.

Sentences that begin with *there* or *here,* or even with a prepositional phrase, may also invert the order of subject and verb. There are many ways to cook chicken or Here is a delicious recipe for *coq au vin* or In the kitchen were three hot and tired chicken-loving chefs. The initial *there* or *here* will never be the subject, but it alerts us to the change in word order: the verb will come next or shortly thereafter and will be followed by the subject. In the first example, the verb is *are,* followed by the subject *ways.* In the second, the verb is *is,* followed by the subject *recipe.* Occasionally a prepositional phrase will perform the same function of inverting the word order. The third example above begins with the prepositional phrase *in the kitchen,* followed by the verb *were,* and then by the subject *chefs.*

## EXERCISE 22.9 Inverted Word Order

Read each sentence below, and identify the subject and the verb. If the verb does not agree, write the correct form on a separate sheet of paper.

Example: __has__ There have been a series of changes to the menu.

________ 1 There is few things more delicious than hot chocolate on a cold day.

________ 2 Fortunately, there are now drugs available to combat the disease.

________ 3 Do the news ever tell the story of such complications?

________ 4 Was the pants too expensive?

________ 5 Here are some other pairs of pants.

________ 6 Only after we have solved many sample problems do mathematics become clear.

## RECIPE FOR REVIEW Subject/Verb Agreement

As you edit your essays, you will want to check each sentence for correct subject-verb agreement. **First, find the simple grammatical subject**, keeping in mind the various confusions and exceptions:

1. Nouns in prepositional phrases or other clauses that look like the subject
2. Compound subjects connected by *and* that are treated as plural
3. Compound subjects connected by *or* or *nor* whose verb must agree with the nearest subject
4. Indefinite pronouns—the large group that is always singular, the four words that are always plural, and the group that can go either way depending on what the pronoun refers to

5. Subjects that appear plural but are treated as singular, such as collective nouns, titles, and measured amounts

6. Irregular nouns ending in *–s*

7. Inverted word order in questions and in statements that begin with *there* or *here*

**Second, check that the verb agrees** with the simple grammatical subject. Singular subjects take verbs that end in *–s*. Plural subjects take verbs without an *–s*. Remember that only verbs in the present tense change form, with the exception of the past tense of *to be* (*was* and *were*).

## ? END OF CHAPTER QUIZ Correcting Subject/Verb Agreement

**DIRECTIONS: Part I.** *Read each sentence below,* * *and identify the subject and verb. If the verb does not agree, write the correct form on a separate sheet of paper.*

_______ 1. Each of the kitchens and restaurants has different menus and assignments.

_______ 2. The chefs from each class has to put their orders in to the storeroom three days in advance.

_______ 3. Without the storeroom, the kitchens cannot operate, and neither the chefs nor the students have anything to do.

_______ 4. In the storeroom, the bright colors and earthy scents of the produce create an inviting atmosphere.

_______ 5. Do the school use many different vendors for its food supply?

_______ 6. Some of the walk-ins has an indescribable cold smell

_______ 7. Freshly picked grapes and a ripe pineapple sits on a small shelf near the door.

_______ 8. Jerusalem artichokes and Daikon radishes are stored in cardboard boxes

*Sentences adapted from various student essays. Errors were introduced to create this quiz.

_______ **9.** There is sixty to seventy varieties of cheese in the refrigerator.

_______ **10.** Like a beating heart, the storeroom pumps out food to all the kitchens at the college.

**DIRECTIONS: Part II.** *Read each sentence in the passage below, and identify the subject and verb. If the verb does not agree, write the corrected sentence on a separate sheet of paper.*

In the television series *The Closer,* C.I.A.-trained interrogator Brenda Leigh Johnson takes a new job as head of the Priority Homicide Division of the Los Angeles Police Department. The transition isn't easy, however. Many of the team members resent her taking over from their former boss, Captain Taylor. Others feel that she won't understand the unique environment of L.A. since she's from Atlanta. There's other difficulties as well, such as the demands she makes on her team and her personal relationship with the Chief of Police.

Brenda's first case is that of an unidentified woman who has been found in the home of a successful computer programmer. Captain Taylor and Lieutenant Flynn considers Brenda's presence at the crime scene especially annoying because she proceeds to explain what they've been doing wrong. In addition, the medical examiner resents being summoned to confirm that the seriously decomposed body is, in fact, dead. By the end of her second day on the job, all of the members of Brenda's team has requested transfer to another division.

The hostility begins to abate, however, when the team observes Brenda's interrogation of the victim's innocent-looking secretary. It turns out that the victim is actually the computer programmer, a woman who has been living as a man in order to avoid being arrested on a murder charge. She and the naïve secretary, who believes she is working for a man, has become romantically involved. But when "his" secretary discovers his true identity, she bashes him over the head, then shoot him in the face. Through a combination of cunning and compassion, Brenda elicits a confession from the distraught woman, and her team members begin to reevaluate "Miss Atlanta."

# Plating II

## Editing for Style and Usage

UNIT 7

chapter

23

# More about Verbs

In the last chapter, we studied the forms of the present tense and the rules of subject–verb agreement. We're not done with verbs yet, though! In this chapter, we will look at the principal parts or forms of verbs, the six basic tenses, and the properties of voice and mood. Information on the progressive tenses (for example, *are reading*) can be found in Chapter 31.

## THE PRINCIPAL PARTS OF VERBS

As we saw in Chapter 22, verbs take different forms in order to reflect the time of the action or description and to agree with the subject in person and number. All these forms are built from the verb's **principal parts:** the base form (remove the word *to* from the infinitive), past tense, past participle, and present participle. Regular verbs form the present tense with the base (*they cook*) and add *–s* in the third person singular (*she cooks*). Regular verbs form both the past tense and the past participle by adding *–d* or *–ed*. The base plus *–ing* creates the present participle. See Figure 23.1 for a summary of the principal parts of regular verbs.

Irregular verbs take unpredictable forms. For example, study the frequently used but highly irregular forms of *to be* in Figure 23.2.

**FIGURE 23-1** **Principal Parts of Regular Verbs.**

| VERB FORM | FORMATION | EXAMPLE |
|---|---|---|
| infinitive | *to* + base form | *to cook* |
| present tense | base form | *cook* |
| | (+ *s* in 3rd person singular) | *(cooks)* |
| past tense | base form + *ed* | *cooked* |
| present participle | base form + *ing* | *cooking* |
| past participle | base form + *ed* | *(have) cooked* |

**FIGURE 23-2** **Principal Parts of the Irregular Verb *to be*.**

| VERB FORM | EXAMPLE |
|---|---|
| infinitive | *to be* |
| present tense | *am, is, are* |
| past tense | *was, were* |
| present participle | *being* |
| past participle | *been* |

## EXERCISE 23.1 Using the Verb *to be*

Rewrite the sentences below, inserting the correct form of the verb *to be.*

1. Last week Rick and Jenny ________ having dinner at a Chinese restaurant.
2. They had ________ dining at a different restaurant each Friday.
3. Jenny ________ a culinary student and plans to graduate in three months.
4. ________ a culinary student gives Jenny a unique perspective on their restaurant experiences.
5. She hopes ________ a restaurant owner herself at some point.

We noted in Chapter 22 that *have* and *do* are also irregular verbs and, like *to be*, they have irregular forms in the present tense (for example, I *have* but she *has*). Irregular verbs may also form the past tense and the past participle in irregular ways. They may replace the internal vowel(s) of the base form, such as *freeze* changing to *froze*; or they may add a *t*, such as *bend* changing to *bent*; or they may make more sweeping changes in the base form, such as *buy* changing to *bought*. Note that while most verbs use the same form for all persons in the past tense, the verb *to be* is an important exception. However, all verbs form the present participle by adding *–ing*, though they may drop a final *e* first. Study the irregular verbs listed in Figure 23.3.

**FIGURE 23-3 Selected Irregular Verbs.**

| BASE | PAST TENSE | PAST PARTICIPLE |
|---|---|---|
| bear | bore | (have) borne |
| beat | beat | (have) beaten *or* beat |
| become | became | (have) become |
| begin | began | (have) begun |
| bend | bent | (have) bent |
| bite | bit | (have) bitten |
| bleed | bled | (have) bled |
| blow | blew | (have) blown |
| break | broke | (have) broken |
| bring | brought | (have) brought |
| build | built | (have) built |
| burn | burned *or* burnt | (have) burned *or* burnt |
| burst | burst | (have) burst |
| buy | bought | (have) bought |
| catch | caught | (have) caught |
| choose | chose | (have) chosen |
| come | came | (have) come |
| cost | cost | (have) cost |
| creep | crept | (have) crept |
| cut | cut | (have) cut |

*(continues)*

**FIGURE 23-3** *(continued)*

| BASE | PAST TENSE | PAST PARTICIPLE |
|---|---|---|
| deal | dealt | (have) dealt |
| dig | dug | (have) dug |
| dive | dived *or* dove | (have) dived |
| do | did | (have) done |
| draw | drew | (have) drawn |
| drink | drank | (have) drunk |
| drive | drove | (have) driven |
| eat | ate | (have) eaten |
| fall | fell | (have) fallen |
| feed | fed | (have) fed |
| feel | felt | (have) felt |
| fight | fought | (have) fought |
| fly | flew | (have) flown |
| forbid | forbade *or* forbad | (have) forbidden |
| forget | forgot | (have) forgotten *or* forgot |
| freeze | froze | (have) frozen |
| get | got | (have) got *or* gotten |
| give | gave | (have) given |
| go | went | (have) gone |
| grind | ground | (have) ground |
| grow | grew | (have) grown |
| hang (a picture) | hung | (have) hung |
| hang (a person) | hanged | (have) hanged |
| have | had | (have) had |
| hear | heard | (have) heard |
| hide | hid | (have) hidden |
| hold | held | (have) held |
| hurt | hurt | (have) hurt |
| keep | kept | (have) kept |
| knit | knit *or* knitted | (have) knit *or* knitted |
| know | knew | (have) known |
| lay | laid | (have) laid |
| lead | led | (have) led |

*(continues)*

**FIGURE 23-3** *(continued)*

| BASE | PAST TENSE | PAST PARTICIPLE |
|---|---|---|
| leave | left | (have) left |
| lend | lent | (have) lent |
| let | let | (have) let |
| lie | lay | (have) lain |
| light | lighted *or* lit | (have) lighted *or* lit |
| lose | lost | (have) lost |
| make | made | (have) made |
| mean | meant | (have) meant |
| meet | met | (have) met |
| mistake | mistook | (have) mistaken |
| pay | paid | (have) paid |
| prove | proved | (have) proved *or* proven |
| put | put | (have) put |
| quit | quit | (have) quit *or* quitted |
| read | read | (have) read |
| ride | rode | (have) ridden |
| ring | rang | (have) rung |
| rise | rose | (have) risen |
| run | ran | (have) run |
| say | said | (have) said |
| see | saw | (have) seen |
| sell | sold | (have) sold |
| send | sent | (have) sent |
| set | set | (have) set |
| sew | sewed | (have) sewn *or* sewed |
| shake | shook | (have) shaken |
| shine | shone *or* shined | (have) shone *or* shined |
| shoot | shot | (have) shot |
| show | showed | (have) shown *or* showed |
| shut | shut | (have) shut |
| sing | sang | (have) sung |
| sink | sank *or* sunk | (have) sunk |
| sit | sat | (have) sat |

*(continues)*

**FIGURE 23-3** *(continued)*

| BASE | PAST TENSE | PAST PARTICIPLE |
|---|---|---|
| sleep | slept | (have) slept |
| slide | slid | (have) slid |
| sow | sowed | (have) sown *or* sowed |
| speak | spoke | (have) spoken |
| speed | sped *or* speeded | (have) sped *or* speeded |
| spend | spent | (have) spent |
| stand | stood | (have) stood |
| steal | stole | (have) stolen |
| stick | stuck | (have) stuck |
| sting | stung | (have) stung |
| stink | stank *or* stunk | (have) stunk |
| strike | struck | (have) struck *or* stricken |
| swear | swore | (have) sworn |
| swim | swam | (have) swum |
| swing | swung | (have) swung |
| take | took | (have) taken |
| teach | taught | (have) taught |
| tear | tore | (have) torn |
| tell | told | (have) told |
| think | thought | (have) thought |
| throw | threw | (have) thrown |
| understand | understood | (have) understood |
| wake | woke or waked | (have) woken/waked/woke |
| wear | wore | (have) worn |
| weave | wove or weaved | (have) woven *or* weaved |
| weep | wept | (have) wept |
| win | won | (have) won |
| wind | wound | (have) wound |
| write | wrote | (have) written |

**EXERCISE 23.2** **Irregular Verbs**

On a separate sheet of paper, write the correct form of the verb base in parentheses.

Example: Susan has driven (drive) across the United States several times.

1. Susan's husband Paul has _________ (come) to enjoy these cross-country trips; he _________ (come) on the trip to Bermuda two years ago.
2. Last year they _________ (choose) to travel during the winter; they had _________ (choose) to travel in the summer the year before.
3. Paul's hands _________ (freeze) one night when he had to change a tire, kneeling on the mud that had _________ (freeze) earlier.
4. To warm up, they _________ (go) into a diner where they had _________ (go) two years earlier.
5. Jessica _________ (drink) a cup of hot chocolate flavored with peppermint, a beverage she has _________ (drink) in a number of different restaurants.

## THE SIX BASIC TENSES

Verb tense has to do with time—past, present, and future—and can be understood in part by using a time line. The present tense is in the middle of the line, the past on the left, and the future on the right.

| past | present | future |
|---|---|---|
| *cooked* | *cook/cooks* | *will cook* |

The **present tense** describes an action that is happening now, in the present. For example, Annabella *cooks* a risotto for dinner. The present tense also describes an often repeated action: Annabella *cooks* a risotto every Sunday. When writing about a book or a film, we typically describe the action in the present tense: Sam Gamgee *cooks* a pair of rabbits for himself and Frodo in *The Lord of the Rings*. Or, in *To Kill A Mockingbird* Scout and Jem *are* attacked as they walk home through the woods. It's as if the story happens for the first time each time we read or watch it; therefore, the present tense seems appropriate.

The **past tense** is formed by adding *–d* or *–ed* [*cooked* or *basted*] to the verb base and describes an action that occurred in the past or a condition that existed in the past. Annabella *cooked* risotto for her grandmother, or Annabella's kitchen *was filled* with mouthwatering aromas. The **future tense** is formed with *will* + base and describes an action that is to take place at some time in the future. For example, Annabella *will cook* risotto for her grandmother again next week, or Annabella's kitchen *will be filled* with mouthwatering aromas when she cooks on Sunday.

The other three tenses are **perfect**, a characteristic that has to do with whether an action has been completed or whether a condition still exists. The three simple and three perfect tenses are pictured below:

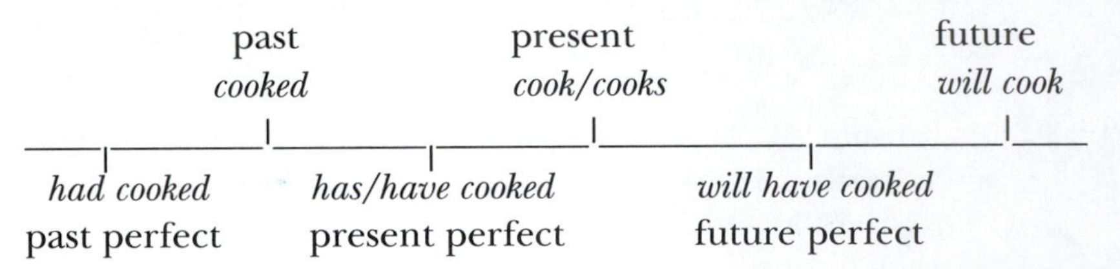


Note that the "perfect" action precedes the "simple" action in each time frame. When two actions occurred in the past but one was completed before the other, we use the **past perfect tense**, which is formed with *had* + *past participle.* For example, Annabella *had cooked* risotto for her grandmother long before she *worked* at the restaurant. Both actions occurred in the past, but one (cooking the risotto) was completed before the other (working at the restaurant). Avoid using the past perfect in place of the simple past. Review the examples below:

**had picked/picked**

Yesterday I bought a steak and mushrooms for dinner. I *had picked* some tomatoes in the garden the day before. [I picked the tomatoes *before* I bought the steak.]

| *had picked* | *bought* |
|---|---|
| past perfect | past |

Yesterday I bought a steak and mushrooms for dinner. Then I *picked* some tomatoes in the garden. [I picked the tomatoes *after* I bought the steak.]

| *bought* | *picked* |
|---|---|
| past | past |

**had bought/bought**

There was a long line for the movie. Fortunately I *had* already *bought* tickets. [I bought the tickets *before* I saw the line for the movie.]

There was a long line for the movie. Eventually I *bought* tickets. [I bought the tickets *after* waiting in line for the movie.]

A common error is to add *had* where it isn't needed. If one action follows another, use the simple past. Use the past perfect only for an action that occurred *before* the other actions described in the simple past.

The **future perfect tense** is formed with *will have* + *past participle* and expresses an action or condition that will be completed before another action in the future. For example, Annabella *will have cooked* risotto many times before she begins her new job at the restaurant. Finally, the **present perfect tense** describes an action that occurred or a condition that existed at some indefinite time in the past. For example, Annabella *has cooked* risotto on many occasions. The present perfect may also describe an action or condition that began in the past and continues up until or into the present: Annabella *has begun* to make risotto at work [and may be doing so still].

The **progressive tenses** are discussed in Chapter 31.

### EXERCISE 23.3 Choosing the Appropriate Tense

Choose the correct tense in the examples below.

1. Yesterday's basketball game, the last of the season, (was/has been) an exciting one.
2. Joe, who (will be/has been) an inconsistent player for several weeks now, finally began to make good decisions.
3. He (stole/had stolen) the ball three times and then hit a series of free throws.
4. Mike made a beautiful pass to Bill, who then unfortunately (missed/will have missed) the shot.
5. However, Bill (has missed/will miss) less than 40% of his shots all season.

## Using Tense Consistently

Another common problem with verb tense that must be corrected during editing is the inconsistent use of the past and present. Often a story or an essay could be written in either tense. However, once we've made that basic decision, logic demands that we stay with it and not move randomly back and forth between the two time frames. Study the following example:

> Yesterday Mark *went* to the store. He *bought* a dozen eggs and a gallon of milk. He *picks* up a loaf of bread. Then Mark *paid* the cashier.

Within the flow of these sentences, the present tense *picks* stands out like a cornstalk in a row of cabbages. It's as if *went, bought,* and *paid* are on one side of a brick wall and *picks* is on the other. In fact, it can be helpful to think of consistent tense in relation to this brick wall. Before you begin to write, you choose the past or present tense, that is, you choose the left or the right side of the wall (Figure 23.4).

Once you've made that first choice, you must continue to select verbs from the same side of the wall. Occasionally, though, an event will fall *outside* the main stream of the story's timeline. Let's go back to Mark and his trip to the store.

> Yesterday Mark *went* to the store. He *bought* a dozen eggs and a gallon of milk. In fact, he *buys* a dozen eggs every week. Then Mark *paid* the cashier.

In this second scenario, the present tense *buys* is correct. The fact that Mark buys a dozen eggs every week is not in the same time sequence as yesterday's specific purchases. There's a door in the brick wall, and on occasions like Mark's weekly purchase of a dozen eggs, we have the key to open that door.

**FIGURE 23-4** Hitting the Wall.

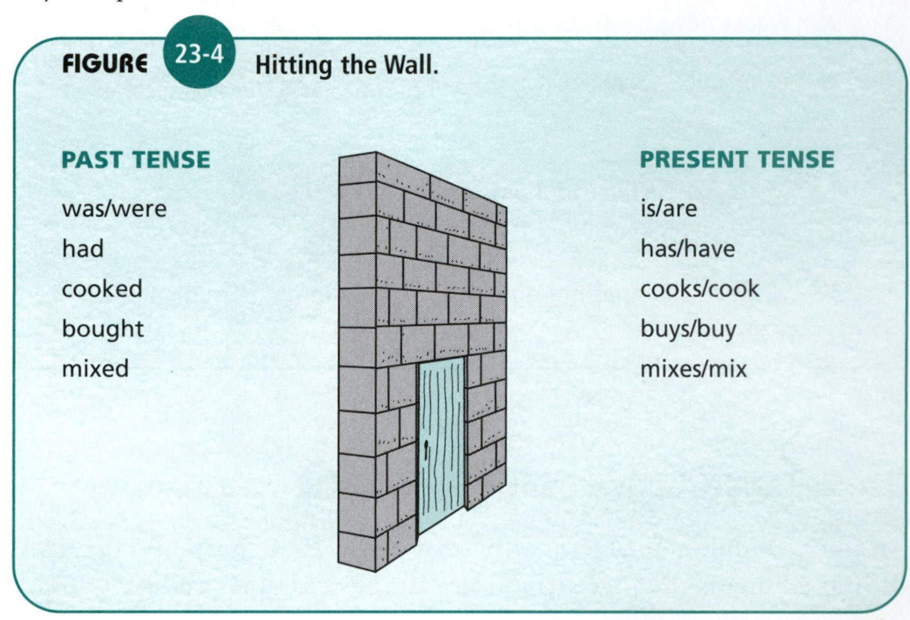

**EXERCISE 23.4** **Maintaining Consistent Tense**

Rewrite the paragraph below, changing verbs when necessary to maintain consistent tense. You may use either the past or present for this exercise.

We walk into Roth Hall and were able to notice right away that the atmosphere from outside to inside is completely different. The light was soft and very dim. I automatically felt comfortable and warm when I step inside. As we walked further down the hall, it becomes less active and more subtle. I notice the smell of breads and other oven-baked products from the kitchen. The front of the hall didn't really have a smell. The floor was made for walking in any type of shoes that you wanted to. The walls were made of brick and give us a very comfortable feeling.

—Yolanda Dillard*.

## CONFUSING VERBS

*Lie* and *lay* are two words that are often confused with one another. They look and sound like each other and even share the form *lay*. The confusion is further heightened by the words' sexual connotations. Perhaps we can address *lie* and *lay* more comfortably with some ordinary examples. *Lie* means to be in or to assume a prone position, as in *to lie down*. For example, The cloth *lies* smoothly on the table, or The baby *lies* in its crib. *Lie* does not take a direct object. *Lay*, on the other hand, does take a direct object; it means to place or put something, as in *to lay something down*. For example, The server *lays* the cloth smoothly on the table, or The babysitter *lays* the baby in its crib. The past tense and participle of each verb add to the potential for misunderstanding. Study Figure 23.5.

Most errors in the use of *lie* and *lay* occur when forms of *lay* are used with the meaning of *lie*. Note that while *lie*'s past tense is identical to *lay*'s present tense, no form of *lie* ends in *–d*. Do not use *laid* unless something is being put down: I *laid* the book on the table.

Another pair of confusing verbs is *rise* and *raise*. *Rise* is used for items that move up on their own, like the sun; *rise* doesn't take an object. For example, The sun *rises* every morning. *Raise*, on the other hand, is used with direct objects and means to bring up. The cadet *raises* the flag every morning. Study the forms and examples in Figure 23.6.

*Adapted from an essay by Yolanda Dillard. The errors were added to create this exercise.

FIGURE 23-5 *Lie* and *lay*.

| Infinitive | **To lie** (down) | **To lay** (something down) |
|---|---|---|
| Present Tense | lie/lies<br>*The cloth lies smoothly on the table.* | lay<br>*The server lays the cloth smoothly on the table.* |
| Past Tense | lay<br>*The cloth lay on the table yesterday.* | laid<br>*The server laid the cloth on the table yesterday.* |
| Past Participle | lain<br>*The cloth has lain smoothly on the table many times.* | laid<br>*The server has laid the cloth smoothly on the table many times.* |
| Present Participle | lying<br>*The cloth is lying smoothly on the table.* | laying<br>*The server was laying the cloth smoothly on the table.* |

A third pair of bewildering verbs is *sit* and *set. Sit* means to assume a sitting position and does not take an object. For example, The students *sit* at narrow desks. In contrast, *set* means "to put or place something"

FIGURE 23-6 *Rise* and *raise*.

| Infinitive | **To rise** (up) | **To raise** (something) |
|---|---|---|
| Present Tense | rise/rises<br>*The internal temperature rises as the chicken cooks.* | raise/raises<br>*You raise the temperature of the chicken by cooking it.* |
| Past Tense | rose<br>*The internal temperature rose as the chicken cooked.* | raised<br>*You raised the temperature of the chicken by cooking it.* |
| Past Participle | risen<br>*The internal temperature has risen as the chicken cooked.* | raised<br>*You have raised the temperature of the chicken by cooking it.* |
| Present Participle | rising<br>*The internal temperature is rising as the chicken cooks.* | raising<br>*You are raising the temperature of the chicken by cooking it.* |

and often takes a direct object. For example, The students *set* their books down on their desk. Because *set* has only one form for the present tense, past tense, and past participle, be sure to use it with the meaning "to set something down." Otherwise, use *sit* or *sat*. See Figure 23.7.

Note that *set* can also mean "to harden or solidify," as in *Let the jello set in the refrigerator overnight.* In this context, *set* does not take a direct object.

**EXERCISE 23.5** **Choosing the Correct Verb**

Choose the correct verb in each sentence below.

1. The terrier circled his bed three times and (lay/laid) down on the soft flannel.
2. After bustling about the dining room all evening, the server (sat/set) down and drank a glass of water.
3. The temperature typically (rises/raises) after the morning fog disappears.
4. The secret to the movie's success (lies/lays) in its exotic locations and thrilling action sequences.
5. When she (lay/laid) out the good china, she accidentally broke a plate.

## ACTIVE AND PASSIVE VOICE

The concept of **voice** has to do with who or what is *performing* the action of the verb and who or what is *receiving* the action. In the **active voice**, the subject of the sentence is performing the action of the verb.

> The chef *whisked* the eggs for the Greek omelet. [The subject, *chef*, is doing the whisking.]

In the **passive voice**, which is formed with *to be* and a participle, the subject of the sentence is *receiving* the action:

> The eggs *were whisked* for the Greek omelet. [The subject, *eggs*, is receiving the whisking.]

If we wished to include the information that the chef performed the action, we could say "The eggs were whisked *by the chef.*"

FIGURE 23-7 *Sit* and *set.*

| Infinitive | **To sit** (on something) | **To set** (something down) |
|---|---|---|
| Present Tense | sit<br>*Every day the diners sit at the wooden table.* | set<br>*Every day the servers set the table for dinner.* |
| Past Tense | sat<br>*Yesterday the diners sat at the wooden table.* | set<br>*Yesterday the servers set the table for dinner.* |
| Past Participle | sat<br>*The diners have sat at the wooden table many times before.* | set<br>*The servers have set the table for dinner many times before.* |
| Present Participle | sitting<br>*The diners are sitting at the wooden table.* | setting<br>*The servers are setting the table for dinner.* |

In general, the active voice cuts out unnecessary words and adds life to your writing. Look at the difference between these two sentences:

PASSIVE: The baguettes were taken out of the oven by Katie.

ACTIVE: Katie took the baguettes out of the oven.

While it isn't incorrect to use the passive voice, we want to be sure it is nicely balanced with the active. Long strings of passive sentences can drain the energy from your writing. However, there are times when the passive voice is preferred. You may wish to highlight the recipient of the action or you may not know who or what performed the action.

The delighted customer was offered a choice between two complimentary desserts. [The emphasis on the recipient of the action; we don't care who made the offer.]

The expensive laptop was stolen from his car. [We don't know who performed the action.]

Sometimes we know who performed the action, but we'd like to avoid naming the person. For example, "A mistake was made in the Accounting Department."

"Katie took the baguettes out of the oven."

## EXERCISE 23.6 The Passive Voice

Each of the sentences below is in the passive voice. Rewrite each one in the active voice, and then explain which one you prefer, and why.

1. The delicate wine glass was broken by the inexperienced server.
2. A three-point shot was made at the buzzer by the phenomenal freshman.
3. The lead guitar in Pink Floyd is played by David Gilmour.
4. A copy of the keys will be made by each new tenant.
5. An essay about the perils of drunk driving was written by the man convicted three times of driving while intoxicated.

## UNDERSTANDING THE MOOD OF VERBS

We probably don't think of grammar as having any emotions—yet it does have moods! The **indicative mood** reflects an ordinary, everyday mood; verbs in the indicative mood tell or ask without suggesting any hidden meaning. The large majority of our sentences are in the indicative mood. Second, the **imperative mood** is used to give orders or commands, for example, "Sharpen your knife" or "Add the egg whites now." Note that the subject of a command is understood to be *you*. You are the one to add the egg whites. The imperative mood occurs frequently in spoken English and is used extensively in a most familiar publication—the cookbook!

Finally, there is the **subjunctive mood**, which is used mostly in formal situations to talk about a wish or to make a statement that is not factual. The subjunctive is typically used with *were* and *be*. Study the difference between these two sentences:

> If the bell *rings* before we finish the story, we will finish it tomorrow. [Indicative mood: it is possible for the bell to ring.]

> If the bell *were to ring* now, it would interrupt the test. [Subjunctive mood: the bell is unlikely to ring.]

This form of the subjunctive mood uses *were* in all cases: I were (not *was*), you were, he/she/it were (not *was*), we were, you were, they were.

> If I *were* you, I wouldn't stir that sauce so vigorously. [But I am not you.]

Another form of the subjunctive uses *be* in all cases: I be, you be, he/she/it be, we be, you be, they be. *Be* is used in certain formal structures, for example, in clauses following words such as *advise, ask, recommend, request*, and *suggest*.

> Jack's supervisor recommended that he *be* given another chance.

> Jill advises that the bell *be* rung ten minutes before closing.

## EXERCISE 23.7 The Subjunctive Mood

Read each sentence carefully. If the subjunctive mood is required, select the appropriate form (*be* or *were*).

_____ 1 The doctor recommended that the x-ray *was* repeated.

_____ 2 If it *rains*, Grandma will take down the clothes that are hanging out to dry.

_____ 3 If I *was* you, I would take those cookies out of the oven.

_____ 4 Jane would be thrilled if Bingley *was* to propose to her.

_____ 5 Jane's mother suggested that Bingley *be* invited to dinner.

## RECIPE FOR REVIEW More about Verbs

### Verb Forms

1. Regular verbs use predictable forms for the tenses and participles (Figure 23.1).
2. Irregular verb forms are unpredictable and must be memorized (Figures 23.2 and 23.3).

### Choosing the Appropriate Tense

1. Use the **past tense** for action that happened or a condition that existed in the past.
2. Use the **present tense** for action that is happening now or is often repeated, as well as to describe the action in a book or film.
3. Use the **future** for action that is to take place in the future.
4. Use the **past perfect** when one action in the past was completed before another.
5. Use the **present perfect** for an action that occurred or a condition that existed at some indefinite time in the past.

6. Use the **future perfect** for an action that will be completed before another time in the future.

**Using the Six Tenses.**

| TENSE | EXAMPLE |
|---|---|
| past | Annabella *cooked* risotto for her grandmother last week. |
| present | Annabella *cooks* a risotto every Sunday. |
| future | Annabella *will cook* risotto for her grandmother again next week. |
| past perfect | Annabella *had cooked* risotto for her grandmother long before she worked at the restaurant. |
| present perfect | Annabella *has cooked* risotto on many occasions. |
| future perfect | Annabella *will have cooked* risotto many time before she begins her new job at the restaurant. |

## Maintaining Consistent Tense

If you begin a story in the past tense, keep the rest of your verbs in the past. Similarly, if you begin in the present tense, stay there. However, if a single action takes place out of the flow of events, choose the appropriate tense.

## Confusing Verbs

Study the confusing verb pairs in Figures 23.5, 23.6, and 23.7.

## Active and Passive Voice

Do not allow long strings of **passive** sentences to creep into your writing. The **active voice** is often clearer and livelier.

## Mood

The **indicative** and **imperative moods** usually give students very little trouble. Use the **subjunctive** *were* to talk about a wish or make a statement that is not factual. Use *be* in clauses following *advise, ask, recommend, request, suggest.*

## ? END OF CHAPTER QUIZ More about Verbs

**DIRECTIONS:** *Correct the fifteen (15) verb errors in the following passage.*

I remember my first casserole well. I arrived home from school knowing it is my night for dinner. I wanted to surprise my mom and has dinner ready by the time she got home from work. My mother had had all of her recipes on note cards in a little plastic box. I remember getting the recipe out of the box and going to work. I put all the ingredients into the bowl. I preheat the oven to 350°. I was looking good on time. The casserole needs 20 minutes to bake, and Mom would arrive in 30 minutes. Now, I had had everything in the oven and am setting the table. Mom came through the door, and I could not waited to tell her that I had dinner ready.

It was a night my dad had to work late. My mother, my brother, and I set down to have dinner. I brought out the tuna noodle casserole and a large serving spoon, and lay them on the table. The casserole looks great until I tried to spooning it out. The casserole turns into a tuna noodle brick. Mom instantly knew what had happened. She asked me if I had looked on the back of the recipe card, knowing I had not. I got the card and look on the back. It all made sense when I read that the noodles should be boiled first. That night we decide take out would be better than the tuna noodle brick.

—Tom Parsil[†].

7

[†]Adapted from an essay by Tom Parsil. The errors were added to create this quiz.

chapter 24

# Pronouns and Point of View

**Pronouns** are words that can substitute for nouns, often to avoid awkward repetition. For example, the following sentence sounds silly because *Elisa* is repeated so often: "After *Elisa* received *Elisa's* instructions from the sous chef, *Elisa* placed the grilled vegetables on *Elisa's* tray." The sentence is so bogged down with *Elisa*'s that we don't even know what Elisa has been doing. By substituting appropriate pronouns for *Elisa,* we can produce a sentence that sounds better and makes its point more clearly: "After Elisa received *her* instructions from the sous chef, *she* placed the grilled vegetables on *her* tray."

Although we easily see the advantages of using pronouns, they are words that are frequently *misused* in terms of case, agreement, reference, and point of view.

Let's keep this between you and *I*. [incorrect case]

Everyone was happy to receive *their* desserts. [lack of agreement]

Mike told Jack that *he* was lazy. [unclear reference]

*You* might have trouble with emulsions. [informal point of view]

When you're editing a final draft, pronouns deserve extra attention!

## USING THE CASES

In many languages, such as Spanish, French, and Italian, words change form to indicate the different jobs they have in a sentence. English has

"After Elisa received her instructions from the sous chef, she placed the grilled vegetables on her tray."

lost many of these changes, but they do remain to some degree in **case**, the name given to the different forms of nouns and pronouns. The three cases left in English are the subjective, objective, and possessive. Study Figure 24.1.

## The Subjective Case

The **subjective case** is used for the subject of a sentence. For example, "*He* is intrigued by the concept of fusion cuisine." This usage prevails in both informal conversation and formal writing. We sometimes make mistakes, however, when the pronoun is in a group of words connected by *and* or *or.* Here informal usage often conflicts with formal. For example, you may hear a sentence such as the following:

> Me and her are planning a big night on the town. [incorrect]

Yet you wouldn't say "*Me* is planning a big night" or "*Her* is planning a big night." In formal usage, both pronouns should be in the subjective case.

**FIGURE 24-1** **Pronoun Cases.**

**Cases of Singular Pronouns**

| Subjective | Objective | Possessive |
|---|---|---|
| I | me | my, mine |
| you | you | your, yours |
| he, she, it | him, her, it | his, her, hers, its |

**Cases of Plural Pronouns**

| Subjective | Objective | Possessive |
|---|---|---|
| we | us | our, ours |
| you | you | your, yours |
| they | them | their, theirs |

*She and I* are planning a big night on the town. [correct]

When used alone, both pronouns sound "right": "*She* is planning a big night" or "*I* am planning a big night."

Similarly, the subjective case is used with subjects of dependent clauses.

Harry is thrilled because *he* [not *him*] and his brother are vacationing in Hawaii this year.

In the sentence above, *he and his brother* is the compound subject of the dependent clause *because he and his brother are vacationing in Hawaii*. We would not say "*Him* is vacationing in Hawaii."

Finally, the subjective case is used with a pronoun that follows a linking verb.

I knew it was *they*. [not *them*]

Linking verbs do not have objects, so the objective case *them* (described in the next paragraph) is incorrect.

## The Objective Case

The **objective case** is used with the objects of verbs (both direct and indirect) and prepositions. The **direct object** of a verb receives the action; it answers the question "Who or what received the action of the verb?" For example, in the sentence "Derek Jeter threw him the ball," *ball* is the direct object of the verb *threw. Him* is the **indirect object**, explaining *to whom* the ball was thrown. Again, complications arise when a pronoun is part of a *compound* object, as in the example below:

Harriet gave Drew and me [not I] the tickets.

To check the usage, drop the noun and consider the pronoun only. It sounds better when we say "Harriet gave *me* the tickets" than "Harriet gave *I* the tickets." Similarly, test the objects of prepositions with the pronouns only. We wouldn't say "Pass the salt to *I*"; therefore, we shouldn't write "Pass the salt to *Tim and I*." Look at the following examples:

Keep this information between you and *me* [not *I*].

Culinary school was the best choice for Andy and *me* [not *I*].

Because we hear this mistake very often in speech, we need to think carefully as we're editing our writing.

Finally, the objective case is used for both the subject and object of an infinitive. For example:

Paul told *him* to serve the first course.

*Him* is the subject of the infinitive *to serve,* as well as the object of the verb *told.* In the next sentence,

Paul told Victor to serve *her*.

In this example, *her* is the object of the infinitive *to serve.*

## Choosing the Correct Case

Choosing the correct pronoun case is made more difficult because we often use one case in informal speech and another in formal writing. The choice is especially confusing when a pronoun and a noun are used together. In the two examples below, check the case of the pronoun by mentally crossing out the noun.

*Us students* often need to work on the weekends. [Incorrect; you wouldn't say "*Us* need to work."]

*We students* often need to work on the weekends. [Correct; think "*We* need to work."]

Sometimes it's difficult for *we students* to make time to study. [Incorrect; you wouldn't say "It's difficult for *we* to make time."]

Sometimes it's difficult for *us students* to make time to study. [Correct; think "It's difficult for *us*."]

A choice is also required after *than* and *as*. Choose the subjective or objective case depending on your meaning, and check the usage by completing the clause. Study the following examples:

Jeannette likes chocolate better *than I*.

Jeannette likes chocolate better *than I* [do]. [complete the clause]

In other words, I don't like chocolate as much as Jeannette likes chocolate.

Jeannette likes chocolate better *than me*.

Jeannette likes chocolate better *than* [she likes] *me*.

In other words, Jeannette doesn't like me as much as she likes chocolate. Study the similar examples with the word *as* below.

Jeannette does not like anchovies *as much as I* [do].

Jeannette does not like anchovies *as much as* [she likes] *me*.

## The Possessive Case

The **possessive case** is used to modify nouns (for example, *my* brother) and words that act like nouns, such as gerunds (verbs plus *–ing*, such as *keeping*). Study the following examples:

> Paul was pleased by *their* neatness. [*their* modifies the noun *neatness*]

> Paul was pleased by *their* keeping the walk-in neat. [*their* modifies the gerund *keeping*]

In these sentences, *keeping* is a gerund used as a noun and is modified by the possessive form *their*. Remember that a verb plus *–ing* may also be a present participle used as an adjective, as in the following sentence:

> The police caught *him* climbing out the window.

Here the phrase "climbing out the window" modifies the pronoun *him*, which is the object of the verb *caught*.

In some situations, you may choose either the objective case to emphasize the pronoun or the possessive case to emphasize the verb plus *–ing*. The first example below emphasizes the pronoun:

> The police observed *him* climbing.

The police were looking at *him;* we don't know whether or not the *climbing* was important. In the second example, however, the police are focused on the *climbing* itself (Figure 24.2).

> The police observed *his* climbing.

**FIGURE 24-2 Using the Objective and Possessive Cases.**

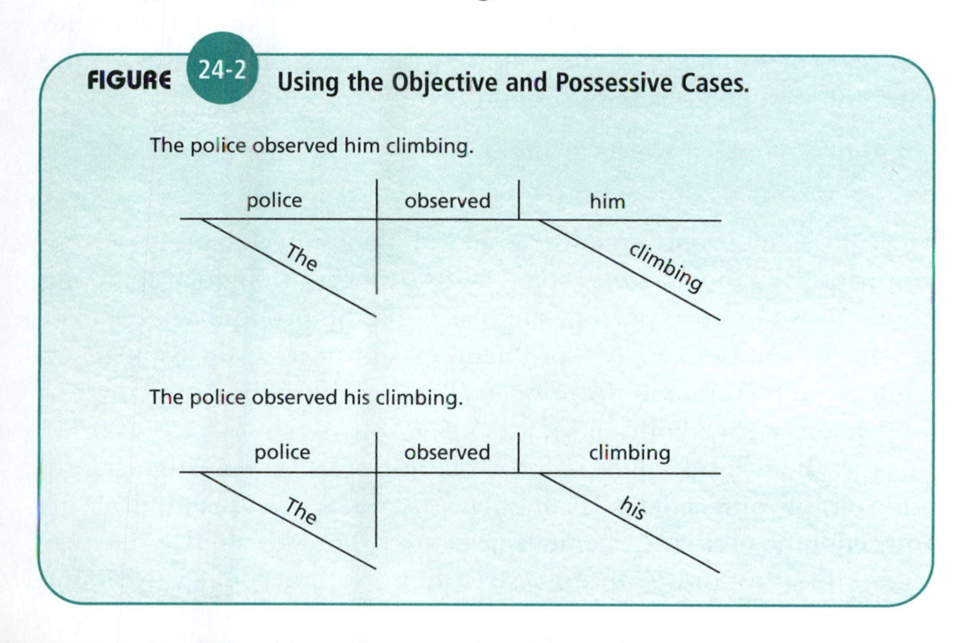


7

**EXERCISE 24.1** **Choosing the Correct Case**

Identify the correct pronoun in each sentence.

1. George knew that Fred and (he/him) would have a busy night on Friday.
2. It was (they/them) who came in on their day off.
3. Between you and (I/me), I'm glad I won't be working that night!
4. I like relaxing at home better than (they/them). [They don't like to relax at home.]
5. Not everyone likes to work hard on (his or her, their) day off.

Additional types of pronouns are discussed in various other chapters. For **interrogative pronouns**, see Chapter 19; for **relative pronouns**, see Chapter 20. For *who* and *whom,* see Appendix III.

## PRONOUN—ANTECEDENT AGREEMENT

Another set of issues regarding pronouns concerns agreement between a pronoun and its **antecedent**, that is, "something that comes before." For example, in the following sentence, *Aunt Frances* is the antecedent while *her* is the pronoun that agrees with it.

> Aunt Frances traveled a great deal to visit her nieces and nephews.

The rule is that pronouns must agree with their antecedents in person, number, and gender. *Aunt Frances* is a third person, singular feminine noun; *her* is the third person, singular feminine pronoun. Agreement in person is not generally a problem, except in terms of consistency. Students will often move from *you* to *I* to *they* within the rough draft of a single essay. (See Point of View below.)

Agreement in number seems straightforward: use a singular pronoun if its antecedent is singular; use a plural pronoun if its antecedent is plural. Agreement in gender follows a similar rule: use a feminine pronoun (*she, her, hers*) when its antecedent is feminine

**Noun–Pronoun Agreement.**

Aunt Frances went to Paris last spring.

She loved the Eiffel Tower.

Aunt Frances took her sister to Paris.

They both filled their suitcases with new clothes.

(*Mrs. Perez, Joyce, girl*); use a masculine pronoun (*he, him, his*) when its antecedent is masculine (*Uncle John, Edward, boy*); and use a neutral pronoun (*it, its*) when its antecedent has no gender (*book, cake, rose*). All third person plural nouns take the pronoun *they* (*them, their*). Study the following examples:

> *Mrs. Perez* is an excellent cook; *she* sometimes shares *her* recipes with friends.
>
> I asked *Uncle John* about *his* trip to Costa Rica when we visited *him* at Thanksgiving.
>
> Don't judge a *book* by *its* cover. (Don't judge *books* by *their* covers.)
>
> Do you know the *Johnsons*? *They* moved into *their* new house across the street last week.

Difficulties can arise, however, when the antecedent is an indefinite pronoun, such as *anyone, anybody, each, either, everyone, neither, no one, nobody, someone,* and *somebody*. In casual speech, these words are sometimes assigned a plural pronoun; for example, "Everyone is ready for *their* quiz tomorrow." However, in formal English these words are always singular and take both singular verbs and singular pronouns (see also Chapter 22). A second question may arise as to which pronoun to use: *his* or *her*? Since writers try to avoid sexist language, we don't want to use simply *his* (see also Chapter 11). We might rewrite the sentence as follows:

> Everyone is ready for *his or her* quiz tomorrow.

The sentence is correct; however, if many repetitions are required, the writing may become awkward and wordy. In a longer work like this textbook, we can alternate between *he* in one chapter and *she* in the next, or, in general, we can rewrite sentences so as to sidestep the agreement issue completely.

Everyone is ready for tomorrow's quiz.

The students are ready for their quiz tomorrow.

Another complication occurs when the antecedent is a **generic noun**, that is, when it names a typical member of a group; for example, "The average server in this restaurant wears a nametag on their uniform." We could substitute *his or her* for *their*; we might also rewrite the sentence more smoothly as follows:

The servers in this restaurant usually wear nametags on their uniforms.

Like agreement between subjects and verbs, agreement between pronouns and antecedents has rules about conjunctions. Though often ignored in informal speaking and writing, these rules apply in formal situations. When antecedents are joined by *and*, a plural pronoun is used.

*The pear and banana* are ripening in *their* skins.

When antecedents are joined by *or* or *nor*, the pronoun should agree with the nearer one.

*The pear or the banana* is ripening in *its* skin.

*The pear or the bananas* are ripening in *their* skins.

Note that sentences will be smoother if the plural antecedent follows the singular.

### EXERCISE 24.2 Pronoun–Antecedent Agreement

On a separate sheet of paper, write the appropriate pronoun for each blank below.

1. The students in the class opened _________ textbooks.
2. Anybody would appreciate such an addition to _________ income.
3. Beth and her mother bought _________ tickets online.
4. Everyone occasionally makes mistakes with _________ pronoun use.
5. Adam or Richard likes sugar in _________ coffee.

## CLEAR PRONOUN REFERENCE

Sometimes we run into trouble when we are not clear about what each pronoun refers to. Look at the sentence below:

> Mike told Jack that *he* was lazy.

Does *he* refer to *Mike* or *Jack?* That is, is Mike himself lazy? Or does Mike think Jack is lazy? This unclear pronoun reference can be corrected by adding a reflexive pronoun (*himself*) or repeating the appropriate noun.

> Mike told Jack that *he himself* was lazy.
>
> Mike told Jack that *Jack* was lazy.

Another problem in pronoun reference arises when we use *you, it,* or *they* without referring to specific persons or things. One of the most common is the use of *you* instead of an indefinite pronoun such as *one* or a common noun such as *people. You* is acceptable only when we intend to address the reader directly, as in a letter or a recipe.

> *You* should baste the turkey frequently to keep it moist.

In most formal writing, avoid *you* altogether.

> Basting the turkey frequently will keep it moist.

The pronoun *it* is sometimes used in conversation without a clear reference. For example, the sentence "In the movie *it* shows how the penguins try to protect their eggs from the bitter Antarctic cold" might be rewritten more clearly as follows:

> *The movie shows* how the penguins try to protect their eggs from the bitter Antarctic cold.

**Who's the lazy one?**

Mike told Jack that he was lazy. So who's that on the couch?

In addition, avoid using *they* when you don't really know who *they* are.

> After a snowfall, *they* work quickly to clear the roads for the school buses.

If you know who *they* are, use the specific noun.

> After a snowfall, *the town* works quickly to clear the roads for the school buses.

If you don't, rewrite the sentence in the passive voice.

> After a snowfall, *the roads are quickly cleared* for the school buses.

Finally, use specific words instead of the pronouns *it, that, this,* or *which* to refer to whole sentences or concepts.

> The episode of *NYPD Blue* in which Bobby Simone died was particularly effective. *This* helps explain why some viewers prefer the relatively impersonal *Law & Order*.

The weak use of *this* might be replaced with a specific noun.

> *The emotional stress of that episode* helps explain why some viewers prefer the relatively impersonal *Law & Order*.

## EXERCISE 24.3 Clear Pronoun Reference

Rewrite each sentence below to correct unclear pronoun reference.

1. Cecilia and her mother thought that *she* needed a vacation.
2. George told his friends that *they* had tickets for the Duke game.
3. The movie was suspenseful, and *you* were on the edge of your seat.
4. In the movie *it* showed the hero dangling from a gutter six stories above the ground!
5. The gutter scene is traumatic for the hero, who is afterwards afraid of heights, and enormously suspenseful for the audience. Later, this prevents him from following the girl up the tower stairs.

**FIGURE 24-3 Reflexive and Intensive Pronouns.**

| Singular | Plural |
|---|---|
| myself | ourselves |
| yourself | yourselves |
| himself, herself, itself | themselves |

## REFLEXIVE AND INTENSIVE PRONOUNS

**Reflexive pronouns** are formed by adding *–self* to the personal pronouns (Figure 24.3) and are used when the subject of the sentence is also the object.

Andrew told *himself* to work more quickly.

**Intensive pronouns**, which have the same forms, are used to add emphasis to their antecedents.

The executive chef *himself* made the rounds of the dining room.

Be sure to use the correct forms of these pronouns: *himself*, not *hisself*; *themselves*, not *theirselves*. In addition, avoid using *myself* or *ourselves* as the subject of a sentence.

The Ortegas and *ourselves* had a late dinner at Bistro Urbano. [incorrect; we wouldn't say "*Ourselves* had a later dinner"]

We and the Ortegas had a late dinner at Bistro Urbano. [correct]

Janet and *myself* enjoyed the fresh Greek salad with oodles of feta cheese. [incorrect; we wouldn't say "*Myself* enjoyed the salad"]

Janet and *I* enjoyed the fresh Greek salad with oodles of feta cheese. [correct; think "*I* enjoyed the salad"]

## POINT OF VIEW

**Point of view**, the narrator's position with regard to the story, can be dramatically illustrated in thrillers like *Psycho* and *Halloween*. Both of these films are disturbing because the camera (a kind of narrator) often takes the villain's perspective. The audience is looking through

7

the eyes of the killer, leading the audience to identify with *him* rather than with the innocent victim!

**EXERCISE 24.4** **Point of View**

Study Figure 24.4. Write a paragraph from the *camera's* point of view describing what is happening in the picture. Then write a second paragraph from the point of view of the student in the foreground, imagining and describing what she sees. Finally, write a third paragraph from the point of view of the student on the far left. What does she see? Check that the use of pronouns in each paragraph is correct and consistent.

In writing, point of view is established partly through the choice of pronouns, for example, *I, we, you, they,* and *one.* Use the first person, *I* and *we,* for writing that emphasizes your personal experience, including letters and essays. Use the second person, *you,* for letters, directions, and recipes. (Note that letters will most likely use both *I* and *you.*) Use the third person—*he or she, they,* or *one*—in very formal academic or professional writing. Follow the guidelines of your instructor or publisher.

**FIGURE 24-4** Each person in the photograph has a different point of view.

Sometimes inexperienced writers tend to shift from one point of view to another within the same piece of writing. Such shifts are confusing to the reader. Study the examples below.

*The audience* clapped enthusiastically when the penguin parents were reunited with their chicks. *You* often feel emotional when nature films involve family groups. [shift from third to second person]

*The audience* clapped enthusiastically when the penguin parents were reunited with their chicks. *Viewers* often feel emotional when nature films involve family groups. [both sentences now in third person]

*I* visited an aquarium in Boston that had a colony of penguins. *You* could see the young ones learning to swim. [shift from first to second person]

*I* visited an aquarium in Boston that had a colony of penguins. *I* could see the young ones learning to swim. [both sentences now in first person]

Like shifts in verb tense, shifts in point of view are distracting and confusing to the reader. Try to stay with the same point of view throughout each essay.

## EXERCISE 24.5 Maintaining Consistent Point of View

Rewrite the passage below, changing five of the pronouns to maintain a consistent point of view.

Food competitions are time-demanding and stressful, but try being the team captain looking over four other students and having to deal with the teachers and judges. First I picked the students that I wanted to be on the team; then you put them in a place they would succeed in. After two long days of brainstorming you finally came to agreement on the menu. If we forgot to pack anything, it will throw you off completely.

Now we get to the meat of the competition: the cooking. I started to work on the salad as Sarah, my assistant, started on the appetizer. This is when things began to go wrong: there was too much liquid in the polenta. Luckily, I found some cheese from another group, and you thickened the polenta with it. We cooked eight portions of the plate. Three of them one presented to teachers with our wine choice, and the other five we got to eat ourselves.

—Joey Jacobsen*.

*Adapted from an essay by Joey Jacobsen. Errors were added to create this exercise.

7

## RECIPE FOR REVIEW Pronouns and Point of View

### Using Pronoun Case

1. Study the case forms in Figure 24.1.
2. Use the **subjective case** for the subjects of independent and dependent clauses, and with subject complements (It is *I* at the door).
3. Use the **objective case** for the direct and indirect objects of verbs, the objects of prepositions, and the subjects and objects of infinitives.
4. Use the **possessive case** to modify nouns and gerunds (words that act like nouns).

### Ensuring Pronoun-Antecedent Agreement

1. Pronouns must agree with their antecedents in person, number, and gender.
2. Avoid *they, them,* and *their* with singular indefinite pronouns such as *everyone* and with singular generic nouns such as the *typical student.*
3. Use a plural pronoun with two or more antecedents joined by *and.* When antecedents are joined by *or* or *nor,* the pronoun should agree with the nearer antecedent.

### Maintaining Clear Pronoun Reference

1. Repeat the antecedent or use a reflexive pronoun to avoid unclear pronoun reference.
2. Avoid using *it, they,* or *you* without a clear, specific antecedent.
3. Avoid using *it, that, this,* or *which* to refer to a whole sentence or concept.

### Using Reflexive and Intensive Pronouns

1. Study the reflexive and intensive forms in Figure 24.3.
2. Use reflexive pronouns when the subject of the sentence is also the object.

3. Use intensive pronouns to add emphasis to their antecedents.
4. Avoid using *myself* or *ourselves* as the subject of a sentence.

## Appropriate and Consistent Point of View

1. In formal essays, avoid addressing the reader directly with *you.*
2. Use point of view consistently; for example, avoid shifting from they to you to one.

## END OF CHAPTER QUIZ Pronouns and Point of View

**DIRECTIONS:** *Rewrite the sentences on a separate sheet of paper, correcting any pronouns that are used incorrectly.*

Example: You and me need to talk.
Correction: You and **I** need to talk.

_______ **1.** Beth thought that her and her husband would enjoy a bottle of chardonnay on their picnic.

_______ **2.** Let's keep that information between you and I.

_______ **3.** We students don't always get enough sleep.

_______ **4.** Everyone needs to have his day of rest!

_______ **5.** The typical Law & Order attorney speaks passionately on their client's behalf.

_______ **6.** The orange or the tangerine has a sticker on its peel.

_______ **7.** Penelope and her sister thought she should apply to culinary school.

_______ **8.** Jim made toast with butter, but it was too cold.

_______ **9.** Karen and myself saw the movie three times.

_______ **10.** When the characters collided on the football field, people in the theater laughed so hard you cried.

*chapter* 25

# Modifiers

A **modifier** is a word that describes, explains, or limits another word in some way. Modifiers can be single words like adjectives and adverbs or groups of words like phrases or clauses. Both the form and the placement of modifiers are important in making our points clear to the reader and in demonstrating our knowledge of standard written English.

## USING ADJECTIVES AND ADVERBS

**Adverbs** are generally recognizable by their *–ly* ending and are used to modify verbs, adjectives, and other adverbs.

The water in the saucepan boiled rapidly. [The adverb *rapidly* modifies the verb *boiled*.]

The water in the saucepan was at a very rapid boil. [The adverb *very* modifies the adjective *rapid*.]

The water was boiling very rapidly. [The adverb *very* modifies the adverb *rapidly*.]

Most **adjectives**, on the other hand, do not end in *–ly* and are used to modify nouns, sometimes following a linking verb.

The rapid boil stopped once the saucepan was removed from the flame. [The adjective *rapid* modifies the noun *boil*.]

The saucepan was still hot, though. [The adjective *hot* follows the linking verb *was*.]

The saucepan was still hot, though.

However, problems do arise when adjectives and adverbs have unexpected forms or are used in place of one another.

While most adverbs end in *–ly,* there are some that do not, such as *fast, hard, late,* and *straight.* In fact, these adverbs have the same form as their adjective counterparts (Figure 25.1).

There are also some adjectives that end in *–ly,* such as *daily, friendly, early,* and *lively.*

Have you put the daily special up on the board?

Friendly wait staff make a big difference to customer satisfaction.

Bruce preferred to work an early shift.

The child's lively behavior made restaurant dining difficult for his parents.

Two of these words can also be used as adverbs: *daily* and *early.*

Jocelyn wrote the specials on the board daily. [modifies the verb *wrote*]

Bruce likes to come in early.

**FIGURE 25-1 Adjectives and Adverbs with Identical Forms.**

| ADJECTIVE | ADVERB |
| --- | --- |
| The Olympic athlete was fast. Compare: The Olympic athlete was quick. | The Olympic athlete ran fast. Compare: The Olympic athlete ran quickly. |
| We studied for the hard test. | We studied hard for the test. |
| Late papers will not be accepted. | Don't hand your paper in late. |
| The road was wide and straight. | Drive straight after the second light. |

*Friendly* and *lively*, however, do not have adverb forms. They must be used in a phrase such as "in a friendly way" or "in a lively manner."

## Well and Good

Special problems are posed by *well* and *good* because they are often used differently in casual speech than they are used in formal written English. While *well* can be used as either an adjective or an adverb, *good* is used only as an adjective. *Good* may follow a linking verb, but it should not modify a verb.

The lasagna tastes good. [correct; the lasagna *is* good]

Tony cooks good. [incorrect]

In the first example, the adjective *good* correctly modifies the noun *lasagna;* the verb *tastes* is a linking verb in this sentence. Remember that linking verbs include *to be* and *to seem*, as well as some meanings of *to feel, to smell*, and *to taste*. In general, if a verb can be replaced with *is* or *are*, it is a linking verb. In the second example above, the word *good* is used incorrectly to modify the verb *cooks* and should be changed to the adverb *well.*

Tony cooks well. [correct]

*Good* is an adjective. While many adverbs are formed by adding *–ly* to the adjective, *good* is an exception. However, note that *good* is used correctly in the following sentence:

The coffee smells good. [correct]

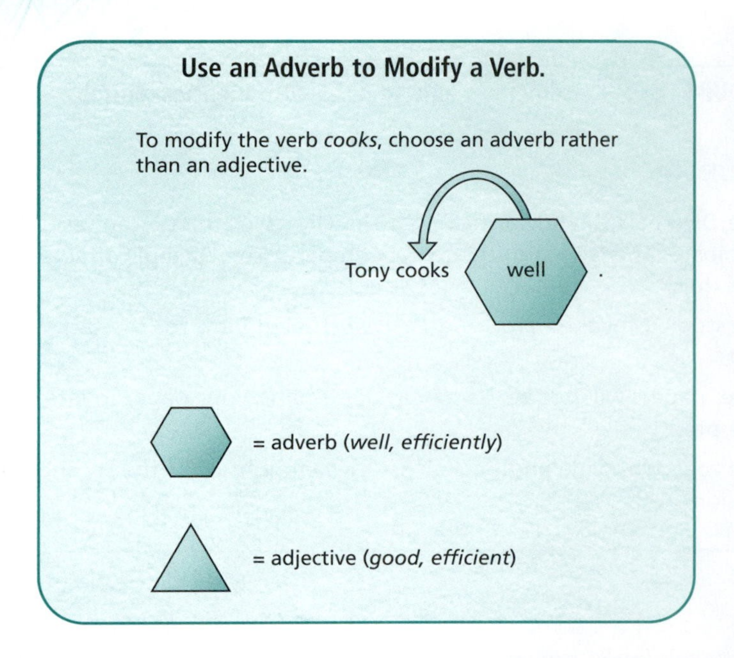


In the sentence *good* is modifying the noun *coffee*, not the verb *smells*. There *smells* is a linking verb; it could be replaced with *is*. However, if *smell* is used with a direct object, it must be followed with an *adverb*.

> The curious dog smelled the strange cat *thoroughly*. [not *good*]

Some other pairs of words that create confusion are *bad/badly*, *real/really*, and *slow/slowly*. In the sentence below, an adjective (*real*) is used informally in place of an adverb.

> The risotto was real tasty. [informal; *real* is an adjective and cannot modify another adjective, *tasty*]

*Real* forms the adverb by adding *–ly*, and the sentence may be corrected as follows:

> The risotto was *really* tasty. [formal; the adverb *really* correctly modifies the adjective *tasty*]

Study the following pairs of correct sentences:

> The performance went badly. [The adverb *badly* modifies the verb *went*.]
>
> The moldy cheese tasted bad. [The adjective *bad* follows the linking verb and modifies *cheese*.]

> The first date went well. [The adverb *well* modifies the verb *went*.]
>
> The first date was good. [The adjective *good* follows the linking verb and modifies *date*.]

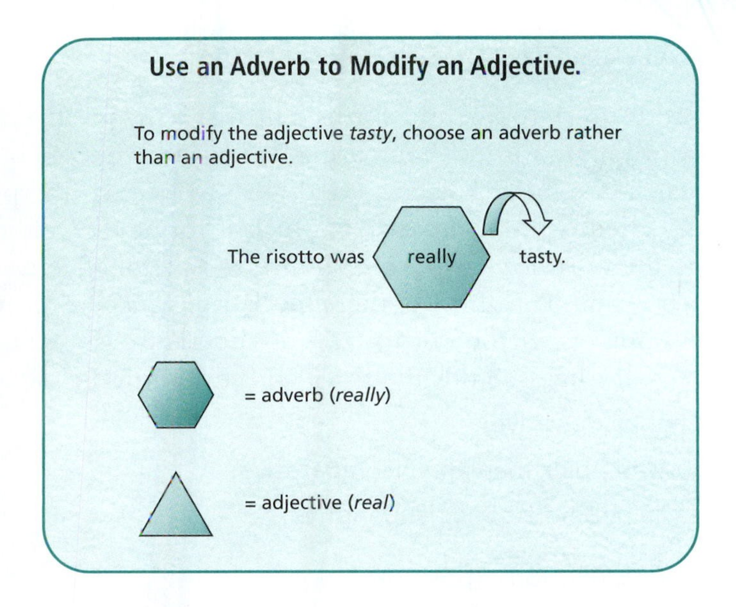


*Slow* is sometimes used as an adverb in the common expressions "drive slow" or "go slow." To be strictly correct, however, use the formal adverb *slowly*.

The bus was traveling slowly through the busy city streets.

Beware of similar confusions with the pairs *awful/awfully, poor/ poorly, quick/quickly*, and *quiet/quietly*. Once you decide whether you need an adjective or adverb, be sure to use the correct form.

## EXERCISE 25.1 Using Adjectives and Adverbs

Rewrite each sentence below, correcting any errors in the use of adjectives and adverbs.

EXAMPLE: Seth wrote *poor* with his left hand.
REWRITE: Seth wrote *poorly* with his left hand.

1. The strange-looking drink nevertheless tasted good.
2. Tom told Maria that she danced good.
3. Leo was real careful not to spill a drop as he poured the coffee.
4. The stage set was so detailed and colorful that it looked real.
5. Jerry reminded his classmates to beat the egg whites real good.

7

## Comparative and Superlative Forms

Adjectives and adverbs have three forms. The first is the **positive**, a form that describes a noun without comparing it to another one. In our preceding discussion, all the adjectives and adverbs were in the positive form. The second is the **comparative**, which is formed by adding the suffix *–er* or the word *more.* The comparative is used to judge *two* items against each other. The third is the **superlative**, which is formed by adding the suffix *–est* or the word *most.* The superlative is used to compare *more than two* items. Study the three adjective forms below:

Trey is happy. [positive]

Bennett is happier than Trey. [comparative]

Trish is the happiest of all. [superlative]

Trey's cake is delicious. [positive]

Bennett's cake is more delicious than Trey's. [comparative]

Trish's cake is the most delicious of all. [superlative]

The adverb forms are similar:

Trey works fast. [positive]

Bennett works faster than Trey. [comparative]

Trish works the fastest of all. [superlative]

Trey washes plates quickly. [positive]

Bennett washes plates more quickly. [comparative]

Trish washes plates the most quickly of all. [superlative]

Unfortunately, it is not always possible to predict which adjectives and adverbs will add a suffix to form the comparative and superlative and which will add the words *more* and *most.* The majority of one-syllable adjectives and adverbs add a suffix, *–er* or *–est.* Those with two syllables use the suffix or the words *more* or *most.* Adjectives and adverbs with more than two syllables typically use the words *more* and *most.* Check a dictionary to be certain.

In addition to the two common variations discussed above, some adjectives and adverbs have irregular forms (Figure 25.2). While it is useful to memorize these forms, remember that they can also be found in a dictionary. Finally, note that only one form of the comparative and superlative may be used. The following sentence is incorrect:

Bennett is more friendlier than Trey. [incorrect]

Bennett is friendlier than Trey. [correct]

Bennett is more friendly than Trey. [correct]

**FIGURE 25-2** **Irregular Forms of the Comparative and Superlative.**

| POSITIVE | COMPARATIVE | SUPERLATIVE |
|---|---|---|
| bad (adjective)<br>badly (adverb) | worse<br>worse | worst<br>worst |
| good (adjective)<br>well (adverb) | better<br>better | best<br>best |
| little (adjective)<br>little (adverb) | littler<br>less | littlest<br>least |
| many (adjective with number)<br>much (adjective with amount) | more<br>more | most<br>most |

## EXERCISE 25.2 Using Comparatives and Superlatives

On a sheet of paper, write the appropriate form of the adjective or adverb in parentheses.

EXAMPLE: The chef searched for the ______ (*good*/superlative) recipe.

WRITE: best

1. Alyssa needed __________ (*many*/comparative) jalapeño peppers for the chili sauce.
2. This restaurant has the __________ (*large*/superlative) selection of wines in the city.
3. I have never seen a __________ (*funny*/comparative) movie than *Tootsie.*
4. Dan diced the onions __________ (*efficiently*/comparative) than Tom did.
5. While everyone felt badly about the abandoned dog, Cindy seemed to feel __________ (*badly*/superlative).

## MISPLACED MODIFIERS

The role of a modifier is to describe or explain one of the words in a sentence. A **misplaced modifier** is too far away from the word it describes, and the reader may be confused. Look at the sentence below:

> The chef tosses the pizza dough into the air in the immaculate white jacket.

This sentence is confusing because it seems as if the *air* is wearing an *immaculate white jacket.* The sentence is clearer when the phrase is placed next to the noun it truly modifies, *chef.*

> The chef in the immaculate white jacket tosses the pizza dough into the air.

Such errors in placement are not uncommon. With all the various phrases and descriptions floating around in our minds, we sometimes write them down in the wrong order.

Be particularly careful with words such as *almost, even, just,* and *only.* They must be placed right next to the words they modify in order to avoid confusion such as the following:

> Javier only eats dark chocolate. [He doesn't eat any other food!]

These students don't need coffee to enjoy a taste of dessert.

Since the word *only* modifies *chocolate,* not *eats,* it should be placed next to it.

Javier eats only dark chocolate. [He doesn't eat any other kind of chocolate.]

Another example:

Javier only eats dessert when he has a cup of coffee.

*Only* does not modify *eats.* Instead, it modifies the clause *when he has a cup of coffee* and should be placed next to it.

Javier eats dessert only when he has a cup of coffee.

### EXERCISE 25.3 Correcting Misplaced Modifiers

Rewrite the sentences below to correct the misplaced modifiers.

EXAMPLE: The goal was to make tall sugar sculptures *of the contest.*
The goal *of the contest* was to make tall sugar sculptures. [corrected]

1. *Pastry Daredevils* followed a competition in sugar sculptures, a series on the Food Channel.
2. One of the competitors almost attempted a sculpture nine feet in height.
3. The sculptures reflected the chefs' imaginative interpretations of a fairy tale airbrushed with food coloring.
4. The *Meilleur Ouvrier de France* (M.O.F.) walked quickly over the obstacle course considered a top craftsman in France.
5. The winner created a six-foot sculpture who received a check for ten thousand dollars.

## DANGLING MODIFIERS

The problem with a **dangling modifier**, which is often an initial participial phrase, is that it does not refer to an actual word in the sentence. Without that connection, the modifier dangles! Look at the example below:

After waiting over an hour, their table was finally ready.

*Who* has been waiting over an hour? The reader expects the phrase *after waiting over an hour* to modify the nearest noun. Logically, though, the *table* hasn't been waiting; *they* have. Yet the word *they* does not even appear in the sentence. To correct the "dangle," we might rewrite the sentence in a couple of different ways.

> After they had waited for over an hour, their table was finally ready.

Here we changed the dangling participial phrase into a solid dependent clause.

> After waiting over an hour, they were finally seated.

In this second rewrite, we changed the subject of the sentence to *they*. Now the phrase *after waiting over an hour* correctly modifies the nearest noun, *they*.

## EXERCISE 25.4 Correcting Dangling Modifiers

Rewrite the sentences below to correct the dangling modifiers.

EXAMPLE: Leaning dangerously to one side, the audience watched the chef navigate the obstacle course. [It's not the *audience* or the *chef* that is leaning; it's the unnamed sugar sculptures.]

The audience watched as the competitor navigated the obstacle course, his sugar sculpture leaning dangerously to one side. [clear]

1. Required to use five methods of preparing the sugar (poured, pulled, blown, *pastiage*, and pressed), the sculptures also reflect the competitors' imaginations.
2. Hoping to avoid comparison with an actual fairy tale, the sculpture was an abstract portrayal of a French harlequin.
3. Using a blow torch, the sculptures were melted and molded into the proper shapes.
4. Balancing the sculptures carefully, they were carried over an "obstacle course" to see if they would break.
5. Judged not only on degree of difficulty but on the enormity of the risk, members of the audience held their breaths during the obstacle course.

## RECIPE FOR REVIEW Modifiers

### Using Adjectives and Adverbs

1. **Adverbs** are generally recognizable by their *–ly* ending; they modify verbs, adjectives, and other adverbs.
2. Most **adjectives**, on the other hand, do not end in *–ly;* they modify nouns, sometimes following a linking verb.
3. Use *good* only as an adjective: The meal was *good* [adjective modifying *meal*], but the meal was cooked *well* [adverb modifying *was cooked*].

### Comparative and Superlative Forms

1. Use the **comparative form** to compare *two* items. Adjectives and adverbs form the comparative with *–er* or *more*. Check a dictionary.
2. Use the **superlative form** to compare *more than two* items. Adjectives and adverbs form the superlative with *–est* or *most*. Check a dictionary.
3. Study the irregular forms of the comparative and superlative in Figure 25.2.
4. Do not use both forms of the comparative (*–er* and *more*) or both forms of the superlative (*–est* and *most*) to modify a single word. Study the examples below.

   Leroy was more friendlier than Alan. [incorrect]

   Leroy was friendlier than Alan, or Leroy was more friendly than Alan. [correct]

### Misplaced Modifiers

Ensure that adjectives and adverbs—whether they are words and phrases—are placed immediately next to the words they modify.

The goal was to make tall sugar sculptures of *the contest*. [The prepositional phrase modifies *goal* and should be placed next to it.]

The goal *of the contest* was to make tall sugar sculptures. [corrected]

## Dangling Modifiers

Ensure that modifiers refer to a specific word in the sentence; don't let them dangle.

After waiting over an hour, their table was finally ready.

After waiting over an hour, they were finally seated. [corrected]

## END OF CHAPTER QUIZ MODIFIERS

**DIRECTIONS**: *Rewrite each sentence below, making corrections to modifiers if necessary.*

_______ **1.** On their trip to Hawaii, Christine noticed that her boyfriend surfed good.

_______ **2.** She also noticed that the fresh pineapples tasted especially good.

_______ **3.** Christine walked quick downstairs every night for a piña colada.

_______ **4.** Her boyfriend is fonder of a beer before dinner.

_______ **5.** Rhoda liked the fresh pineapple who was another friend on the trip.

_______ **6.** She only ate pineapple slices, however.

_______ **7.** Rhoda could surf more better than Christine's boyfriend.

_______ **8.** After spending a week on the beach, her tan was the deeper of the three friends'.

_______ **9.** Watching her friends in the water, Christine wished she had learned to surf, too.

_______ **10.** Burned by the sun, the umbrella kept her in the shade.

chapter 26

# Understanding Parallel Structure

You may remember having studied parallel lines in high school. The lines run along the same distance apart into infinity. It's important to maintain that parallelism in real-life situations such as train tracks. If one track went off in a different direction, the train might jump the rails! Parallelism is also important in the culinary world. A restaurant wouldn't use different patterns of china and silver on the same table because the place settings wouldn't be *parallel*.

Maintaining parallelism in writing is about meeting the reader's expectations, making your meaning clear, and creating smooth and pleasing rhythms in your sentences. If the *ideas* in a sentence are somehow equivalent, the *structure* that expresses them should be similar: that is **parallelism**. For example, notice the parallel structure of the second, third, and fourth sentences in the passage below:

**STUDENT WRITING** Payson S. Cushman

When one lists the responsibilities of a chef, it usually does not include addressing food-related issues or politics at all. It would, of course, include serving quality food that is safe. It would include providing an enjoyable setting to experience the food. It would include making a meal at their establishment satisfying and valuable.

After the first sentence, which states what a chef's responsibilities do *not* include, the passage adds three parallel sentences that outline what these responsibilities *do* include. Each sentence begins with *It would include* followed by a gerund.

> *It would include* + *serving* + quality food that is safe.
>
> *It would include* + *providing* + an enjoyable setting to experience the food.
>
> *It would include* + *making* + a meal at their establishment satisfying and valuable.

Parallel structures are also used to add emphasis. The repetition of the same structure—"it would include"—builds to the final important phrase "satisfying and valuable."

## PARALLEL STRUCTURE IN A SERIES

When we have two or more equivalent items in a series, we want to be sure they are constructed using the same forms or parts of speech, for example, all nouns or all adjectives. Consider the example below:

> The steak was thick, juicy, and *it had a good flavor.*

Suppose a triangle represents an adjective and a rectangle represents a clause. Note how the sentence in Figure 26.1 does not have a parallel structure. Instead, we have a series of three comments on the steak in two different grammatical forms: two adjectives and one clause.

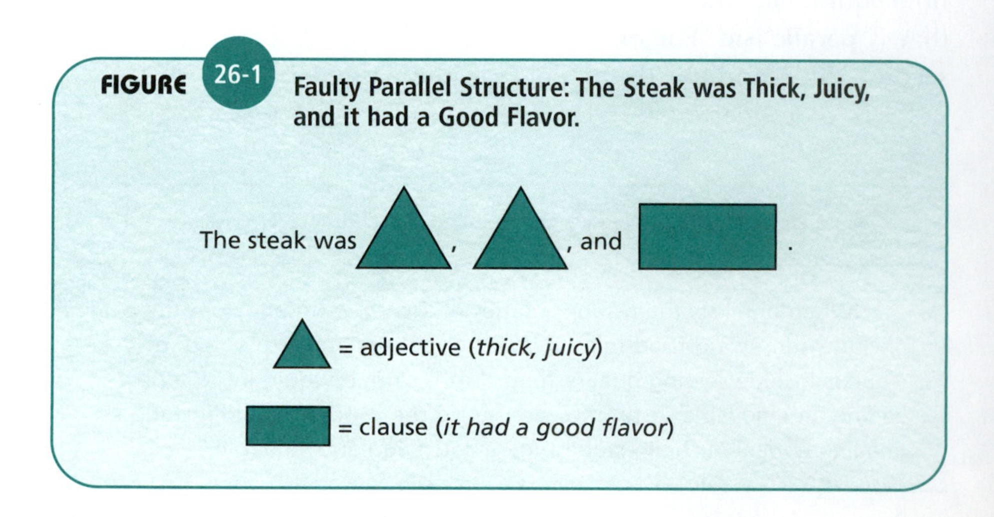


**FIGURE 26-1** Faulty Parallel Structure: The Steak was Thick, Juicy, and it had a Good Flavor.

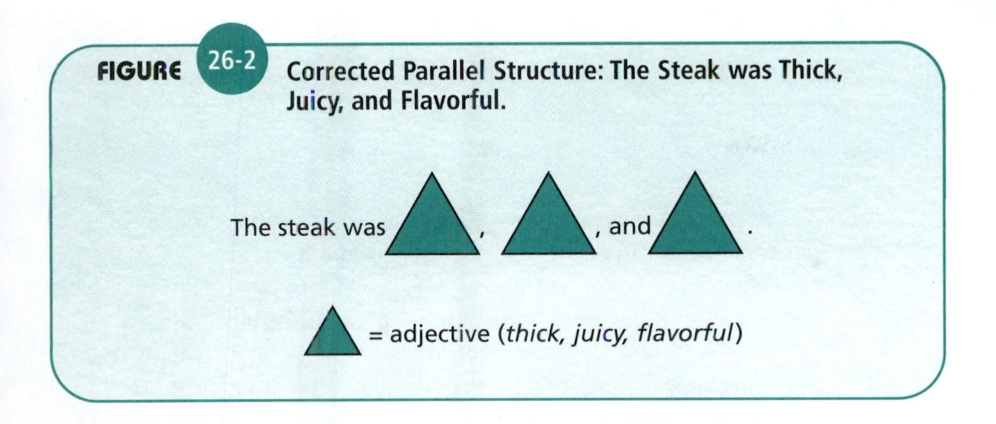


FIGURE 26-2 Corrected Parallel Structure: The Steak was Thick, Juicy, and Flavorful.

To correct the problem, we need to put all three comments in the same form, for example, by changing the clause to an adjective—

The steak was thick, juicy, and *flavorful*.

—that is, by changing the rectangle to a triangle (see Figure 26.2).

In the next example, the three objects of the preposition *for* are not parallel in structure. The first and second are ordinary nouns, but the third is a verbal.

Dr. House is known for his biting sarcasm, his intensely blue eyes, and *making clever deductions* about a patient's illness.

In the second sentence, all three objects are ordinary nouns (*sarcasm, eyes, deductions*), and the structure *is* parallel.

Dr. House is known for his biting sarcasm, his intensely blue eyes, and *his clever deductions* about a patient's illness.

Phrases should also be in the same form. In the sentence below, however, the first phrase is a gerund, the verb + *-ing*, while the second is an infinitive, *to* + verb.

The large tiger cat loved *sleeping* on pillows and to *annoy* the dog.

To correct the parallelism, you must either begin both phrases with a gerund or begin both with an infinitive. It doesn't matter which structure you choose as long as both phrases have the same structure.

The large tiger cat loved *sleeping* on pillows and *annoying* the dog. (gerunds)

The large tiger cat loved to *sleep* on pillows and to *annoy* the dog. (infinitives)

7

## EXERCISE 26.1 Maintaining Parallelism in a Series

Rewrite the sentences below and correct the parallelism.

1. *The Wedding Crashers* is a movie about two friends who like to crash weddings and eating fun finger foods.
2. Owen Wilson plays John, with Jeremy played by Vince Vaughn.
3. The film's humor comes from its witty dialogue and how it often approaches the content from a politically incorrect viewpoint.
4. Vince Vaughn is an especially delightful "straight man," who delivers his lines with dexterity and his 6'5" frame is surprisingly agile.
5. The supporting cast is marvelous, from Christopher Walken's crazed-looking but supportive father of the bride and Isla Fisher plays his spoiled but charming daughter.

## PARALLEL STRUCTURE IN A COMPARISON

You're probably familiar with the expression "You can't compare apples and oranges." Thus when we compare or contrast two ideas—for example, with the words *as* and *than*—we must use the same species of fruit, the same grammatical structure. We *can,* however, compare Red Delicious *apples* and Granny Smith *apples.*

Is it more important *to serve* healthy food than *keeping up* with food fashions?

The first phrase is an infinitive, *to serve,* while the second is a gerund, *keeping.* To maintain parallelism, put both in the same form. Just as we can compare two apples, we can compare two infinitives (*to serve, to keep*), or two gerunds (*serving, keeping*).

Is it more important *to serve* healthy food than *to keep* up with food fashion?

Is *serving* healthy food more important than *keeping* up with food fashion?

Look at another example:

Julius doesn't like *basketball* as much as *grilling a steak.*

Here a noun is paired incorrectly with a verbal. Instead, a noun should be paired with another noun or a verbal with another verbal.

Julius doesn't like *playing basketball* as much as *grilling a steak*.

See also Chapter 24 for information on using pronouns with *than* and *as*.

**EXERCISE 26.2** **Maintaining Parallelism in a Comparison**

Rewrite the sentences below, correcting any problems with parallel structure.

1. Did you enjoy Vince Vaughn's dancing as much as how he made balloon animals?
2. Owen Wilson's character was initially more romantic than the way his friend was.
3. During the touch football session on the lawn, Wilson was more interested in flirting with another of Walken's daughters than he was focused on playing the game.
4. Is Vaughn's slapstick humor during the game as funny as he has a deadpan expression?
5. Walken's scowling son Todd, hunched over his painter's palette, is more appealing than his daughter has a manic fiancé.

7

## PARALLEL STRUCTURE WITH *BOTH/AND, EITHER/OR, NEITHER/NOR,* AND *NOT ONLY/BUT ALSO*

When ideas are joined with the conjunctions *both/and, either/or, neither/nor*, or *not only/but also*, they must be expressed using the same forms or structures.

Not only was the food delicious but we also got good service. [not parallel]

The sentence could be rewritten as follows:

Not only did we receive delicious food but also good service. [parallel]

Group work requires both an exchange of ideas and a sense of humor.

We received two things: *food* and *service*. Let's look at another example:

> Brenda Johnson is both smart and has a lot of courage. [not parallel]

The sentence might be rewritten with two adjectives (*smart, courageous*) or two nouns (*intelligence, courage*).

> Brenda Johnson is both smart and courageous. [parallel]

> Brenda Johnson has both intelligence and courage. [parallel]

Place the initial word of the pair (*both, either, neither*, or *not only*) carefully so that the two structures that follow are identical. For example, in the sentence below, the word *either* should immediately precede the word *Russian*.

> The actress either was of Russian or Ukrainian descent. [poor placement]

> The actress was of either Russian or Ukrainian descent. [clear placement]

**EXERCISE 26.3** **Maintaining Parallelism with *Both/And***

Rewrite each sentence below to correct the parallel structure.

1. Police chief Brenda Johnson is both an astute interrogator and she is also poor at housekeeping.
2. Her subordinates not only admire her but also they think she's a bit odd.
3. Her relationship with her boss is complicated not only because they were once lovers but also he left her for another woman.
4. Sergeant Gabriel was the first officer at her new job who began either to like her and he respects her now, too.
5. Unable to find her way around Los Angeles, Chief Johnson needs him both as moral support and she needs a guide so she doesn't get lost.

## MAKING PARALLEL STRUCTURES COMPLETE

Sometimes we have the beginnings of parallel structure between two elements or between items in a series, but we lack certain words that would make the structure completely parallel. Study the following example:

> Edoardo liked the fresh herbs at the farmer's market better than the store.

Because the second item in this sentence is incomplete, the sentence seems to be comparing *herbs* and *store*. By adding a pronoun and a preposition, we make the second item complete and the meaning clear.

> Edoardo liked the fresh herbs at the farmer's market better than *those* at the store.

In this revised sentence, we are clearly comparing the two locations for the fresh herbs, the *market* and the *store*. Here's another example:

> Before taking the job, SooJin met with the manager and sous chef.

In this sentence, it is not clear whether SooJin met with one person or two. Add the article *the* to clarify that the manager and the sous chef are two separate people.

7

**Completing parallel structures.**

**Incomplete parallel: See what's missing?**

Edoardo liked the  at the farmer's market better than the 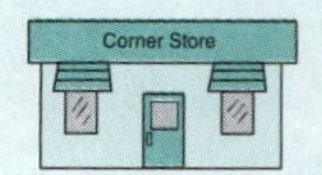

**Completed parallel:**

Edoardo liked the fresh **herbs**  at the farmer's market better than **those**  at the store.

> Before taking the job, SooJin met with the manager and *the* sous chef.

Finally, with a series of clauses that begin with *that*, it is often clearer to repeat the pronoun.

> She told me that she wanted to see a comedy, Tracy wanted to see an action film, and George wanted to stay home.

As written, this sentence could be a comma splice with the second independent clause beginning at *Tracy*. Note how the addition of *that* makes the structure clear.

> She told me that she wanted to see a comedy, *that* Tracy wanted to see an action film, and *that* George wanted to stay home.

## EXERCISE 26.4 Completing Parallel Elements

Rewrite the sentences below, adding the appropriate words to complete the parallel elements in the sentences below.

1. In the classic 1980s comedy *Tootsie,* Dustin Hoffman had more trouble finding acting work as a man than a woman.
2. His agent told him that he was too much of a perfectionist, he argued too often with directors, and no one in New York would work with him.
3. Hoffman's character found that he became both a better person and actor when he dressed as a woman.
4. He liked himself better as Dorothy than Michael.
5. His friends at home liked him more than at work.

## RECIPE FOR REVIEW Understanding Parallel Structure

1. Put two or more equivalent items in a series in the same form or part of speech, for example, all nouns or all adjectives, all gerunds or all infinitives.

   The steak was thick, juicy, and it had a good flavor. [not parallel]

   The steak was thick, juicy, and flavorful. [parallel]

2. In a comparison using *as* or *than*, put equivalent items in the same form or part of speech, for example, both would be nouns, both adjectives, or both infinitives.

   Julius doesn't like basketball as much as grilling a steak. [not parallel]

   Julius doesn't like playing basketball as much as grilling a steak. [parallel]

7

3. When ideas are joined with the conjunctions *both/ and, either/ or, neither/ nor, or not only/ but also,* put them in the same form.

 Brenda Johnson is both smart and has a lot of courage. [not parallel]

 Brenda Johnson is both smart and courageous. [parallel]

 Brenda Johnson has both intelligence and courage. [parallel]

4. Add words where necessary to complete a parallel structure.

 Edoardo liked the fresh herbs at the farmer's market better than the store. [incomplete]

 Edoardo liked the fresh herbs at the farmer's market better than ***those at*** the store.[complete]

## END OF CHAPTER QUIZ Understanding Parallel Structure

**DIRECTIONS:** *Rewrite each sentence below, correcting the errors in parallel structure.*

1. In *Crash*, Shaniqua dislikes Officer Ryan because he is disrespectful, manipulative, and he's in an irritable mood.
2. Jean realizes that she likes her maid more than talking to her friend Carol.
3. Officer Hanson is not only upset with his partner's racism but also he feels that his boss is being unfair.
4. Anthony is kinder to the illegal immigrants in the back of the van than Jean and Rick.
5. Among the various characters, Daniel the locksmith is one who both maintains his dignity in the face of prejudice and he also encourages others to rediscover their self-respect.

# Cleanup

## Proofreading the Final Draft

UNIT 8

**Chapter 27**

Punctuation I: End Marks, Capitalization, Apostrophes, and Quotation Marks

**Chapter 28**

Punctuation II: Commas, Colons, and Semicolons

**Chapter 29**

Punctuation III: Abbreviations, Numerals, Italics, Underlining, Parentheses, Brackets, Hyphens, Dashes

**Chapter 30**

Proofreading the Final Draft

smoky
raw
brackish
savory
ripe
tart
saccharin
tangy
briny
honeyed
pungent
spicy
curdled
gamey
juicy

chapter 27

# Punctuation I: End Marks, Capitalization, Apostrophes, and Quotation Marks

**Punctuation marks**, such as those that come at the end of a sentence, provide visual, nonverbal guides to the structures and meanings of written language. They function like road signs, directing the reader's movement through the essay. With capital letters, punctuation marks frame our ideas, just as an elegant table setting frames the culinary delicacies from the kitchen.

## END MARKS

**End marks** are those punctuation marks that indicate the *end* of a sentence. They're like stop signs that tell us we've reached the end of a thought, like the end of a city block. They also tell us something about the *kind* of sentence, whether it is a statement, a question, or an exclamation.

1. A statement, like this one, is followed by a **period**.

   Punctuation is an important part of clear writing.

2. A question is followed by a **question mark**.

   What is the most common type of punctuation?

3. However, an **indirect question** would be followed by a period rather than a question mark.

   The students asked what the most common type of punctuation is.

Capital letters and punctuation marks frame our ideas, just as this elegant table setting frames the culinary delicacies from the kitchen.

4. An **exclamation point** follows an exclamation or emphasizes a command.

What a delicious dessert!

Don't touch that pan!

## EXERCISE 27.1 End Marks

Punctuate the following paragraph with the appropriate end marks.

Clint Eastwood's career has taken some interesting turns in the last decade From *Unforgiven* to *Mystic River*, he has revisited the genres of the cowboy movie and the police thriller that were his bread and butter twenty and thirty years before He has also, as a director, coaxed some superb acting from his stars Sean Penn's emotional portrayal of a vengeful father in *Mystic River* was recognized with an Oscar Do you remember the scene outside the park gates when Penn realizes his daughter has been murdered What a brilliant performance

## CAPITALIZATION

Like punctuation marks, **capital letters** provide visual reinforcement of the structure and meaning of the sentence. Specifically, capital letters indicate certain categories of words, just as menus are organized into categories such as appetizers, beverages, or desserts. Capital letters are used for two main purposes: to mark the beginning of a sentence or a direct quote and to indicate a proper noun. Just as in this sentence, the first word of every English sentence begins with a capital letter. The capital letter that begins the sentence and the period or question mark that ends it indicate the conclusion of a complete thought.

1. Capitalize the **first word of a sentence**.

   The movie was released in December, just in time to be considered for the Academy Awards.

2. Capitalize the **first word of a direct quote**.

   The food safety instructor says, "Wash your hands for twenty seconds in the hottest water you can tolerate."

3. Do ***not*** capitalize the first word of a quoted *phrase.*

   Dominique paid close attention to the instructions regarding washing hands "in the hottest water you can tolerate."

4. Capitalize the first word of a statement or a question that forms part of a sentence, even when it is not set off by quotation marks, as in the example below.

   We asked ourselves, What are the differences between the cuisines of Greece and those of other Mediterranean countries?

5. **Do *not* capitalize the first word that follows a semicolon**, unless that word would be capitalized for another reason.

   Captain Ahab is obsessed with the idea of killing the white whale who took off his leg; his story is told in the novel *Moby Dick*. [Do not capitalize *his*.]

   The narrator of the story is a young man called Ishmael; Queequeg, a cannibal from the South Seas, is his friend. [Capitalize the proper noun *Queequeg*.]

### EXERCISE 27.2 Capitalization and End Marks

Correct the capitalization, and add appropriate end marks as you rewrite the sentences below.

1. one of my favorite restaurants is chez marcel
2. the newspapers say it offers "The best french cuisine outside of paris"
3. as we arrive, the hostess exclaims, "welcome to chez marcel"
4. my friends want to try something new; They order escargot
5. i ask myself, how daring do i feel tonight

## More Capitalization

In addition to marking the beginning of a sentence, capitalization points out the difference between a general category and a specific individual. It is the difference between *the teacher* and *Professor Fromm,* between *restaurant* and *Chez Marcel,* between *movie* and *Babette's Feast* (see Figure 27.1).

6. **Capitalize proper nouns**, which include the following:

   - The names of specific persons and animals, such as *John Hancock, Sojourner Truth,* and *Seabiscuit;* capitalize their titles when these precede the name, such as *Professor Malik, Dr. Garcia,* and *Chef Bocuse,* and when the title is used in place of the name, such as *Senator, Coach,* or *Chef.*

   - Words showing family relationships when used with the person's name or when used in place of the person's name, such as *Aunt Sally* or *Grandma;* do not capitalize these titles when they are preceded by a possessive, such as *my aunt Sally* or *my grandmother.*

   - The names of religions, races, and nationalities, such as *Islam, Judaism, Hispanic, Caucasian, Brazilian, Thai.*

**FIGURE 27-1 Proper and Common Nouns.**

| PROPER NOUNS | COMMON NOUNS |
|---|---|
| Kazuo Shiguro | a novelist |
| Julia Child | a famous female chef |
| Man o' War | the horse that won the race |
| Dr. Kildare | the doctor on television |
| Inspector Clousseau | a police detective |
| Aunt Debbie | my aunt, my aunt Debbie |
| Dad | my father |
| Ibiza | an island in the Mediterranean |
| Danube River | a river |
| Fifth Avenue | a street in Manhattan |
| Wednesday | one day last week |
| Passover | a religious holiday |
| Snickers | a candy bar, a Snickers bar |
| the Civil War | a civil war |
| the National Restaurant Association | an organization of restaurants |
| Kitchen Confidential | a book about the food industry |
| The London Times | a newspaper |
| a Spanish dialect | a local dialect |

- The names of specific places, including specific cities, states, countries, continents, such as *Seoul, Alaska, Mexico,* and *Australia;* specific islands, mountains, oceans, lakes, and rivers, such as *Ellis Island, Mount Kiliminjaro, the Indian Ocean, Lake Michigan,* and *the Yangtze River*; and specific streets, such as *Fifth Avenue, Mulberry Street,* and *Sunset Boulevard.*
- The days of the week, months, special dates—but not the seasons—such as *Monday, July, Valentine's Day,* and *summer.*
- The names of other specific things—ships, planes, awards, etc.—such as *the Titanic, Air Force One,* and *the Academy Awards.*

8

- The names of specific brands, but not the common nouns that may follow: *Cheerios, Sprite, Heinz ketchup, Dodge minivan.*

7. Capitalize the titles of specific organizations, events, and historical periods, such as *the League of Women Voters, the Boston Tea Party,* and *the Colonial Era.*

8. Capitalize the first word and other important words (such as nouns, adjectives, verbs) in the titles of books, newspapers, stories, poems, movies, and other works of art, such as *The Joy of Cooking, The Washington Post,* "A Rose for Emily," "Sailing to Byzantium," *Men in Black,* and the *Mona Lisa.* Do *not* capitalize articles and prepositions *within* the title.

9. Capitalize the adjectives formed from proper nouns, for example, *the English language, Roman numerals.*

10. The pronoun *I* is always capitalized in English. The other pronouns are not capitalized, unless they begin a sentence.

## EXERCISE 27.3 Capitalization

Rewrite the phrases below, correcting the capitalization where necessary. Some items are already correct.

EXAMPLE: New York city

REWRITE: New York City

1. mexico city
2. Cape Town, South Africa
3. mr. kim
4. professor Lauria

5. chez panisse
6. aunt Winifred
7. my cousin Tony
8. a vietnamese dish
9. a hotel on lake Victoria
10. the Rocky mountains
11. a place in the country
12. a park avenue address
13. new year's eve
14. a Winter sport
15. on tuesday night
16. the ivy award
17. "sink the Bismarck!"
18. rice krispies
19. a General Mills cereal
20. the screen actors guild
21. the Middle ages
22. the bill of rights
23. *the Lord of the rings*
24. a Mediterranean diet
25. my father and Me

**EXERCISE 27.4** **End Marks and Capitalization**

Correct the capitalization, and add appropriate end marks in the sentences below.

1. although brooke has never visited the campus of duke university, she is a big fan of its basketball team
2. like many fans, she considers mike krzyzewski—more often known as coach k—to be one of the top college coaches in the united states
3. one of her favorite players was bobby hurley, who was born in jersey city, new jersey
4. do you remember the blue devils' back-to-back wins of the ncaa tournament in 1991 and 1992
5. what a triumph

## APOSTROPHES

Like capital letters, **apostrophes** have specific meanings and must be used correctly in your academic writing. Apostrophes are used to form the possessive of nouns and to make contractions, which are combinations of two words in which some letters are omitted. Apostrophes are also used to form the plural of letters of the alphabet.

Dot your *i*'s and cross your *t*'s.

In general, the possessive of a singular noun is formed by adding an apostrophe plus the letter *s*. For example, instead of writing *the story of the iron chef,* we could write *the iron chef's story;* instead of *the reputation of Chicago for deep-dish pizza,* we could write *Chicago's reputation for deep-dish pizza.* For singular nouns that already end in *s,* we still add an apostrophe plus *s* to form the possessive: *the bus's route, Henry James's book.* However, if the noun has more than one syllable, some writers prefer to add only an apostrophe. Either is correct: *Chef Stevens's recipe* or *Chef Stevens' recipe.* Writers tend to add the *s* if they pronounce the word in three syllables, *Steven-ses.*

For most plural nouns, add just an apostrophe to form the possessive: *the girls' team, the Andersons' restaurant.* For nouns with irregular plural forms, add apostrophe plus s: *the children's team, the men's team.* Be careful to distinguish the simple plural, the singular possessive and the plural possessive.

The chefs = more than one chef

The chef's = belonging to one chef

The chefs' = belonging to more than one chef

The chefs were unpacking a mystery basket.

The first chef's basket contained an assortment of tropical fruit.

The competition judged the chefs' creativity.

## EXERCISE 27.5 Forming the Singular Possessive with Apostrophes

Rewrite each phrase below, changing the italicized singular noun to its possessive form.

EXAMPLE: the *muffin* aroma-the muffin's aroma

1. the *cake* ingredients ______________
2. the *woman* promotion ______________
3. the *dress* color ______________
4. *Julia recipe* ______________
5. Mr. Damon *travel plans* ______________

## EXERCISE 27.6 Forming the Plural Possessive with Apostrophes

Rewrite each phrase below, changing the italicized plural noun to its possessive form.

EXAMPLE: the *muffins* aroma-the muffins' aroma

1. the *cake* ingredients ______________
2. the *woman* promotion ______________

8

3. the *dress* colors ________________
4. Julia's *daughters* recipes ________________
5. the *Damons* travel plans ________________

Apostrophes are *not* used in constructing the possessive forms of the personal pronouns: *my, mine, our, ours, your, yours, his, her, hers, its.* Be careful not to confuse the possessive it's with the contraction *it's* (see also Appendix III). To form the possessive of singular indefinite pronouns, however, add an apostrophe plus *s: someone's glove, nobody's fool.*

To form the possessive of hyphenated words, add apostrophe plus *s* to the last word: *brother-in-law's car, passer-by's expression.* (Note that the plural of hyphenated words is formed differently: *brothers-in-law, passers-by.* Note also that *passer-by* may be spelled without the hyphen.) The plural possessive of hyphenated words contains two *s*'s*: brothers-in-laws' cars.* The names of organizations also form the possessive by adding apostrophe *s* to the last word: *Johnson & Johnson's, Simon & Schuster's.*

When an item is owned in common, the possessive is formed by adding apostrophe *s* to the last name: *Jack and Jill's rhyme,* Bill and Ted's Excellent Adventure. When ownership is not shared, each name forms the possessive with apostrophe *s: Anna's and Carla's hats.* Each woman has her own hat.

## EXERCISE 27.7 More Possessives

Correct the italicized words below.

1. *Jim and Judy* recipe for cherry pie (joint ownership).
2. *Jim and Judy* recipes for pumpkin pie (separate ownership)
3. her *mother-in-law* recipe for apple pie
4. *Ben and Jerry* ice cream
5. *everybody* favorite chocolate chip recipe

Another use of the apostrophe is to indicate that one or more letters have been left out, as in the following contractions:

| | |
|---|---|
| I'm = I am | you're = you are |
| isn't = is not | he's = he is |
| hasn't = has not | they're = they are |
| haven't = have not | didn't = did not |

You may wish to avoid such contractions altogether in formal academic writing. If you do use them, however, be sure to use them correctly.

### EXERCISE 27.8 Forming Contractions

Write the appropriate contraction for the italicized words below.

1. In Pieces of April, April *has not* seen her family since she moved to New York City.
2. *She* is planning a special Thanksgiving dinner for them.
3. Unfortunately, April *can not* cook.
4. She *does not* even know how to light her oven.
5. *There is* something very funny about that scene.

Finally, the apostrophe may be used to form the plurals of letters.

When Debbie was a little girl, she made hearts over her *i*'s.

The rules concerning apostrophes continue to evolve. For example, dates used to require an apostrophe (the 1980's); now they do not (1980s). Follow the guidelines of your instructor, and make the reader's understanding your priority.

8

## QUOTATION MARKS

**Quotation marks** are used to **set off the titles** of songs, poems, short stories, individual episodes in a television series, articles, and chapters. Note that the titles of books, magazines, and movies are underlined or italicized (see Chapter 29), although many newspapers put movie titles in quotation marks rather than italics.

"Happy Birthday" is one of the world's most popular songs.

Edgar Allan Poe's poem "The Raven" has been memorized by scores of schoolchildren.

Poe also wrote short stories, such as "The Tell-Tale Heart."

"Chowder" is the title of a chapter in Herman Melville's nineteenth century novel *Moby Dick*.

Note that the period in the third example falls *inside* the end quote.

Quotation marks are also used to ***set off words and phrases that might be unfamiliar*** to the reader, such as certain slang expressions or quoted phrases. If the expression or phrase is commonly used, it does not require quotation marks.

Joanie told Angela to "bring it."

Angela advised her friend to let sleeping dogs lie.

Finally, quotation marks set off **direct speech**, that is, the words exactly as they are spoken. Put the quotation marks before the first and after the last word. Don't forget the final set of quotation marks.

Angela said, "I read *Moby Dick* in college."

Quotation marks are ***not*** used to mark **indirect speech**, which does not quote the exact words spoken.

Angela said that she had read *Moby Dick* in college.

[She actually said, "I read *Moby Dick* in college."]

Use a comma, a question mark, or an exclamation point to separate a direct quote from the rest of the sentence.

"Did you like *Moby Dick?*" asked Jennifer.

Do not use more than one punctuation mark at the end of a quote.

"Yes, I did!," replied Angela. [incorrect]

"Yes, I did!" replied Angela. [correct]

When a sentence in the quote is interrupted, do not capitalize the next word.

"When we go out to dinner on Friday nights," said Jamal, "we like to try new restaurants." [The *we* after the comma is not capitalized.]

Use single quotation marks to set off one quoted phrase within another.

"Sometimes the waiters sing 'Happy Birthday,'" said Jamal.

While commas and periods fall within quotation marks, semicolons and colons fall outside. Question marks and exclamation points belong inside if the quote itself is a question or an exclamation; otherwise they too go outside.

Diane asked, "Do you like to sing along?"

Did the other customers sing "Happy Birthday"?

If you are quoting a conversation, begin a new paragraph with each change of speaker, as in the following example:

"What's your favorite television show?" asked Cheryl.

"I'm torn between *House* and *Grey's Anatomy,*" Sandra replied. "What about you?"

"I love *The Closer.*"

## EXERCISE 27.9 Using Quotation Marks

Rewrite the sentences below, adding quotation marks where appropriate.

1. A classic Motown hit of the 1960's was I Want You Back by the youthful Jackson Five.
2. One of Ernest Hemingway's most famous short stories is Hills Like White Elephants.
3. I read an article in *Food and Wine* called Grilled in Japan.
4. Have you ever been to a Japanese restaurant? they asked.
5. Yes, she replied. I'm a big fan of sake.

## RECIPE FOR REVIEW Punctuation I

1. **End marks**, such as periods, question marks, and exclamation marks, are used to indicate the end of a sentence and to show whether the sentence makes a statement, asks a question, or receives special emphasis.

"I'm doing the blood oranges. Who's got the cranberries?"

The cranberry relish was made with blood oranges.

Did it taste sweeter because of that?

It was absolutely delicious!

2. **Capital letters** are used to mark the beginning of a sentence, a direct quote, or a statement that forms part of a sentence. Do *not* capitalize the word that follows a semicolon.

The hostess checked the reservation book.

She asked, "May I seat your party now, sir?"

The customer thought to himself, This is a good restaurant.

He was also pleased with his waitress; she was very prompt in bringing out the drinks and appetizers.

Capital letters are used to indicate a proper noun, that is, the name of a specific person, place, or thing. See Figure 27.1.

3. Use **apostrophes** for the following:

   Possessives: *the student's schedule, the students' schedules*

   Contractions: *isn't, they're*

   Plurals of letters: *x's* and *y's*

4. Use **quotation marks** for the following:

   Songs, poems, short stories, articles, and chapters

   Unfamiliar words and phrases

   Direct speech

## END OF CHAPTER QUIZ Punctuation I

**DIRECTIONS**: *Rewrite the following passage, correcting the capitalization, and adding appropriate end marks, apostrophes, and quotation marks.*

alex and his friend milo often talked about music
alex was a big fan of green day and especially liked such political songs as american idiot his friends taste was a bit different
weird al yankovich is a genius alex youve got to admit it milo frequently remarked
yes hes funny alex would reply but whats he doing to wake people up
what about canadian idiot milo suggested now theres a song with a political message
sometimes i think youre the idiot milo exclaimed alex

chapter 28

# Punctuation II: Commas, Colons, and Semicolons

Like other punctuation marks, commas, colons, and semicolons direct the reader's attention to meaningful groups of words. They are like the hands of a traffic cop telling cars to stop or go, turn right or left. They are like the hands of a waiter guiding the customer through a meal, from pointing out the table and offering the menu to serving the dishes and sweeping up the crumbs.

## THE COMMA

The **comma** is one of the most misused punctuation marks; in fact, it is often *over* used. There are certain occasions when commas must be used in standard written English, and there are occasions when they must not be used. There are also times when the comma *may* be used to add emphasis or to make the meaning clearer. Since the trend in American English is to use fewer commas than we used to, the rather vague advice to put a comma wherever you pause in the sentence is not always a reliable guide. Rather, it makes sense to consult a list of the rules and to use commas only when these rules apply. When in doubt, leave commas out.

**Add a comma and a coordinating conjunction to separate two independent clauses.**

One train stopped, and the other kept going.

Joaquin grilled two steaks, and Dexter added a garnish of mushrooms.

**The comma is like a hand giving a short wave to the reader as if to say, okay, here's where the next clause begins.**

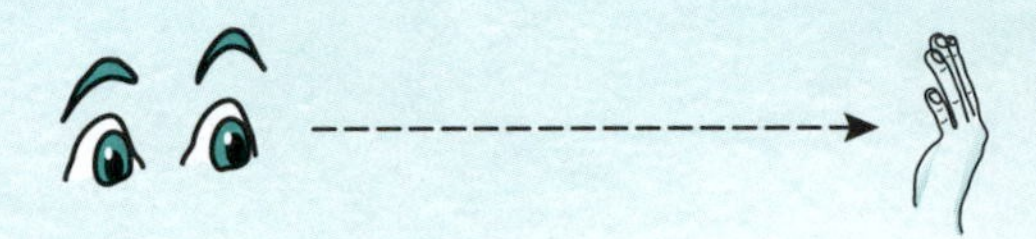

Joaquin grilled two steaks, and Dexter added a garnish of mushrooms.

In these examples the comma is like a hand giving a short wave to the reader as if to say, okay, here's where the next clause begins.

## EXERCISE 28.1 Using Commas in Compound Sentences

Rewrite the following sentences, adding commas where needed.

1. The movie *Emma* is based on Jane Austen's novel of the same title and it takes place in England in the late eighteenth century.
2. Emma Woodhouse is smart and pretty but she thinks a little too highly of her own abilities.
3. Emma has long been jealous of Jane Fairfax for Jane is a very beautiful and accomplished musician.
4. Mr. Knightley lives in the same village and he admires both women.
5. Frank Churchill admires them as well yet he is already secretly engaged to Jane.

**Add a comma after an introductory word or phrase**. The purpose of punctuation is to map out the meaning of the sentence so that the reader is able to follow it clearly. Since the main clause and its subject are especially important to the meaning of a sentence, commas are often used after an introductory word or phrase as an indication that these are going to be *followed* by this main clause.

At the beginning of the meal, the server asked for their drink orders.

After ordering their drinks, the customers opened the menus.

Finally, they made their decisions.

Many writers place a comma after four or more introductory words, or after two or more prepositional phrases. Note that if the introductory word or phrase is short and the meaning is clear, the comma may be omitted.

Yesterday we went to a new restaurant downtown.

In the window there was a photograph of the award-winning chef.

## EXERCISE 28.2 Using Commas with Introductory Words and Phrases

Rewrite the sentences below, adding commas where necessary.

1. In one popular television show about doctors the main character's name is House.
2. As in a murder mystery the cause of a patient's illnesses must be investigated.
3. Indeed House is famous for diagnosing rare and difficult conditions.
4. During one memorable episode he found that a woman's psychotic symptoms were caused by a vitamin K deficiency.
5. Despite Dr. House's undeniable intelligence he often has difficulty with personal relationships.

**Add a comma after an introductory subordinate clause.** Remember, a subordinate clause is a group of words that contains a subject, a verb, and a subordinating conjunction ("coat hanger").

While one train stopped, the other kept going.

After Joaquin grilled two steaks, Dexter added a garnish of mushrooms.

Again, the comma is like a hand drawing the reader's eye to the subject of the main clause, Dexter.

**EXERCISE 28.3** **Using Commas with Subordinate Clauses**

Rewrite the following sentences, adding commas where necessary.

1. When Emma first meets Frank Churchill she thinks he is very handsome and well mannered.
2. As she gets to know him better she does not completely trust him.
3. Although Frank is not perfect Emma would like him to marry her best friend, Harriet.
4. While Emma is busy match-making she fails to notice that both Frank and Harriet are in love with other people.
5. Despite Emma's normally acute perceptions she also doesn't realize that she herself is in love with Mr. Knightley.

If the subordinate clause *follows* the main clause, the comma is omitted when the subordinate clause is **restrictive**, that is, when it limits the meaning of a word in the main clause or is otherwise essential to the meaning of the sentence. Note that a subordinate clause that begins with the pronoun *that* is always restrictive; therefore it is not preceded by a comma.

Restrictive: The crowd cheered for the player who hit the home run.

We came home early because we were tired.

I went to the new restaurant that everyone was talking about.

If the clause is **nonrestrictive**, that is, if it could be omitted without altering the fundamental meaning of the sentence, it is preceded by a comma.

| | |
|---|---|
| Nonrestrictive: | The Boston fans greatly admired David Ortiz, who had hit several home runs during the season. |
| | We came home early, although the party was not over. |
| | I went to the new restaurant, which we learned later had received good reviews. |

**Add commas between three or more items in a series.** Commas are used to separate items in a series, as in the following example:

The apple was green, round, and juicy.

Although it is more formal to include it, some writers choose to omit the comma right before the *and.* Find out which style your audience prefers, and be sure to use a comma if it clears up a potential misunderstanding in the *meaning* of the sentence. Whichever option you choose—comma or no comma—you should be consistent throughout any particular essay.

Note that if you use a conjunction between each item, you do not need commas.

The apple was green and round and juicy.

You may choose to add commas where they are not required, however, as in the example that follows, if each item is rather long and the sentence would be clearer with punctuation.

The fruit salad consisted of Granny Smith apples bought from the grocery store and used with their bright green skins intact, and blueberries and raspberries picked that morning at a local farm, and walnut pieces that were left over from holiday baking.

When you have a series of adjectives describing a noun, you insert a comma between those that modify the noun directly.

The café was painted a bright, cheerful color.

Both *bright* and *cheerful* describe *color*, and the sentence might have been written as follows: *The café was painted a bright and cheerful*

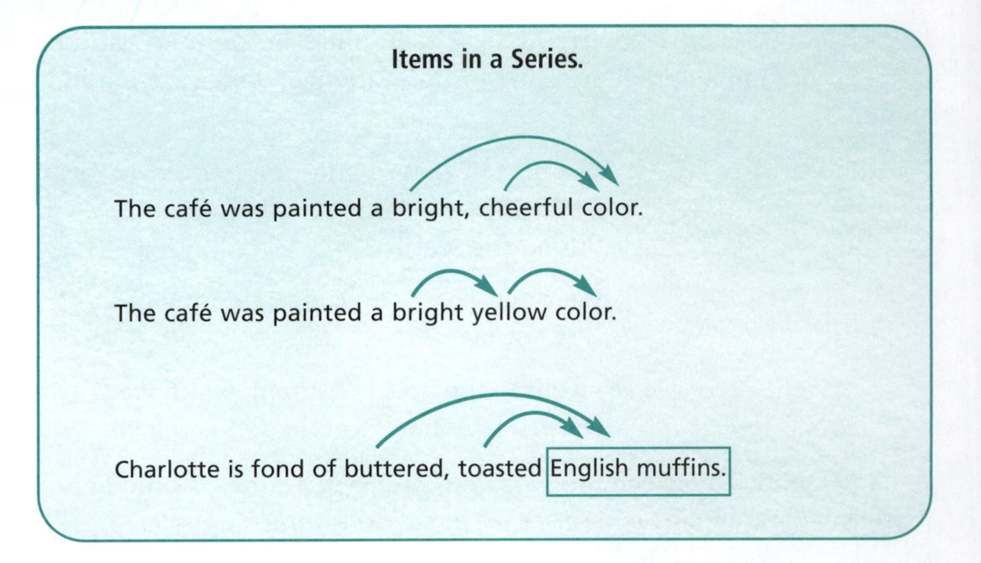


*color.* When adjectives modifying the same noun could be joined by *and* (*bright and cheerful color*), they are separated by commas if *and* is removed (*bright, cheerful color*). In the next example, however, *bright* is now an adverb modifying *yellow* and is not separated by a comma.

> The café was painted a bright yellow color.

We wouldn't write *a bright and yellow color*; thus, we do not place a comma after *bright.* Finally, do not add a comma between an adjective and a noun if the adjective limits or qualifies the meaning of the noun. For example, there is no comma between *toasted* and *English* in the sentence below because the adjective *English* limits the muffins to a certain variety. They're not *corn* muffins, for example.

> Charlotte is fond of buttered, toasted English muffins.
>
> [*not* Charlotte is fond of buttered, toasted, English muffins.]

One common qualifier is the season of the year. In the sentence below, we wouldn't speak of the mild *and* summer evening; therefore, there is no comma between them.

> The children watched the fireflies throughout the mild summer evening.
>
> [*not* The children watched the fireflies throughout the mild, summer evening.]

## EXERCISE 28.4 Commas with Items in a Series

Rewrite the following sentences, adding commas where necessary.

1. The two business partners bought a small attractive restaurant on a side street.
2. The newly opened café had fresh seasonal vegetables on the menu.
3. The customers especially enjoyed the varied tender salad greens.
4. Another popular item was the thick nutritious vegetable soup.
5. For dessert many customers ordered the very rich chocolate cake.

**Add commas on both ends of words, phrases, or clauses that interrupt the sentence**. Use *two* commas—like two hands holding a tray—unless the phrase begins or ends the sentence.

> Dr. House, *however,* is not always popular with his coworkers.
>
> *However,* Dr. House is not always popular with his coworkers.
>
> Dr. House is not always popular with his coworkers, *however.*

In the second two examples, the capital letter and the period, respectively, act as the other "hand" that helps the comma hold the tray.

**Appositives**, which are words or phrases that sit next to a noun and rename it, are usually separated from the rest of the sentence with commas, unless they are very short or unless they restrict the meaning of the noun.

> The new show, a hospital drama, premiered on Tuesday night. [commas set off *a hospital drama,* which renames *show*]
>
> My aunt Sally loves the show. [*Sally* restricts the meaning; it's not my aunt *Betty.*]

Use *two* commas—like two hands holding a tray—to enclose a word or phrase that interrupts a sentence.

8

Words or phrases used in **direct address** are also separated from the rest of the sentence by commas.

I am sorry, Dr. House, but your license has been suspended.

## EXERCISE 28.5 Using Commas with Interrupters

Rewrite the following sentences, adding commas where necessary.

1. Another popular medical series this one with a large ensemble cast is *Grey's Anatomy*.
2. Meredith Grey from whom the series derives its name is an intern at a hospital.
3. She shares a house with two other interns Izzy and George.
4. The interns' supervisor Dr. Bailey runs a tight ship.
5. In addition the show is enlivened by the many romantic entanglements.

**Add a comma in numbers larger than four digits**, such as 150,000 or 23,671. The comma is optional in four-digit numbers:

Marlene made 1,129 phone calls at work last month. [correct]

Marlene made 1129 phone calls at work last month. [correct]

At that rate, she will have made 13548 by the end of the year. [incorrect]

At that rate, she will have made 13,458 by the end of the year. [correct]

Do *not* add commas to telephone numbers, years, or zip codes.

**Add a comma between elements of a date**; add a comma after the date if the sentence continues, as in the example below.

Julia Child was born on August 15, 1912.

Julia Child was born on August 15, 1912, and was an influential figure in the American culinary world.

However, when the date is given in inverted order—day, month, year—no commas are used: 15 August 1912. Do *not* use a comma when you write just the month and day or just the month and year.

Julia Child was born on August 15.

Julia Child was born in August 1912.

**Add a comma between elements of an address, except between the state and the zip code.** Use a comma after the address if the sentence continues.

The President of the United States lives at 1600 Pennsylvania Avenue, Washington, D.C., 20500.

**Add a comma between a name and abbreviations** such as *M.D.* or *Jr.* Use a comma after the abbreviation if the sentence continues.

T. Berry Brazelton, M.D., is a widely respected author of parenting books.

**Add a comma after the greeting of a personal letter and after the closing of all letters**. Note that the greeting of a *business* letter is followed by a colon.

Dear John,
Sincerely yours,
*BUT*
Dear Superintendent Yang:

## THE COLON

**Colons** are like a pair of eyes tipped on its side, like the emoticon :) without the smile. They mean "look at this." Colons are often used **to introduce a list or a quotation.**

**Colon "Eyes."**

Dexter dislikes all vegetables except the following: broccoli, carrots, and corn.

Dexter dislikes all vegetables except the following: broccoli, carrots, and corn.

Perhaps one of the most well-known sentences in the English language comes from *Hamlet:* "To be or not to be, that is the question."

Colons may be used **between independent clauses**, if the first clause is explained or summarized by the second.

Dexter applies this same selectivity to ice cream flavors: he prefers pure chocolate or vanilla, with an occasional scoop of cookies and cream to add texture.

Colons are also used **following the greeting in a business letter**—

Dear Sales Associate:

—and in **expressions of time**.

Dinner was served at 7:00 p.m.

Finally, colons are used in certain parts of **bibliographic citations**. Study the examples below:

Chicago: University of Chicago Press [between city and publisher]

*Food and Wine* 18:63-65 [between volume and page numbers]

Some style manuals require a colon between biblical chapter and verse; the Modern Language Association prefers a period. Follow the preference of your instructor or publisher.

Do not use a colon after the expressions *for example, including,* and *such as*; between a verb and the rest of the sentence; or between a preposition and its object.

Dexter enjoys many types of potatoes, such as: mashed, baked, and fried. [incorrect]

Dexter enjoys many types of potatoes, such as mashed, baked, and fried. [correct]

Dexter's favorite vegetables are: broccoli, carrots, and corn. [incorrect]

Dexter's favorite vegetables are broccoli, carrots, and corn. [correct]

Dexter's mother made his favorite meal of: lasagna and garlic bread. [incorrect]

Dexter's mother made his favorite meal of lasagna and garlic bread. [correct]

**EXERCISE 28.6** Using the Colon

Add or delete a colon where necessary in the sentences below.

1. Each episode of *Law & Order* follows a specific sequence the discovery of the body, the police investigation, and the jury trial.
2. Over the years, several different actresses have played Jack McCoy's assistant, for example: Jill Hennessey, Carey Lowell, and Angie Harmon.
3. Detective Briscoe's partner has also been played by different actors, including: Benjamin Bratt and Jesse L. Martin.
4. Jennifer is a huge fan of the show she has at least ten seasons on DVD.
5. Did you watch the episode last night at 1000 p.m.?

## THE SEMICOLON

A **semicolon**, as we saw in Chapter 21, is used to **join two independent clauses** that are very closely related in meaning. It's as if our friend Comma Man got into a semi.

Raw eggs may contain a dangerous bacterium called *Salmonella;* they should always be stored at cool temperatures.

A **conjunctive adverb**, such as *consequently* or *however*, may be used to describe the relationship between the two clauses.

Raw eggs may contain a dangerous bacterium called *Salmonella;* consequently, they should always be stored at cool temperatures.

A "Semi" Colon.

Such conjunctive adverbs are typically followed by a comma, particularly when they interrupt the sentence, as in the example above.

Semicolons may also be used instead of commas **to separate independent clauses that are joined by a coordinating conjunction and already contain a number of commas.**

> On the following day, dressed in clean chef whites, the new students learned to prepare vegetable stock, beef stock, and chicken stock; and their instructor, who was fair but strict, was pleased with their progress.

In addition, semicolons may be used instead of commas **to separate items in a series when these already contain a number of commas.**

> *Mystic River* is the story of three friends: Sean Devine, a quiet child, now a state trooper and the only one of the three to go to college; Jimmy Markham, the leader of the group, a small-time crook who went straight after serving two years in prison; and Dave Boyle, perhaps the central figure, the boy who was kidnapped, the man who couldn't escape.

See Exercises 21.4 and 21.5 for additional practice with semicolons.

## RECIPE FOR REVIEW Punctuation II

### Commas (the Helping Hands)

Add a comma in the following situations:

1. Before a coordinating conjunction to separate two independent clauses
2. After an introductory word or phrase in a sentence
3. After an introductory subordinate clause
4. Before a nonrestrictive clause or phrase
5. Between three or more items in a series
6. On both sides of words, phrases, or clauses that interrupt the sentence, including words or phrases used in direct address
7. To mark the thousandth place in numbers *larger* than four digits; the comma is optional in four-digit numbers
8. Between elements of a date and after the date if the sentence continues

9. Between elements of an address, except between the state and the zip code, and after the address if the sentence continues
10. Between a name and abbreviations such as *M.D.* or *Jr.* and after the abbreviation if the sentence continues
11. After the greeting of a personal letter and after the closing of all letters. Note that the greeting of a *business* letter is followed by a colon.

## Colons (a Pair of Eyes)

Add a colon in the following situations:

1. Introducing a list or quotation
2. Following the greeting in a business letter and in expressions of time
3. Indicating certain parts of bibliographic citations

Do *not* use a colon in these situations:

1. After the expressions *for example*, *including*, and *such as*
2. Between a verb and the rest of the sentence
3. Between a preposition and its object

## Semicolons (Comma Man in a Semi)

Add a semicolon in the following situations:

1. Between two independent clauses
2. Between two independent clauses that contain several commas and are joined by a coordinating conjunction
3. Between items in a series when these already contain a number of commas

## END OF CHAPTER QUIZ Commas, Colons, and Semicolons

**DIRECTIONS**: *Rewrite the following sentences, adding or deleting commas, colons, and semicolons as necessary in the sentences below. Assume all other punctuation is correct.*

1. While some characters in *Mystic River* are timid and reserved others are intense and fearless.

2. My personal favorite is, Jimmy Markham played by Sean Penn: who happens to be one of my favorite actors.

3. Jimmy is a loving compassionate family man but he is forced to backpedal into a life that had been forgotten a life in organized crime.

4. Markham captained a crew that included the Savage brothers and "Just Ray" Harris, it landed him directly in federal prison.

5. These painful years mold the character into what the film depicts; a powerful rugged leader that commands loyalty.

6. However with the birth of his eldest daughter the elements of love compromise and family are created as well.

7. As the film continues to unfold; Jimmy Markham rises to a climactic implosion and his wrath is felt by all.

8. Jimmy's sight once transparent and clear is now opaque with rage and vengeance.

9. In the end the teachings of Katie's birth are ironically in vain Markham is forced to honor his slain daughter, by dishonorable means and actions.

10. His past is now the present and the dragon's slumber is permanently disturbed.

—Adam McGlone

*Sentences adapted from an essay by Adam McGlone. Errors were introduced to create this quiz.

chapter 29

# Punctuation III: Abbreviations, Numerals, Italics, Underlining, Parentheses, Brackets, Hyphens, Dashes

Like the punctuation marks discussed in the two preceding chapters, those discussed below are used to make your meaning clear. For example, abbreviations are often spelled out to help the reader understand what we mean, while parentheses may indicate that some information is relatively less important than that in the rest of the sentence. Other uses of punctuation—such as underlining a book title—enhance the presentation of an essay to meet the expectations of academic writing.

## ABBREVIATIONS

In general, *avoid abbreviations* in your academic writing; for example, do not use an ampersand or plus sign for "and."

We ordered the macaroni & cheese. [incorrect]

We ordered the macaroni and cheese. [correct]

Titles are an exception to this rule. Titles such as Mrs. and Dr. are typically written as abbreviations when used before a name: *Mrs. Robinson, Dr. Seuss, St. Nicholas.* When they are not used with a name, however, such titles are spelled out, for example, *Is there a doctor in the house?* The abbreviations in Figure 29.1 are always acceptable.

Just as cooks wash their pots and pans, writers "clean up" their sentences by correcting the punctuation.

## NUMERALS

As with abbreviations, there are rules for deciding whether to use numerals or words to represent numbers. Rules about numbers vary from style to style. The Modern Language Association suggests spelling out numbers that can be written as one or two words. Thus, *fifty* and *ninety-eight* would be spelled out, while *150* and *125th* would be represented by numerals.

The new restaurant did 150 covers per night.

The owner bought fifty new chairs.

We got off the train at 125th Street.

Bonnie scored in the ninety-eighth percentile on the math test.

**FIGURE 29-1 Acceptable Abbreviations.**

| | EXAMPLES | MEANING |
|---|---|---|
| Titles | Mr., Mrs., Ms. | man, married woman, woman |
| | Dr., St. | doctor, saint |
| Expressions of Time | AM, a.m. | before noon |
| | PM, p.m. | after noon |
| College Degrees | A.O.S. | Associate of Occupational Studies |
| | B.P.S. | Bachelor of Professional Studies |
| | B.A. | Bachelor of Arts |
| | M.S. | Master of Science |
| | M.D. | Doctor of Medicine |
| Latin Expressions | etc. | and so on |
| | i.e. | that is |
| | e.g. | for example |
| Government Agencies | CDC | Centers for Disease Control |
| | FDA | Food and Drug Administration |
| | FEMA | Federal Emergency Management Agency |

However, do not begin a sentence with a numeral.

150 covers were expected on a Friday night. [incorrect]

One hundred fifty covers were expected on a Friday night. [correct]

Further, round numbers are typically written as words rather than numerals.

They estimated that the renovation would cost over a million dollars.

I told you that for the hundredth time.

Numerals should be used in dates, addresses, and page numbers, and with symbols or abbreviations. Some style guides prefer that numbers over ten—including ages—be written as numerals.

Finally, it is important to *be consistent.* If the rules require a numeral for one of the numbers in a sentence or paragraph, use numerals for all the numbers in that category.

There were only three croissants and 100 munchkins on the table. [inconsistent]

There were only 3 croissants and 100 munchkins on the table. [consistent] *OR*

There were only three croissants and one hundred munchkins on the table. [consistent]

### EXERCISE 29.1 Abbreviations and Numerals

Rewrite the sentences below, spelling out the abbreviations and numerals where necessary. Abbreviate any words from Figure 29.1.

1. I went to the Dr.'s office yesterday & got 3 Rx for the flu.
2. Did you know that mister León has a B.P.S. in cul. arts?
3. 2 oz. of cheese won't be enough; we need at least four.
4. Have you been to that new restaurant on Second Ave.?
5. 100 years ago there were no televisions, computers, cell phones, etc.

## ITALICS AND UNDERLINING

**Italics** and **underlining** are generally used in the same way, though italics are used only with word processors. Italics are used for the titles of books, periodicals, movies, television series, works of art, and ships. Chapters in a book, articles in a periodical, and episodes in a television series are enclosed in quotation marks, however. Look at the following examples:

For further information on quotation marks, see Chapter 27, "Punctuation I: End Marks, Question Marks, Apostrophes, and Quotation Marks," in *A Taste for Writing: Composition for Culinarians.*

Brian and Patrick's favorite episode of *The Office* was "Dwight's Speech."

When writing by hand, it is correct to underline the titles of books and movies, though students and newspapers frequently use quotation marks instead. In the example below, the two film titles should be underlined or italicized in formal situations. Check with your instructor or publisher.

**STUDENT WRITING** Matthew Berkowski

There is actually one more similarity between cooks and writers. There is the odd occasion where the cook is a writer. Take Anthony Bourdain, author of <u>Kitchen Confidential</u>, for example. He is a recognized chef, yet he wrote about the food industry from a chef's perspective. Other authors use food in a symbolic way to represent people's emotions, for example, "Chocolat" and "Like Water for Chocolate."

Italics are also used when referring to a letter or word and when using words in a foreign language.

Some students add an extra stroke to the letters *m* and *n*.

The cooks prepared their *mise en place.*

## EXERCISE 29.2 Italics and Underlining

Rewrite the sentences below, underlining words and phrases where necessary.

1. Have you read The Lord of the Rings or just seen the movie?
2. Crash won the Academy Award for Best Picture.
3. Fernando enjoyed that recent article in Food and Wine.
4. Large ensemble casts have been popular on television with such series as Lost and Desperate Housewives.
5. The students in beginning French quickly learned to say bonjour.

## PARENTHESES

**Parentheses** are used to set off words and phrases that explain or refer to something within the main sentence. The parentheses indicate that this information is less important and that the structure and meaning of the sentence are not affected. Look at the examples:

> The Food and Drug Administration (FDA) does not regulate the sale of dietary supplements.

> Parentheses may be used like commas to set off explanatory information (see Chapter 28).

Parentheses are like "stage whispers" or "asides" in a play. The actor turns from the other cast members and speaks directly to the audience, often in an exaggerated whisper. While the stage whisper is intended to be heard, and sometimes contains such information as explanations of the actor's motives or of elements in the story, it is not part of the regular dialogue. Think about Romeo in the garden, listening raptly to Juliet's outpouring of desire. In an aside to the audience he wonders, "Shall I hear more, or shall I speak at this?"

Punctuation marks generally fall outside the parentheses, unless they are part of the phrase inside the parentheses.

> After Barrett saw *Ray* (What a great movie!), he bought the soundtrack.

> After Barrett saw *Ray* (the biography of Ray Charles), he bought the soundtrack.

## BRACKETS

**Brackets** are used within quotations to set off words, phrases, or explanations that were not in the original text.

> "After I saw *Ray* [the biography of Ray Charles]," said Barrett, "I bought the soundtrack."

Brackets are also used to set off explanatory words and phrases within parentheses.

**EXERCISE 29.3** **Parentheses and Brackets**

Rewrite the following sentences, adding parentheses or brackets where necessary.

1. The students developed a Hazard Analysis and Critical Control Point HACCP plan for the restaurant in their case study.
2. "My mom's the chef at Bistro Urbano a trendy downtown restaurant," the boy bragged to his friends.
3. Commas are also important in setting off words and phrases from the rest of the sentence see Chapter 28.
4. *American Idol* a reality series drew more viewers than the Olympic Games that night.
5. The company's new CEO Chief Executive Officer made it a policy to visit each department once a week.

## HYPHENS

**Hyphens** are the short lines used within single words to form **compound numbers**, such as "thirty-three" or "ninety-one," and with the prefixes *all-*, *ex-*, and *self-*, for example, "ex-boyfriend" or "self-esteem." Hyphens are also used with **compound adjectives**, that is, adjectives that are modified by an adverb and directly precede a noun. For example,

> The servers appreciated the well-behaved children. [hyphen required]

However, no hyphen is required if the adverb ends in *–ly*.

> The servers appreciated the surprisingly polite children [no hyphen]

When writing by hand, use hyphens to break words at the end of a line in order to maintain an orderly margin. When using a word processor, set the margin to align left and press enter only at the conclusion of the paragraphs; the computer will "wrap" the text automatically at the end of each line.

## DASHES

**Dashes**—like commas—are used between words and phrases to set off a thought that interrupts the rest of the sentence.

> The new bakery—the one around the corner—featured an assortment of muffins.

Dashes may also be used like a colon to introduce additional material or to explain or rename a word or phrase in the sentence.

> In a single year Jamie Foxx was nominated for two Oscars for two separate films—*Ray* and *Collateral.*

## RECIPE FOR REVIEW Punctuation III

### Abbreviations

Avoid abbreviations in your academic writing, except for those described in Figure 29.1.

### Numerals

1. Spell out numbers less than one hundred.
2. Do not begin a sentence with a numeral.
3. Write round numbers as words.
4. Be consistent.

### Italics

1. Use italics for the titles of books, periodicals, movies, television series, etc.
2. Use italics when referring to a letter or word and when using words in a foreign language.

### Underlining

1. Use underlining in the same ways as you use italics.
2. If you're writing on a computer, italics is generally preferable to underlining. Check with your instructor or publisher.

## Parentheses

1. Use parentheses to set off information that explains or refers to something within the sentence.
2. In general, place punctuation marks outside the parentheses.

## Brackets

Use brackets within quotations to set off words, phrases, or explanations that were not in the original text.

## Hyphens

1. Use hyphens to form compound numbers and with the prefixes *all-*, *ex-*, and *self-*.
2. Use hyphens with compound adjectives before a noun, for example, *a well-known recipe.*
3. Use hyphens to break words at the end of a line (except on a word processor).

## Dashes

1. Use dashes to set off a thought that interrupts the rest of the sentence.
2. Use dashes to introduce additional material or to explain or rename a word or phrase in the sentence.

## ? END OF CHAPTER QUIZ Punctuation III

**DIRECTIONS**: *Rewrite the following sentences, correcting the punctuation (abbreviations, numerals, italics, underlining, parentheses, brackets, hyphens, and dashes) as necessary.*

1. 2 servers + a line cook were still needed at the new restaurant on Park Ave.
2. Doctor Casey had already made a reservation for 5 for Thurs. night.
3. 1 of Jeremy's favorite dishes is liver & onions.

4. The case of Jane Doe the label given to the unidentified victim depended on DNA evidence.

5. Reggie was still confused about the difference between affect and effect.

6. Mister Jones often ordered 2 helpings of mac & cheese when he ate lunch at the local diner.

7. 100 people attended the gallery opening on Oct. third.

8. Amber saw Pirates of the Caribbean in the theater 3 times.

9. Ben checked the walk-in and found 5 bags of carrots, 12 heads of lettuce, and two boxes of cremini mushrooms.

10 After reading Kitchen Confidential, the culinary students applied at 3 or 4 restaurants on Cape Cod.

chapter

# 30

# Proofreading the Final Draft

Once the final draft of your essay, letter, e-mail, or résumé is complete and polished in terms of content, organization, grammar, and usage, you must then proofread it carefully one or more times before submitting or mailing it. Proofreading is like the final wiping of the plate just before you serve it to the customer. Proofreading shows that you are proud of your work and respectful toward your audience.

## THE PROOFREADING PROCESS

When you **proofread** a piece of writing, look at each sentence individually, from the capital letter to the period, in order to check that each sentence follows the rules of grammar and punctuation outlined in Units 6, 7, and 8. (See Figure 30.1.)

You must be careful to look at each word as it is written. If we go too quickly, we tend to "read" the sentence as it exists in our minds, not on the page. Since the brain's job is to make sense of the words, we may correct errors automatically inside our heads and so miss them on the paper. One way of short-circuiting the brain's autocorrect feature is to read the essay *backwards*. Start with the last sentence, then move to the sentence before it, then the sentence before that. The idea is to separate the meaning of the passage from its mechanics, spelling, and punctuation. Having someone else check the essay is also useful. In any case, you would be wise to proofread it yourself at least twice and perhaps a third time.

FIGURE 30-1 Proofreading Checklist.

| CHECK | PROOFREADING STEPS | REFERENCE |
|---|---|---|
| | Check that each sentence is complete. | Chapter 20 |
| | Eliminate run-on sentences and comma splices. | Chapter 21 |
| | Check that each verb agrees with its subject. | Chapter 22 |
| | Check that verb forms are correct and that verb tense is appropriate and consistent. | Chapter 23 |
| | Check that pronouns are used correctly in terms of case, agreement, reference, and point of view. | Chapter 24 |
| | Check that modifiers are used correctly. | Chapter 25 |
| | Correct any errors in parallel structure. | Chapter 26 |
| | Check that each sentence begins with a capital letter and ends with the appropriate punctuation mark. | Chapter 27 |
| | Check overall use of capital letters and apostrophes. | Chapter 27 |
| | Check internal punctuation. | Chapters 27-29 |
| | Use spell check; then check the list of commonly misspelled words and/or your personal list of commonly misspelled words. | Chapter 30<br>Appendix I<br>Appendix II |
| | Check that commonly misused words are used correctly. | Appendix III |

## SPELLING

An important part of proofreading, of course, is to check the spelling. Spelling is interesting. Some writers seem to spell correctly without much difficulty; others make mistakes even with the simplest words. Further, while some people place a great value on correct spelling, others do not see what the big deal is. The purpose of spelling, like that of punctuation, is to make sure the reader knows what you mean; that's the bottom line. However, spelling also reveals how much care you have taken with your writing, even how much respect you have for your audience! You certainly don't want bad spelling to show up in a review of your restaurant:

> "Many of the Italian ingredients and cooking terms—arugula, bruschetta, mascarpone, carbonara—are misspelled on the menu."[33]

Poor spelling may seem relatively unimportant to a restaurant's success, but the next paragraph adds that the restaurant's "casual approach to service can be off-putting," and the reviewer's overall assessment is negative. A misspelled word is like a dirty water glass—its lack of clarity makes a bad impression.

So, what can you do if you're one of those people who just can't spell? First of all, don't feel bad! Just because you can't spell doesn't mean you can't *write* or *think*. Spelling seems to be a separate skill or ability. If you don't have it, it's okay—but you will have to compensate. Many professional writers can't spell, but they have editors to help them. Shakespeare spelled the same word three different ways on a single page, and he is one of the most admired authors in the English language. In other words, let's not get emotional or judgmental about spelling; let's just get the job done.

Sometimes we've been told to check spelling in a dictionary. The difficulty there is that if you have no idea how to spell a word, you don't know where to look. If the word begins with the *s*-sound, for example, will you look under *s* or *c*, or even *p*? *Certain, scenery,* and *psychology* all begin with the same sound but definitely not with the same letters. Difficulties exist also with spell check. Although useful and convenient, especially when you're using a word processor, spell check does not contain all the words or names you may use and so cannot check their spelling.

Further, although spell check can identify a correct spelling—and even suggest possible corrections for some misspellings, a particularly helpful feature—it cannot tell you whether words have been *used* correctly. For example, it cannot tell you that *chose* is the incorrect form in the infinitive phrase *to choose.* Spell check doesn't even realize that *dinning* is not the correct spelling of *dining.* Does that mean you shouldn't use spell check? No, of course not. *Always* check your writing with spell check before giving it to anyone to read. Spell check is also available in a handheld form, like a calculator.

English is a difficult language to spell. It has borrowed words from different languages and gone through various stages of pronunciation and

**FIGURE 30-2 Spelling with *ie* and *ei*.**

| SPELLING RULE | EXAMPLES | EXCEPTIONS |
| --- | --- | --- |
| Use ie (not ei) for the -e-sound | brief, niece, retrieve | |
| BUT after c use ei | ceiling, deceive, receive | either, leisure, weird |
| Use ei when the sound is not e, especially when it is a | neighbor, weigh, height | friend, mischief |

spelling rules. The end result is kind of crazy. However, there are certain rules that govern spelling, and it is useful to know them. Another helpful practice is to keep a list of the words that you yourself often misspell. You can use this list in two ways. First, you can memorize the correct spellings on your list. Second, you can look for these words as you proofread your writing, and then use the list to check whether they are spelled correctly. If you find that you are misspelling words because you're confusing the meaning, write the definition as well as the spelling on your list, and see the list of commonly misspelled words in Appendix I and the selected culinary terms in Appendix II.

## Spelling Rules

Do you remember the rhyme "*i* before *e* except after *c*, and when sounded like *a* as in *neighbor* and *weigh*"? Thus we write *receive* and *ceiling* because of the *c*, but *believe* without it (see Figure 30.2). Of course there are exceptions, such as *seize, either, weird, leisure, neither.*

Another type of confusion concerns words ending in the sound "seed." There are three different spellings: *-cede, -ceed, -sede* (see Figure 30.3).

**FIGURE 30-3 Words with *-cede, -ceed,* and *-sede*.**

| SUFFIX | EXAMPLES |
| --- | --- |
| -cede (several words) | concede, precede, recede |
| -ceed (three words) | exceed, proceed, succeed |
| -sede (one word) | supersede |

**FIGURE 30-4 Spelling with Suffixes.**

| SPELLING RULE | EXAMPLES |
|---|---|
| Do not change the spelling when you add –ly or –ness | stubborn + ness = stubbornness<br>EXCEPT: happy + ness = happiness |
| Drop final e if suffix begins with a vowel | dine + ing = dining<br>hope + ing = hoping |
| Keep final e if suffix begins with a consonant | hope + ful = hopeful<br>EXCEPT: argue + ment = argument |
| If word ends in consonant + y, change y to i, unless suffix begins with i | hurry + ed = hurried<br>hurry + ing = hurrying |
| Double final consonant before suffix IF the word ends in one vowel plus one consonant and is accented on the last syllable | run + ing = running<br>occur + ed = occurred<br>BUT cancel + ed = canceled (last syllable is not accented) |

When a **prefix** is added to the beginning of a word, the original spelling does not change; we simply add the two parts together. For example, *mis* + *spell* = *misspell.* However, there are a number of rules that govern the spelling of a word when a **suffix** is added to the end (see Figure 30.4).

8

## EXERCISE 30.1 Tracking Misspelled Words

Start a list of words that you've misspelled on your essays. Look over the commonly misspelled words in Appendix I and the selected culinary terms in Appendix II, and add to your list any whose spelling you are unsure of.

**EXERCISE 30.2** **Identifying Misspelled Words**

Identify the correctly spelled word in each pair.

| | | |
|---|---|---|
| 1 | definate | definite |
| 2 | succeed | suceed |
| 3 | writing | writting |
| 4 | occured | occurred |
| 5 | disappointed | dissappointed |
| 6 | extremely | extreemly |
| 7 | schedual | schedule |
| 8 | probably | probaly |
| 9 | accessible | accessable |
| 10 | recieve | receive |

## Culinary Terms

Since the culinary world uses words from many different languages, spelling can be especially complicated. We must move smoothly from French (*hors d'oeuvre, niçoise*) to Italian (*foccaccia, prosciutto*) to Spanish (*paella, tortilla*). We must also spell words from languages that use a different alphabet (*challah, Szechuan*). And, of course, we must spell all kinds of English words, such as *cocoa, doughnut, leek,* and *Worcestershire sauce!*

**EXERCISE 30.3** **Identifying Misspelled Culinary Terms**

Identify the correctly spelled word in each pair.

| | | |
|---|---|---|
| 1 | restaurant | restaraunt |
| 2 | vegtable | vegetable |

| | | |
|---|---|---|
| 3 | tomatos | tomatoes |
| 4 | license | licence |
| 5 | dining | dinning |
| 6 | guacemole | guacamole |
| 7 | thyme | tyme |
| 8 | aperitif | apperitif |
| 9 | cinamon | cinnamon |
| 10 | vinagrette | vinaigrette |

A tableside *aperitif* creates a welcoming atmosphere.

## COMMONLY MISUSED WORDS

Salt and sugar look very much alike, though salt granules are generally larger. However, the two have quite different tastes and chemical properties. What would happen if you mixed up the quantities in a cookie dough recipe and added a cup and a half of *salt* instead of sugar? Ugh! Similarly, English contains a number of words that resemble one another but that have different spellings and meanings. Skim through Appendix III, and identify any word pairs that you know you find confusing. Read and study the explanations. When you proofread a paper, pay special attention to these words. Look up others as questions arise.

## RECIPE FOR REVIEW Proofreading the Final Draft

1. Use spell check on every piece of writing.
2. Use a dictionary to check the spelling and definition of words highlighted by spell check.
3. Add words and names to spell check as needed.
4. Learn spelling rules.
5. Keep a list of words that you sometimes misspell, and check for them in each piece of writing. Add new words to the spell check feature on your computer.
6. Check lists of commonly misspelled words and culinary terms.
7. To learn the correct definitions of commonly misused words, read through Appendix III and write down five pairs of words that look confusing or that you know have confused you in the past. Study the explanations and examples. Then, for each word, write a sentence that illustrates its correct usage.
8. Look at your essays and note which of the words in the appendix you tend to use or misuse. Study them carefully. Write a sentence for each that illustrates the correct usage.

## END OF CHAPTER QUIZ Proofreading the Final Draft

**Part I. Spelling**. *One word is misspelled in each of the following sentences. Write the correct spelling on a separate sheet of paper.*

_______ **1.** The customers were pleased with the hotel accomodations.

_______ **2.** It was necessary to chose smoking or non-smoking rooms, of course.

_______ **3.** Latter, they went to the hotel's three-star restaurant.

_______ **4.** The dinning room was beautifully decorated with red velvet curtains and gold picture frames.

_______ **5.** The soup du jur was butternut squash with apple slices and sour cream.

_______ **6.** No one was dissappointed with its creamy texture and rich autumn flavors!

_______ **7.** The next course followed imediately, a salad of impossibly fresh mixed greens.

_______ **8.** There were also hot roles and butter, as well as large Calamata olives.

_______ **9.** By the time the main course arrived, the dinners were not even hungry.

_______ **10.** After a last sip of wine, everyone headed threw the doors toward the elevators.

**Part II. Commonly Misused Words.** *Fill in the blanks below with the appropriate word in parentheses.*

**1.** The customers were unable to __________ between the lemon meringue pie and the apricot flan. (choose/chose)

**2.** The daily special was served with a __________ cup of coffee or tea. (complementary/complimentary)

**3.** "Which of the __________ looks good to you?" asked Marianne. (deserts/desserts)

**4.** "What's for __________?" asked Joey. (diner/dinner)

**5.** "__________ important to work with a sharp knife," said the chef-instructor. (Its/It's)

**6.** The knife had a speck of tomato sauce on __________ blade. (its/it's)

**7.** Many Americans resolve to __________ weight after the winter holidays. (loose/lose)

**8.** Andrew and __________ shared a delicious mushroom risotto. (I/myself)

9. We had already driven __________ the restaurant before we saw the sign. (passed/past)

10. We __________ the restaurant before we saw the sign. (passed/past)

11. "May I have a __________ of that New York cheesecake?" (peace/piece)

12. The customers at the local Italian restaurant look at __________ menus. (their/there/they're)

13. __________ are four kinds of homemade pasta available. (Their/There/They're)

14. __________ excited about trying the vodka a la penne. (Their/There/They're)

15. No one is fonder of chicken lo mein __________ Vinnie is. (than/then)

16. You __________ may enjoy Vinnie's favorite some day. (to/too/two)

17. Is this the pistachio pudding cake __________ you were talking about this morning? (that/which)

18. Is Rick the one __________ intends to make the cake this afternoon? (who/which)

19. Rick learned the recipe from Will, __________ had learned it from his mother. (that/who)

20. Jeremy saw a girl __________ he immediately admired. (who/whom)

21. John saw a girl __________ was dancing with her father. (who/whom)

22. "__________ ready to bake cookies?" asked George. (Whose/Who's)

23. "I am," said Derek. "__________ recipe are we going to use?" (Whose/Who's)

24. "We will use __________ recipe, Derek." (your/you're)

25. "__________ going to need lots of brown sugar, then." (Your/You're)

# International Cuisine

## English as a Second Language

UNIT 9

**Chapter 31**

English as a Second Language

chapter 31

# English as a Second Language

Writing in a second language is challenging for many reasons. Not only must we master complex rules of grammar and memorize unpredictable idiomatic expressions, but we also need to understand something about the culture's expectations concerning the content and style of our writing. In other words (as always in a piece of writing), we must consider how best to communicate with a particular audience.

One potential difference between cultures is the directness with which the writer approaches the main point. In general, your American instructors will expect you to make a "straight line" to the topic and its supporting information. Further, they will expect you to "keep to the point" and "stay focused," unlike readers in other cultures that value circuitous discussions and inventive digressions. Although it is appropriate to explore all possible aspects of an issue, American instructors may often expect you to "take a side" or "have an opinion" (see Chapter 14). In addition, in the same way that American instructors value a direct, focused approach to the topic, they also value concise sentences that get right to the point, avoid repetition, and use relatively few adjectives.

Another difference between cultures is the writer's relationship to "authority" and "intellectual property." American instructors place great value on the thoughts and opinions of individuals, including their students. In general, you are expected to express *your* ideas in an essay; you

Studying in a foreign country can be exciting as well as challenging.

are the authority on your own thoughts. Further, your written words become to some extent "property" that you own, just as the written words of others are their property. Therefore, if you borrow ideas or words from another source, you are expected to cite them appropriately (see Chapter 17) or even ask permission to copy them. American professors feel quite strongly that it is unacceptable to use the words or ideas of another person without correctly citing the material. Original thinking is highly prized and should comprise the bulk of your essays. Material from outside sources is used to support or "prove" your points.

However, other cultures do not always share these views of "intellectual property" and "authority." In other cultures, students may be encouraged to learn by closely imitating the thoughts and perhaps even copying the words of respected authors. Indeed, some cultures would look with disfavor on the idea that students' opinions were as valuable as these respected authors' opinions. No approach is right or wrong in an absolute sense; we just want, as writers, to be aware of the expectations of our particular audience. When you're writing for your American professors, remember to cite any words or ideas that you borrow from another source.

A last cultural difference has to do with the relationship between the instructor and the student. American instructors in general welcome free and lively discussion and particularly encourage their students to ask questions. Although in some cultures it would be considered disrespectful, in American classrooms asking questions is valued as a sign of interest and dedication. Therefore, if you do not understand the directions for writing a particular essay, or if some aspect of grammar or usage is unclear, be sure to ask for clarification!

## ORGANIZING YOUR IDEAS INTO PARAGRAPHS

A **paragraph** is a series of sentences that explores or illustrates a single idea. While there is no hard and fast rule about the length of a paragraph, it is typically longer than one or two sentences. As in so many aspects of writing, the purpose of the paragraph is to make the writer's meaning clear. A new paragraph tells the reader that we've moved to a new idea. To make that even clearer, in English we indent the first line of each new paragraph about one half-inch or, if typing, one tab space. Writers also pay attention to the length of paragraphs so that they make an interesting pattern on the page and are not so long that they look unappetizing to the reader. An American essay will have clearly defined introductory and concluding paragraphs, as well as several body paragraphs (see Chapters 3 and 9).

## SENTENCE STRUCTURE

Languages differ in terms of which structural elements in a sentence must be spelled out and which may be implied. In standard written English, each sentence must contain a verb. Avoid sentences like the following in which the **linking verb** is omitted:

> The tiger cat fat because of his long career of catching mice. [incorrect]
>
> The tiger cat is fat because of his long career of catching mice. [correct]

Every English sentence also has a stated subject, as in the sentences below:

> George sat down at the table and prepared to enjoy a large breakfast. [The subject is *George*.]

The students opened their books to page 193 and began to read. [The subject is *the students*.]

However, sentences in the imperative mood—that is, commands—do not need to state the subject. Commands are understood to have the subject *you,* as in the following examples:

Please sit down. [The subject is understood to be *you*.]

Open your books to page 193. [The subject is understood to be *you*.]

English cannot omit the subject *it* in the expression *it is.*

Is important to be on time to class. [incorrect]

*It* is important to be on time to class. [correct]

If *is* is followed by a noun, rather than an adjective, it may be preceded by the word *there.*

Is a noisy dog living in that house on the corner. [incorrect]

*There is* a noisy dog living in that house on the corner. [correct]

A noisy dog lives in that house on the corner. [correct]

*There* is not used when the linking verb is followed by an adjective:

There is foggy outside. [incorrect]

*It* is foggy outside. [correct]

*There* is fog outside. [correct when linking verb is followed by a noun]

English sentences that are missing a subject or a verb (with the exception of commands) are called **sentence fragments** (Chapter 20).

American English—as distinct from other dialects of English—is very particular about avoiding what it calls a **comma splice**, that is, joining two or more independent clauses with a comma (Chapter 21).

Joaquin grilled two steaks, Dexter added a garnish of mushrooms. [comma splice]

Joaquin grilled two steaks; Dexter added a garnish of mushrooms. [correct]

This is an important difference from the rules of punctuation in Great Britain and other parts of the world, where commas and semicolons are often used interchangeably. In American English, the **semicolon** is used *only* to join two independent clauses. However, the **comma** can be used for several different purposes. For example, American writers use a comma between an initial subordinate clause and the main clause, as in the sentence below.

Although Joaquin grilled two steaks, he did not add a garnish of mushrooms.

Note that a semicolon *could not* be used in place of that comma (see Chapter 30).

**EXERCISE 31.1** **Sentence Structure**

Rewrite the sentences below, adding missing words or editing the punctuation. Be prepared to explain your reason for each choice.

1. Makiko visited the United States for the first time at the age of ten; when her parents vacationed in California.
2. She returned after graduating from college in order to study baking and pastry arts, she hopes to open a small bakery when she returns to Tokyo.
3. Makiko's roommate also from Japan.
4. Is a student from Thailand in the same dormitory.
5. Is helpful for the three girls to speak English together.

## NOUNS AND ARTICLES

Nouns can be divided into two categories, count and noncount nouns. **Count nouns** generally refer to concrete things and can be counted: one *car*, two *cars*, three *cars*, or one *banana*, two *bananas*, three *bananas*. Count nouns have a singular and a plural form. In addition, count nouns can be quantified with expressions such as *many* or *few*. **Noncount nouns** include things that cannot be counted, such as *salt* or *cream*. Noncount nouns do not have a plural form.

I added salt to the eggs.

We couldn't answer the question "How *many* salts did you add?" because salt is not counted in the way that eggs are. Noncount nouns are quantified in a different way, with expressions such as *much* or *little.* "How *much* salt did you add?" would be an appropriate question. Noncount nouns also include abstract concepts, such as *courage* or *perseverance,* and some nouns considered in a collective sense, such as *equipment* or *work.*

> The student showed *many perseverances.* [incorrect]
> The student showed *much perseverance.* [correct]

> The restaurant purchased *few new equipments.* [incorrect]
> The restaurant purchased *little new equipment.* [correct]

Many nouns can be used either as count or noncount nouns, as in the following sentence:

> Chefs may choose from a variety of salts, such as kosher salt, Hawaiian sea salt, or *Fleur de Sel.*

In this "count" context, we could talk about *many salts.* Singular count nouns may be quantified with words like *one, each,* or *every.* Plural count nouns may be quantified by words like *many, several,* or *few.* Quantities of noncount nouns (which have no plural form) can be described as *much, less,* or *little.* Finally, both noncount and plural nouns may be quantified with words such as *more, some,* or *any.* In making decisions about count and noncount nouns, remember to consider how the noun is used in a given sentence, as well as the general class to which it belongs. See Figure 31.1 for a list of common noncount nouns in English. Note that the list of noncount nouns is different for different languages. Further, remember that some of the nouns on the list can also be *used* as count nouns.

> We had chicken for dinner. [noncount]

> The Murphys raised their own chickens. [count]

The distinction between count and noncount is not involved in choosing when to use *this, that, these,* or *those.* Use *this* or *that* with singular nouns and *these* or *those* with plural nouns.

> Would you help me move *this furniture*? [noncount—no plural form]

> Would you help me with *this suitcase*? [singular]

> Would you help me with *these suitcases*? [plural]

> *That milk* has not been pasteurized. [noncount—no plural form]

> *That carrot* has not been washed. [singular]

> *Those carrots* have not been washed. [plural]

**FIGURE 31-1** Common Noncount Nouns.

| | |
|---|---|
| **Abstract Nouns** | advice, anger, beauty, confidence, courage, education, fun, happiness, health, honesty, information, knowledge, leisure, love, precision, progress, truth, wealth |
| **Concrete Nouns—Edible*** | bacon, beef, bread, broccoli, butter, cabbage, candy, cauliflower, celery, cereal, cheese, chicken, chocolate, coffee, corn, cream, fish, flour, fruit, ice cream, lettuce, meat, milk, oil, pasta, rice, salt, spinach, sugar, tea, water, wine, yogurt |
| **Concrete Nouns—Inedible** | air, dirt, gasoline, gold, paper, plastic, rain, silver, snow, soap, steel, wood, wool |
| **Additional Examples of Noncount Nouns** | clothing, equipment, furniture, homework, jewelry, machinery, mail, money, news, poetry, research, scenery, traffic, transportation, violence, weather, work |

*Note, however, that quite a few of these nouns may also be used in a "count" sense.

However, the distinction between count and noncount nouns *is* important when choosing the appropriate articles. These words—*a*, *an*, and *the*—give certain kinds of information about the nouns that follow them. Not every language has articles, however, and their usage can therefore be quite confusing. If possible, check your use of articles with a fluent English speaker when you edit a piece of writing.

The **indefinite articles** *a* and *an* often have the meaning "one." They are used with singular count nouns when the noun is first introduced to the reader or is otherwise unknown. *A* is used before words beginning with a consonant sound (*a car, a horse*) while *an* is used before words beginning with a vowel sound (*an earthquake, an honor*). Note that despite its spelling, *honor* begins with a vowel sound. Further, since one or more adjectives may intervene between an article and the noun it modifies, the decision to use *a* or *an* may not involve the noun itself, but rather the word that immediately follows the article. Study the following examples:

Flicka is the name of *a horse* in *a* popular children's book.

Flicka is the name of *an intelligent horse* in a popular children's book.

Flicka is the name of *a very* intelligent horse in a popular children's book.

Indefinite articles are not used with noncount nouns or with nouns used in a noncount context.

We added *salt* to the eggs. [not *a salt*]
*but*
The chef tried *a different salt* each day. [She tried one of a variety of salts—"count" usage.]
*or*
Would you like *a coffee*? [*a coffee* = *a cup of coffee* in this context]

Some international students incorrectly omit the indefinite article, as in the example below:

One sunny day, our city was hit by strong earthquake.

English requires that an article precede the singular count noun *earthquake.* In general, when a noun is first introduced to the reader, use an indefinite article:

One sunny day, our city was hit by a strong earthquake.

Use the **definite article**, *the*, for subsequent allusions to the same noun:

After *the* earthquake, our family packed some items to prepare for the next one.

In contrast to *a* and *an, the* does not have the meaning *one.* Rather, it indicates that the noun has been mentioned earlier or is otherwise known to the reader.

*The* sun is hiding behind *the* clouds. [The reader is familiar with *sun* and *clouds*.]

*The* sun is hiding behind *a* cloud. [This is the first mention of a *particular* cloud.]

*The* sun is hiding behind *the* cloud that looks like a rabbit. [The sentence refers to a *specific* cloud that is defined by the clause *that looks like a rabbit*.]

Do not use *the* with noncount or plural nouns with the meaning "all" or "in general."

Playing *the music* to young children is said to enhance their math ability. [incorrect; *music* is a noncount noun with the meaning "music in general"]

Playing *music* to young children is said to enhance their math ability. [correct]

But *do* use *the* when the meaning is limited by another word or phrase or when the noun is known to the reader.

Playing *the music of Mozart* to young children is said to enhance their math ability. [correct; *the* indicates the meaning of *music* here is limited to that *of Mozart*]

Here is another example:

The *doctors* are often particularly bright and hard-working. [incorrect when referring to "all" doctors]

*Doctors* are often particularly bright and hard-working. [correct for "all" doctors]

*The doctors at this hospital* are often particularly bright and hard-working. [correct; the word *doctors* is restricted to those "at this hospital"]

*The best doctors* are often particularly bright and hard-working. [correct; the word *doctors* is restricted by *best*]

Do not use *the* with most singular proper nouns—

We asked to speak with Dr. Kildare. [not *the* Dr. Kildare]

Exceptions to this rule are found among proper nouns that refer to newspapers (but not magazines), historical periods, official titles, and organizations, for example, *The London Times,* the Ice Age, the Dean of Students, and the National Restaurant Association. Magazine titles, such as *Newsweek* or *Wine Spectator,* are not preceded by *the.*

Place names can be confusing. Use *the* with names of rivers, canals, oceans, seas, gulfs, peninsulas, and deserts, such as the Mississippi River, the Erie Canal, the Indian Ocean, the Black Sea, the Gulf of Mexico, and the Gobi Desert. Use *the* also with island chains, mountain ranges, geographical regions, and the names of certain countries, for example, the Virgin Islands, the Himalayas, the Pacific Northwest, and the United States of America.

Do *not* use *the* with names of streets, parks, cities, states, counties, countries (with some exceptions, such as the Netherlands or the United States), continents, and individual lakes, mountains, and islands: Mulberry Street, Central Park, Chicago, Illinois, Orange County, Mexico, South America, Lake Michigan, Mount Kiliminjaro, and Ellis Island.

Finally, do use *the* with the names of museums, libraries, and hotels: the Museum of Science and Industry, the Library of Congress, and the Four Seasons.

**EXERCISE 31.2** **Using the Definite and Indefinite Articles**

Add or delete articles as needed in the sentences below. Be prepared to explain your reason for each choice.

1. Brenda was staying on the Park Avenue at Grand Hotel.
2. Adrianne spoke with the Professor Jordan about homework assignment.
3. We are not happy with the freshness of this cabbages.
4. Peter chose restaurant that offered the sushi.
5. Tim was huge fan of Boston Celtics.

## ADJECTIVES AND ADVERBS

Adjectives are words that describe or modify a noun. In English, unlike some other languages, the adjectives are generally placed *before* the noun they describe.

> The fuzzy little puppy had grown into a sleek adult. [*fuzzy* and *little* describe *puppy; sleek* describes *adult*]

When a series of adjectives describes a single noun, the adjectives usually appear in a certain order according to type. (See Figure 31.2; you will find exceptions, however.) Interestingly, many English speakers are unaware of these rules; they simply "hear" that one order "sounds right."

**FIGURE 31-2** Order of Adjectives.

| PLACEMENT | TYPE OF ADJECTIVE | EXAMPLES |
|---|---|---|
| 1 | article/possessive | a, an, the, her, Daniel's |
| 2 | number | two, second, last |
| 3 | opinion | beautiful, dishonest, ridiculous |
| 4 | size | small, little, big |
| 5 | age | old, young, 18-year-old |
| 6 | shape | round, square, pointed |
| 7 | color | green, red, bright blue |
| 8 | origin | French, Mexican, Egyptian |
| 9 | material | wooden, silvery, rubber |
| 10 | qualifier | jazz (music), kitchen (counter), steak (knife) |
| 11 | the noun itself | hen, ball, grandmother |

**the little red hen** = 1 (article) + 4 (size) + 7 (color) + 11 (noun)

**a small, round rubber ball** = 1 + 4 + 6 + 9 + 11

**her old French grandmother** = 1 (possessive) + 5 + 8 + 11)

**Daniel's beautiful big green pool table** = 1 + 3 + 4 + 7 + 10 + 11

In general, American essay writers do not use more than two or three adjectives with any one noun.

Adverbs are words that describe verbs, adjectives, or other adverbs. When adverbs modify a form of the verb *to be*, they are typically placed *after* the verb:

> Desserts *are often* tastier when accompanied by coffee. [*often* follows *are*]

Otherwise, adverbs typically *precede* the verb:

> Desserts *often contain* chocolate. [*often* precedes *contain*]

Adverbs usually *precede* the adjectives or adverbs they modify.

The soup was *very* smooth and flavorful. [*very* precedes the adjective *smooth*]

We *very* often ordered the butternut squash soup. [*very* precedes the adverb *often*]

**EXERCISE 31.3**

### Placing Adjectives and Adverbs Correctly

Rewrite each sentence, inserting the words in parentheses into the correct place. Be prepared to explain your reason for each choice.

1. The restaurant on the corner is owned by couple. (*couple* is modified by *a/an, Italian, old, tiny*)
2. It is decorated with fixtures. (*fixtures: light, wrought iron, elegant*)
3. Each table has a vase. (*vase: deep blue, glass, Venetian*)
4. The tiles were made in Tuscany. (*tiles: floor, square, colorful*)
5. Customers praise the restaurant's scheme. (*praise: frequently; scheme: color, attractive*)

## PREPOSITIONS

Prepositions can be difficult since they often do not translate directly from one language to another. Again, it is helpful to check your sentence with a speaker of American English. Two prepositions that seem particularly troublesome are *in* and *on*. Use *in* to describe one item that is *inside* another.

The coffee is *in* the cup.

Use *on* to describe one item that is *resting on the top of* another.

The cup is *on* the table.

Use *in* with books and magazines and with place names.

I found the answer *in my math textbook*.

Have you read the article about Japanese cuisine *in this month's* "Food Arts?"

Vince grew up *in Lake Forest,* a suburb of Chicago.

*In* and *On*.

Use *on* with television, radio, and the Internet.

*House* was his favorite show *on television*.

Jennifer was hoping to hear some of her favorite songs *on the radio*.

Have you tried to find a map *on the Internet*?

Use *in* with a car, but *on* with a bus, train, plane, or boat.

According to some statistics, Americans eat 19% of their meals *in a car*.

Do you prefer to ride *on a bus* or *on a train*?

Prepositions are also used in expressions of time. Use *in* for a period of time or for a particular month or year.

I will be ready *in ten minutes*.

She is looking forward to her daughter's birthday *in August*.

Use *on* for a particular day or date.

The test will be *on Monday*.

Julia was born *on August 15*.

Use *at* for a particular time.

Class begins *at 8 o'clock*.

The family meets *at breakfast* to go over plans for the day.

Prepositions are also used following nouns, adjectives, and verbs. Prepositions are particularly difficult here because there are no general "rules" to study. You will have to learn each expression as you encounter it. As with vocabulary words, it is helpful to keep your own list of these expressions. While Figure 31.3 lists some examples, it is by no means comprehensive.

**FIGURE 31-3 Prepositions Used with Other Words.**

| WORD + PREPOSITION | SAMPLE SENTENCE |
|---|---|
| accompanied **by** [someone or something] | Martin was **accompanied** by his wife. |
| acquit **of** [a charge] | The defendant was **acquitted of** the charge of murder. |
| admit [*no preposition*] | Beryl **admitted** that she had taken two desserts. |
| admit **to** | Beryl **admitted to** taking two desserts. |
| admit **into** | The celebrities **were admitted into** the VIP room. |
| agree **to** | The contestants **agreed to** the rules of the game. |
| agree **on** [terms] | The city and the workers **agreed on** a 3% pay raise. |
| agree **with** [someone] | I **agree with** you that lighting is a very important part of a restaurant's atmosphere. |
| answer **to** [someone] | Juan had to **answer to** his mother. |
| answer **for** [what was done] | Juan had to **answer for** the broken window. |
| argue **with** [someone] argue **about,** argue **over** [something] | Makiko **argued with** Yuko **about** whose turn it was to wash the dishes. |
| base **on** (not *upon*) [an idea or rule] | Heinrich's decision to use a tourniquet **was based on** his knowledge of first aid. |
| comment **about** [someone not present] | The surgeon **commented about** his new interns. |
| comment **on** [something] | The surgeon **commented on** the unusual nature of the patient's injury. |
| comment **to** [someone present] | The surgeon **commented to** Mrs. Davies that her husband was a strong man. |
| confide **in** [someone] | Mrs. Davies **confided in** her oldest friend, Mrs. Edwards. |
| confide **to** [someone] that | Mrs. Davies **confided to** the surgeon that her husband was actually quite frightened of needles. |
| consist **of** [parts] | The play **consisted of** three acts. |

*(continue)*

| **WORD + PREPOSITION** | **SAMPLE SENTENCE** |
|---|---|
| consist **in** [traits] | The excellence of the play **consisted in** its witty dialogue and tightly constructed plot. |
| depend **on** | You can **depend on** me. |
| different **from** [generally preferred] | Marjoram is quite **different from** oregano. Cooking with butter is somewhat **different from** cooking with safflower oil. |
| differently **than** [may be used with an adverbial clause] | Isabella made the paella **differently than** Julia did. |
| equivalent **in** | The dishes were **equivalent in** cost and preparation time. |
| equivalent **to** | One stick of butter is **equivalent to** 8 teaspoons. |
| forbid [someone] **to** | **I forbid** you **to** speak! |
| independent or independently **of** [something] | Dennis came up with the plan **independently of** his colleagues. |
| inquire **about** [something] | Alexandra politely **inquired about** her grandmother's health. |
| inquire **into** [circumstances] | Alexandra **inquired into** the circumstances of her grandmother's hospitalization. |
| inquire **of** [someone] | Alexandra **inquired of** the doctor whether her grandmother was ready to go home. |
| inquire **after** [someone] | Alexandra **inquired after** you when she visited last week. |
| listen **to** | Alexandra **listened** carefully **to** the doctor. |
| preferable **to** | For Leonard, dining at home is **preferable to** grabbing something at a fast food restaurant. |
| reconcile **to** [circumstances] | On this occasion, however, Leonard was **reconciled to** a cheeseburger on the go. |
| reconcile **with** [someone] | In his particularly compliant mood, Leonard **reconciled with** his former partner. |

*(continues)*

**FIGURE 25-3** *(continued)*

| WORD + PREPOSITION | SAMPLE SENTENCE |
|---|---|
| skillful **at** or **in** [doing something] | The line cook was youthful but nevertheless quite **skillful at** handling his orders. |
| skillful **with** [something] | The youthful line cook was quite **skillful with** a knife. |
| succeed **as** [a job or role] | Rachel **succeeded as** an actress. |
| succeed **in** [an attempt] | Jeremy did not **succeed in** evading Gloria's advances. |
| wait **for** [someone or something] | We **waited for** the train. |
| wait **on** ["serve"] | The young man who **waited on** me was very efficient. |

### EXERCISE 31.4 Using Prepositions

Rewrite the sentences below, filling in the blanks with the appropriate preposition. If no preposition is required, do not write anything in the blank. Be prepared to explain your reason for each choice.

1. William Shakespeare was born ______ England ______ April 23.
2. The actress was accompanied ______ the party ______ her husband.
3. The class is held ______ room 320 and begins ______ 10 a.m.
4. Do you go to work ______ a car or ______ a bus?
5. Will you wait ______ me before you go ______ home?

## VERBS

### The Progressive Tenses

The six tenses outlined in Chapter 23 may also be **progressive**, that is, they may indicate that the action was, is, or will be continuing. Study the examples below:

- The **past progressive** is formed with *was/were* + present participle: *Vanessa* was *chopping nuts for the biscotti.* The chopping was a continuous action in the past, but did not move into the present.
- The **present progressive** is formed with *is/are* + present participle: *Vanessa is* chopping *nuts for the biscotti.* The chopping started earlier and is still going on.
- The future progressive is formed with *will be* + present participle: *Vanessa* will be chopping *nuts for the biscotti.* She hasn't started yet, but the action will continue for an indefinite length of time in the future.

The perfect tenses also have progressive or continuous forms:

- The past perfect progressive is formed with *had been* + present participle: *Vanessa* had been chopping *nuts for the biscotti.* The action continued for an indefinite period of time in the past, but was completed before another action in the past.
- The present perfect progressive is formed with *has/have been* + present participle: *Vanessa* has been chopping *nuts for the biscotti since 10 o'clock this morning.* She was chopping until recently and may even be chopping still.
- The future perfect progressive is formed with *will have been* + present participle: *By the time we arrive at 11 o'clock, Vanessa* will have been chopping *nuts for the biscotti for one hour.* The action begins in the future, continues for an indefinite time, and ends before the start of another action in the future.

There are certain English verbs, however, that are rarely used in the progressive. Most of these refer to a mental activity (*want*) rather than an action in space (*beat*).

Sarah is wanting to bake a perfect angel food cake. [incorrect]

Sarah *wants* to bake a perfect angel food cake. [correct]

Sarah *is beating* the eggs for the cake batter. [correct]

Sarah *beats* the eggs for the cake batter. [correct]

Other verbs that rarely take the progressive form are *appear, believe, belong, contain, have, hear, know, like, need, see, seem, taste, think,* and *understand.*

I was not knowing her one year ago. [incorrect]

I *did not know* her one year ago. [correct]

**EXERCISE 31.5** **Using the Progressive Tenses**

Select the appropriate form of the verb in the sentences below. Be prepared to explain your reason for each choice.

1. Rosanna (was shopping/is shopping) at the mall all day yesterday.
2. She (was wanting/wanted) to find a special dress for the Valentine's Day party.
3. Her friend Lara also (is needing/needs) a new dress.
4. Lara (will be joining/will have been joining) Rosanna tomorrow for another shopping trip.
5. By tomorrow, Rosanna (was looking/will have been looking) for a dress for two weeks.

## Helping Verbs

The **helping verbs** (sometimes called **auxiliary verbs** or **modals**) are used with the main verb to show tense, voice, and mood (see Chapter 23). One pair that sometimes causes difficulty is *can* and *could*, verbs that add the idea of ability or permission to the main verb.

> Todd *can* paint well.

*Could* is the past tense of *can*.

> Todd *could* paint well as a child.

*Could* also adds an element of doubt or uncertainty to the main verb and can be used in the subjunctive mood.

> If Todd were to try singing, though, I'm not sure he *could* do it.

Be careful of using *could* to mean simply "was able to."

> After I *could* arrive at our apartment, I felt relief to see her face. [incorrect]

Since the uncertainty of arrival has passed, *could* does not work in the sentence above. The simple past tense is clear and appropriate:

After I *arrived* at our apartment, I felt relief to see her face. [correct]

The helping verbs *can, could, may, might, must, shall, should, will,* and *would* are followed by the base form of the main verb.

Brittany and Todd *might decide* to stay an extra night on their vacation.

The verb *ought* is followed by *to* + the base form of the main verb.

Brittany and Todd *ought to consider* the expense of an extra night, however.

The helping verbs *do, does,* and *did* are used to ask a question, to emphasize the meaning of the main verb, or to create a negative meaning with *not* or *never.* These helping verbs are also followed by the base form of the main verb.

*Does* Brittany *like* [not *likes*] the food at the hotel?

Todd and Brittany *do enjoy* [not *enjoying*] the beach.

Todd *did not eat* [not *ate*] much seafood on this trip.

He *never did like* [not *liked*] shrimp.

**EXERCISE 31.6** **Using the Helping Verbs**

Rewrite the sentences below, correcting any problems with verbs. Be prepared to explain your reason for each choice.

1. What type of vacation does Melissa enjoys?
2. She might likes to camp out in the mountains.
3. She did not had fun on her trip to the beach, however.
4. Melissa ought remind herself that she can use sunscreen next time.
5. Would Melissa to choose a vacation in Venice?

## More about Verbs

A number of verbs may be followed by an **infinitive** (*to* + the base form of the verb) or by a **gerund** (the base form of the verb + *-ing*). There are no general rules, only accepted usage. The lists below provide only a limited number of examples; again, you will have to keep your own list as you come across new words.

Verbs that use an infinitive rather than a gerund include the following:

| | | | | |
|---|---|---|---|---|
| agree | ask | beg | decide | expect |
| have | hope | manage | mean | offer |
| plan | promise | refuse | want | wish |

For example:

I *hope to complete* my studies next year.

The three students *offered to give* her a pastry.

Verbs that use a gerund rather than an infinitive include the following:

| | | | |
|---|---|---|---|
| appreciate | avoid | deny | discuss |
| enjoy | finish | imagine | miss |
| practice | quit | resist | suggest |

"The three students offered to give her a pastry."

Study these examples:

> I *imagine completing* my studies next year.
>
> Brandon *suggested giving* her a ride home.

Some verbs may use *either* an infinitive *or* a gerund: *begin, continue, hate, like, love,* and *start.*

> Jorge *began cooking* at a very young age.
>
> Jorge *began to cook* at a very young age.

Another set of verbs has one meaning with an infinitive and another with a gerund: *forget, remember, stop,* and *try.*

> I *remember waking* up early. [I remember the event of waking.]
>
> I *remember to wake* up early on schooldays. [I must wake up early and I do.]

Some verbs may be followed by a noun or pronoun and then an infinitive.

| | | | | |
|---|---|---|---|---|
| ask | advise | allow | command | convince |
| encourage | have | instruct | need | persuade |
| remind | require | tell | want | would like |

In addition, *ask, need, want,* and *would like* may be followed by a noun or a pronoun and then the infinitive. Compare the two sentences below:

> The chef would like to add a dessert special to the menu.
>
> The chef would like *you* to add a dessert special to the menu.

A final pair of verbs that causes some confusion is *do* and *make.* Since there is no clear rule about which to use, you must learn each expression separately. Figure 31.4 lists a few common examples.

**FIGURE 31-4 Common Expressions with *Do* and *Make*.**

| DO | MAKE |
|---|---|
| do the dishes | make the bed |
| do a favor [for someone] | make an impression [on someone] |
| do your hair | make a living |
| do your homework | make (cook) a meal |
| do a job | make a mistake |
| do the laundry | make a speech |

**EXERCISE 31.7** **More Verbs**

Select the correct verb or verb form in the sentences below. Be prepared to explain your reason for each choice.

1. On an episode of *Grey's Anatomy,* the chief of surgery told everyone (to evacuate/evacuating) the hospital because of a bomb threat.
2. Dr. Burke persuaded his girlfriend (to leave/leaving) the operating room.
3. However, he himself resisted (to leave/leaving) his patient.
4. At great risk to himself, Dr. Burke finished (operating/to operate) on the patient.
5. If he (did/made) a mistake, the consequences would be fatal.

## RECIPE FOR REVIEW English as a Second Language

It isn't easy to write in a language that you didn't grow up speaking! Be patient with yourself, and don't expect your progress to be either quick or easy.

- Keep lists of vocabulary words and idiomatic phrases.
- Read as much as you can.
- Talk to "native" speakers.
- Listen to the radio, particularly news stations or NPR (National Public Radio).
- Watch movies and DVDs (with English subtitles, if possible, so that you can see the words as well as hear them).
- Tape television shows so that you can listen to the same thing several times.

- Watch television shows with closed captioning so that you can see the words.

See the following chapters for more information on grammar and punctuation.

- Subject/verb agreement—Chapter 22
- Additional tenses, commonly confused verbs, irregular verbs—Chapter 23
- Use of pronouns—Chapter 24
- Use of modifiers (adjectives, adverbs, phrases)—Chapter 25
- Punctuation—Chapters 27–29
- Commonly misused words—Appendix III

## A Taste for Reading

1. As you read the stories, poems, and essays in Appendix VII, add new words to your vocabulary list.
2. Talk to others about the stories in order to practice your speaking and listening skills as well as your reading skills.

## Ideas for Writing

1. When did you first learn English? Was it in the classroom? Was it through speaking and listening? Do you have any difficulty in speaking or understanding English? Explain.
2. Write a paragraph or short essay describing the similarities and differences between writing in English and writing in your first language.
3. What are the three most difficult things about writing in English?
4. What are three things you can do to improve your written English?

appendix II

# Commonly Misspelled Words

Words marked with an asterisk are defined in Appendix III: Commonly Misused Words.

**A**

absence
accept*
acceptable
accessible
accidentally
accommodate
accurate
acknowledge
acquaintance
acquire
address
admission
adolescent
affect*
alumni*
annihilate
anonymous
answer
anxious
apologize
apparatus
apparent
appearance
appreciate
argument
association
attendance

**B**

basically
beautiful
because
beginning
believe
beneficial
business

**C**

calendar
camouflage
candidate
carburetor
catalogue
category
capital*
capitol*
ceiling
cellar
certain
choose*
chose*
cloths*
clothes*
colonel
column
commitment
committed
complement*
compliment*
conscience*
conscientious
conscious*

consciousness
continuous
convenience
cooperation
correspondence
courtesy

### D

decision
definite
dependent
descendant
describe
description
desert*
dessert*
dilemma
dining
disappointment
discipline
discrimination
discussion
dissatisfied
division
doubt

### E

effect*
efficient
eighth
either
embarrassed
emergency
emphasize
entertainment
enthusiastically
environment
especially
exaggerate
excellent
except*
exhausted
experience
extremely

### F

familiar
fascinating
February
fiery
fluorescent
foreign
forth
fourth
fulfill

### G

gauge
generally
government
grammar
grateful
guarantee
guess
guidance

### H

handicapped
handkerchief
happily
harassment
height
hospital
humorous
hygiene

### I

ideally
ignorance
imagination
immediately
incidentally
independent
inevitable
infinite
influential
institution
insurance
intelligent
interference
irrelevant
irresistible
island

### J

jealous
judgment
judgement[34]
judicial

### K

knives
knowledge

### L

laboratory
later
latter
laugh
legitimate
leisure
library
license
lieutenant
lightning
loneliness
loose*
lose*
luxurious

### M

magnificent
maintain

maintenance
manageable
marriage
medicine
mediocre
miniature
miscellaneous
mischievous
mortgage
mosquito
mountain
muscle
mutual
mysterious

## N

necessary
necessity
negotiate
neighbor
neither
nickel
ninety
ninth
nuisance

## O

obedient
occasionally
occurrence
official
often
omit
omitted
opinion
opponent
opportunity
optimistic
original
ourselves

## P

parallel
participate
particularly
perceive
permanent
permissible
perseverance
persistent
personal*
personnel*
persuade
physical
possession
possibility
practically
precede
preference
prejudice
preparation
principal*
principle*
privilege
probably
procedure
proceed
professional
professor
psychology

## Q

quantity
questionnaire
quiet
quite
quiz
quizzes

## R

realize
really
receipt
receive
recognize
recommend
reference
referred
reign
rein
relevant
relieve
remember
repetition
responsibility
rhyme
rhythm
roommate

## S

safety
salary
scary
scenery
schedule
scissors
secretary
seize
sense
separate
sergeant
significant
similar
simultaneously
since
sincerely
sophomore
souvenir
specifically
straight
strategy
strength
subtle

succeed
success
sufficient
surprise
surroundings
suspicious

### T

technical
technique
temperamental
tendency
themselves
thorough
though
through*
tomorrow
toward
tragedy
twelfth
typical

### U

unanimous
unique
unnecessary
until

### V

vacuum
vain*
valuable
vane*
vehicle
vein*
vicinity
village
villain
violence
visibility
volume

### W

weather*
Wednesday
weird
welcome
where*
whether*
which
whisper
whistle
writing
written

### Y

yawn
yesterday
yield
young

appendix II

# Spelling of Selected Culinary Terms

**A**

a la carte
aioli
alcohol
al dente
amandine
ambrosia
anchovy
anglaise
antipasto
aperitif
appetizer
artichoke
arugula
asparagus
aspic
au gratin
au jus
au lait
avocado

**B**

bain marie
baklava
balsamic
banana
barbecue
baste
béarnaise
béchamel
benedict
beurre blanc
beverage
biscotti
biscuit
bisque
bleu (cheese)
blintz
bok choy
bologna
bordelaise
borscht
bouillabaisse
bouillon
bouquet garni
bourbon
braise
breakfast
brie (cheese)
brioche
broccoli
broccoli rabe
brochette
brunoise
bruschetta
Brussels sprouts
buffet
bulgur
bundt
burrito

**C**

cacciatore
Caesar
café
cafeteria
caffeine
Cajun
calamari
calcium
calorie
calzone

Canadian (bacon)
canapé
cannelloni
cantaloupe
caper
cappuccino
carambola
caramel
caraway
carbonara
cardamom
carob
carpaccio
carrot
casserole
cassoulet
casual
catsup
cauliflower
caviar
cayenne
celery
cèpes
cereal
ceviche
challah
chalupas
champagne
charcuterie
chardonnay
charlotte
chateaubriand
chef
chervil
chestnut
chicory
chiffonade
chili con carne
chipotle
chocolate
cholesterol
chorizo
chow mein
chowder
chutney
cilantro
cinnamon
cocoa
coconut
colander
coleslaw
collard greens
compote
condiment
consommé
coq au vin
cordon bleu
coriander
coulis
couscous
crème brûlée
crème caramel
crêpe
cremini, crimini
croissant
croquette
crouton
crudités
crystallized
cucumber
cuisine
cumin
curry
custard
customer

## D

Daikon
daily
daiquiri
decadent
demi-glace
demitasse
dessert
Dijon
diner
dining
dinner
dolmades, dolmas
dough
doughnut
duxelle

## E

employee
emulsify
enchilada
endive
enoki
entrée
escargot
escarole

## F

fagioli
Fahrenheit
fajita
falafel
fennel
feta
fettuccine
filet, fillet
fines herbes
Florentine
foccaccia
fondue
formaggio
french (fries)
fricassée
frisée
fromage

## G
ganache
garde manger
gastronomy
gazpacho
gelatin
gelato
ghee
giblets
gluten
gnocchi
Gorgonzola
goulash
gourmet
granola
gratinée
gratuity
Gruyère
guacamole
gumbo
gyro

## H
halibut
haricot verts
haute cuisine
herring
hollandaise[35]
hors d'oeuvre
hummus

## J
jaggery
jalapeño
jambalaya
jardinière
Jarlsberg
Jerusalem
artichoke
jicama
jigger
juice
julienne

## K
Kahlúa
Kaiser
kale
kasha
kebab, kabob
kedgeree
kernel (of corn)
ketchup
kielbasa, kielbasy
kitchen
kiwi
knead
knife, knives
kohlrabi
kosher
kumquat

## L
lasagna
leek
lentil
license
linguine
liqueur
liquor
loin
lox

## M
mâche
Madeira
mahi mahi
maitre d'hotel
manager
marinade
marinara
marinate
marsala
marzipan
mascarpone
mayonnaise
measure
medallion
Mediterranean
menu
meringue
mesclun
minestrone
mirepoix
mise en place
miso
mocha
morel
mousse
mousseline
mozzarella
mussel

## N
nachos
niçoise
nutrition
nutritious

## O
omelet
oregano
orzo
ouzo

## P
paella
palate
pallet
parfait
parmesan
pasteurized
pâté de foie gras

persillade
pesto
petit four
pico de gallo
pierogi
pilaf
pimiento, pimento
polenta
pomegranate
porcini
portabella
potage
potato
potatoes
poultry
primavera
prix fixe
prosciutto
protein
Provençal
provolone
purée

## Q

quahog
quesadilla
quiche
quinoa

## R

radicchio
ragout
raisin
ramekin
raspberry
ratatouille
ravioli
receipt
recipe
rémoulade
restaurant
restaurateur
rhubarb
ricotta
rigatoni
risotto
roll
Roquefort
rotini
roulade
roux

## S

sachet
saffron
sake
salmon
salsa
salsify
sambuca
sandwich
sashimi
sauerbraten
sausage
sauté
scallion
scallops
scampi
semolina
sesame
shallot
sherbet
shiitake
shish kebab
sommelier
sorbet
soufflé
soup du jour
sous chef
spaetzle
spaghetti
spatula
special
spinach
streusel
strudel
sushi
swordfish
Szechuan, Sichuan

## T

Tabasco
tamale
tandoori
tapas
tapenade
tarragon
tartar sauce
temperature
tempura
teriyaki
terrine
Thai
thermometer
thyme
tiramisu
tofu
tomato
tomatoes
tortellini
tortilla
tostada
tournedos
tuile
turmeric

## U

udon (noodles)
utensil

## V

vanilla
variety

vegetable
velouté
venison
vermicelli
Vidalia
vinaigrette
vinegar
vintage
virgin (olive oil)
vodka

## W

Waldorf
wasabi
weight
whisk
whiskey
wok
wonton
Worcestershire sauce

## Y

yeast
yogurt

## Z

zabaglione
zinfandel
ziti
zucchini

appendix III

# Commonly Misused Words

## accept, except

- *Accept* is a verb that means "to receive," as in "The restaurant manager *accepted* the delivery of the fresh produce."
- When *except* is used as a verb, it means "to exclude." For example, "The manager *excepted* the lettuce from her receipt of the fresh produce delivery." As a preposition *except* means "excluding," as in "The manager accepted all of the produce *except* the lettuce."

## affect, effect

- *Affect* is most often used as a verb that means "to influence the outcome," as in "The poor spelling *affected* the grade on his essay." Less frequently, *affect* may be used as a noun to mean an emotional state. "The doctor noted that the patient's *affect* was good."
- *Effect* is most often used as a noun that means "consequence" or "result." For example, "The most likely *effect* of his improved study habits is a change in his grades." When used as a verb, *effect* means "to cause or achieve," as in "His improved study habits *effected* a change in his grades."

## alumni, alumnae

- *Alumni* is a plural form that means "male graduates of a school" or "both male and female graduates of a school." For example, "The college's *alumni* give generously to the scholarship fund." *Alumnus* is the singular form and may refer to both male and female graduates.

- *Alumnae* is also a plural form, but it refers exclusively to *female* graduates. *Alumna* is the singular form and refers only to female graduates.

## amount, number

- Use the word *amount* for the quantity of a noncount noun, a quantity considered a unit. "The caterer required a large *amount* of dough for dinner rolls."
- Use the word *number* for the quantity of a count noun, a quantity considered as several separate items. "The caterer required a large *number* of dinner rolls."

## because

- The use of the construction "the reason is because" is common but *informal*. "For example, we might *say,* "The reason he likes this restaurant is *because* it has good service.
- In more *formal* situations, "the reason is that" is preferred. "The reason he likes this restaurant is *that* it has good service."
- A more concise wording is the following: "He likes the restaurant because it has good service."

## being as, being that

- The phrases "being as" and "being that" may be used in conversation, but they are too informal and wordy for academic writing. "Being that I was hungry, I fixed myself a sandwich."
- The preferred usage is "since" or "because": *"Since* I was hungry, I fixed myself a sandwich."

## beside, besides

- *Beside* is a preposition that means "next to" to someone or something. For example, "The server laid the spoon *beside* the knife."
- *Besides* is an adverb that means "also" or "furthermore": "*Besides* setting the tables for lunch, the server wiped off the menus."

## between/among

- *Between* is used in formal English when considering two items, even when they are part of a larger group. "How can I choose *between* coffee and espresso?" Note that, like all prepositions, *between* takes the objective case. It's always "between you and *me,*" not "between you and *I.*"

- *Among* is used with a group. "There was agreement *among* us that this was the funniest movie we'd ever seen."

## brake, break

- *Brake* refers to the part of the car (think of the central *a* in both words) that stops the wheels. For example, "Dolores used the *brake* as she approached the intersection."
- *Break* refers to the action of splitting something into two or more pieces: "Dolores used to *break* the eggs into their own bowl before combining them with the rest of the ingredients."

## capital, capitol

- *Capital* can be used as a noun to mean "assets" or "resources" or to mean the first city. "The Johnsons invested a good deal of *capital* in the new restaurant," or "Springfield is the *capital* of Illinois." As an adjective, *capital* often means "punishable by death," as in "Murder is a *capital* crime."
- *Capitol* refers to the building that houses the legislature. "The senators headed for *Capitol* Hill to vote on the energy bill."

## choose, chose

- The difference between these two is time (and pronunciation): *Choose/chooses* are in the present tense; *chose* is the irregular past tense form.
- Present tense: "The Dietrichs often *choose* an Italian restaurant for special occasions."
- Past tense: "On Halloween, however, they *chose* to eat at a Japanese restaurant."
- Infinitive: *Choose* is used with the *infinitive* form: "It was difficult to *choose* between the two restaurants."

## cite, sight, site

- *Cite* means "to quote or refer to": "The food critic often *cited* the works of M. F. K. Fisher."
- *Sight* means "vision" or "view": "The customers were fascinated by the *sight* of the roasted suckling pig."

- *Site* means "place" or "location": "The entrepreneur studied possible *sites* for the new restaurant."

### cloths, clothes

- *Cloth* describes fabric in general or a useful item such a dishcloth or tablecloth. "Cooks often keep a clean *cloth* tucked into their aprons." (Keep that extra *e* tucked out of sight.)
- *Clothes* refers to items that people wear: "Fortunately they had brought dry *clothes* to change into after their visit to the water park."

### complement, compliment

- As a verb, *complement* means to balance or match, particularly in the hospitality industry. "The wine was chosen to *complement* the main course." As a noun, *complement* means quota or amount. "The new restaurant did not yet have a full *complement* of wait staff."
- In contrast, the verb *compliment* means "to flatter or praise": "The baking instructor *complimented* the student on her marzipan." As a noun, *compliment* means "flattery" or "praise": "My *compliments* to the chef!"

### complementary, complimentary

- *Complementary* means "matching" or "balanced": "The couple's personalities were *complementary*."
- *Complimentary* means either "approving" or "free": "The hotel guests were *complimentary* about the *complimentary* Continental breakfast."

### conscience/conscious

- A *conscience* tells us when we're doing wrong: "My conscience is clear," we say, or "My conscience is bothering me."
- We're *conscious* when we're awake and aware. "The patient opened his eyes; he was conscious" or "I was conscious of a smoky odor coming from the kitchen."

### coarse, course

- The definitions of these two words are clear; however, a misspelling can create a problem for the reader. *Coarse* is an adjective meaning "rough" or "untreated": "The sea salt had a *coarse* texture."

- *Course*, on the other hand, refers to a path, direction, academic class, or part of a meal: "The main *course* at the wedding reception was served under the tent."

### could/can

- This pair of helping verbs adds the idea of ability or permission to the main verb: "Todd *can* paint well." *Could* is the past tense of *can:* "Todd *could* paint well as a child."
- *Could* also adds an element of doubt or uncertainty to the main verb and can be used in the subjunctive mood: "If Todd were to try singing, though, I'm not sure he *could* do it."
- Do not use *could* to mean simply "was able to." The simple past tense is clear and appropriate.

  After I *could* arrive at our apartment, I felt relief to see her face. [incorrect]

  After I *arrived* at our apartment, I felt relief to see her face. [correct]

### could have/could of

Do not use *of* in expressions such as *could of, would of,* or *should of*; the correct usage is *could have, would have,* or *should have.* "The customers *could have* skipped dessert, but they yielded to temptation."

### desert/dessert

A *desert* is full of sand—it's dry and uncomfortable. A *dessert* is full of sugar—it's rich and delicious! The difference between them is the extra *s.* Think of it this way: dessert is so sweet.

> Travelers in the dry and sandy *desert* often long for a cool and refreshing *dessert.*
>
> *Desert* can also be used as a verb meaning "to abandon": "The boys *deserted* their playmates when they were called in for dessert."

### diner/dinner

The clue here is in the pronunciation of the two words. You know that *diner* with a long *i* refers to the person who is eating or to a type of restaurant. Remember that a single consonant means that the *i* is long, while a double consonant means that the *i* is short.

- *Diner* with a long *i* refers to the person who is eating ("The *diners* at the new restaurant were enthusiastic about their meal") or to a type of restaurant ("We ate at many different *diners* on our cross-country trip").
- *Dinner* with a short *i* refers to the meal. "The *dinners* at the new restaurant were enjoyed by all the customers."

### few/little

- *Few* is used with count nouns, that is, with individual items. "There were *few* diners in the restaurants after 10 p.m."
- *Little* is used with noncount nouns, that is, with amounts rather than countable items. "There was *little* light in the restaurant after it closed."

### fewer, less

- *Fewer* is used with count nouns (see *few* above): "There were *fewer* diners in the restaurant on Monday than on Thursday."
- *Less* is used with non-count nouns (see *little*) above: "There was *less* trouble with the seating chart when Maria was working."

### has/have got

Rather than writing *has got* or *have got,* write simply *has* or *have*: "That ugly sofa *has* to go" or "They *have* too much time on their hands."

### have, of

Do not use *of* in expressions such as *could of, would of,* or *should of*; the correct usage is *could have, would have,* or *should have.* "The customers *could have* paid by cash or credit card."

### in, into

- In formal English, *in* refers to a static location: "The lion stalked regally back and forth *in* its cage."
- *Into* refers to movement from one location to another: "The children skipped merrily *into* the zoo to visit the lions."

### its, it's

These two words sound the same, that is, they are *homonyms,* but they are spelled differently and have completely different meanings. Try to understand the difference and use the two words correctly.

- *Its*—like the possessive forms of other personal pronouns (for example, *hers, ours,* and *yours*)—does not use an apostrophe. "She noticed that the risotto had begun to burn in *its* pan." (This is confusing because *indefinite* pronouns, as well as nouns, *do* use an apostrophe for the possessive: *nobody's, Cheryl's.)*
- *It's* uses the apostrophe to indicate that this word is a contraction, in other words, that some letters are missing. *It's* is a shorter or contracted way to write *it is.* For example, "*It's* easy to burn risotto if you are not careful."

### lead, led

- *Lead* is the correct spelling of the metallic element pronounced "led," as well as of the present tense of the verb "to lead": "The negative effects of kryptonite on Superman could be blocked by *lead.*"
- *Led* is the correct spelling for the past tense of the verb "to lead": "Superman *led* the way into the dark tunnel."

"Keri uses her knife like a professional."

### leave, let

- *Leave* means "to go away" or "to abandon": "*Leave* your collection of shot glasses behind when you spend a month in the mountains." The expression "*Leave* me alone" means "go away."
- *Let* means "to allow": "*Let* the customers finish their main course before you offer them the dessert menu." The expression "*Let* me alone" means "stop bothering me."

### like, as

- While *like* is a preposition and must be followed by a noun, *as* is typically a subordinating conjunction. "Keri uses her knife *like* a professional" or "Keri uses her knife *as* a professional does."

- In formal English, do not use *like* for *as if* or *as though.* "He looks *like* he would be interesting to talk to" is informal. The better choice would be "He looks *as though* he would be interesting to talk to."

## loose, lose

- *Loose* is most often an adjective that means "relaxed, free, or baggy": "The *loose*-fitting pants allowed the chefs to move freely about the kitchen."
- *Lose* is a verb that means "to misplace, elude, or be defeated": "The vegetables will *lose* a good deal of their nutritional value if they are boiled too long."

## myself

Do not use the reflexive pronoun *myself* in place of *I* or *me.* Write "my friends and I," not "my friends and *myself.*" See Chapter 24.

## of, off

Watch your spelling here.

- *Of* indicates possession: "The top *of* the table was scarred by knife cuts."
- *Off* indicates location: "The knife fell *off* the table." Note that in formal English we do not use *of* with *off*:

  The knife fell *off of* the table. [informal]

  The knife fell *off* the table. [preferred]

## palate, palette, pallet

These three nouns sound alike, but each spelling is associated with a different meaning.

- *Palate* refers to the "roof of the mouth" and also means "taste" or "appreciation": "This sommelier has an excellent *palate.*"
- *Palette* indicates the "range of colors used by a painter" or the "board on which paints are mixed": "The artist added a fresh tube of red paint to her *palette.*"
- *Pallet* means the "large, stackable wooden tray" used in storage: "We unloaded a few *pallets* of lettuce this morning."

### passed, past

- *Passed* is the past tense of the verb "to pass": "Despite the double yellow line, Brendan *passed* the slowly moving car in front of him."
- *Past* can be used as a preposition meaning "beyond": "Brendan drove *past* the car in front of him."
- *Past* can also be used as a noun referring to "an earlier period of time" or as an adjective that means "historical" or "earlier."

  In the *past,* refreshments had been served at these tournaments. [noun]

  Refreshments had always been served at *past* tournaments. [adjective]

### peace, piece

- *Peace* is the opposite of war; note that they both contain the letter *a:* "Make *peace,* not war."
- *Piece* is the spelling for a serving of pie; note that *pie* is a part of *piece:* "I'd like a *piece* of that cherry pie, please."

### personal, personnel

- *Personal* means "individual" or "private": "The chief executive officer made a *personal* decision to retire early."
- *Personnel* refers to employees: "*Personnel* decisions are handled by the human resources department."

### principal, principle

- The noun *principal* means "head" or "leader," or perhaps a sum of money. "The *principal* of the high school declared a snow day." You might remember the spelling by thinking of the high school principal as your *pal*—or as someone who made life difficult for you and your *pals*!
- The noun *principle* means "rule" or "belief": "The class studied the *principles* of English grammar." Note that both *principle* and *rule* end in *-le.*

### shall, will

In very strict formal English, *shall* is used as a helping verb with the first person *I* and *we:* "We *shall* enjoy a visit to the zoo tomorrow." However, in most cases *will* is acceptable: "We *will* enjoy a visit to the zoo tomorrow."

## should have/should of

Do not use *of* in expressions such as *could of, would of,* or *should of*; the correct usage is *could have, would have,* or *should have.* "The customers *should have* skipped dessert, for they were carrying some extra weight."

## stationary, stationery

- *Stationary* means "motionless" or "in one place"; note the repeated *a,* like the letter *a* in *place.* "The crane was *stationary* over the weekend."
- *Stationery* means "writing paper"; think of the *-er* in *paper.* "Matilda is fond of lavender-colored *stationery.*"

## than, then

- *Than* is a conjunction used with comparisons; note the *a*'s in *than* and *compare.* "This chili is much hotter *than* that one."
- *Then* is an adverb having to do with time; note the e's in *then* and *time.* "I'll taste the chili; *then* I'll have a glass of milk."

## there, their, they're

These three words are often confused with one another. Try to remember the difference with associations such as the following:

- *There* has to do with place, even with pointing. It contains the word "here," which also relates to place. "The knife is *there* on the table."
- *Their* is the possessive form of the pronoun *they*; it refers to a person, not a place. Furthermore, it contains the word "heir," which also refers to a person. "The knife is on *their* table."
- *They're* is completely different because it's spelled with an apostrophe. It's a contraction, like *it's,* meaning that some letters are missing. *They're* is shorthand for *they are.* For example, "Chef Trotter owns many knives; *they're* on the table."

## threw, through, thru

- *Threw* is the past tense of the verb "to throw": "Carlotta *threw* the ball to third base and got the runner out."
- *Through* is the preposition that means "from one end to the other" or "because of": "Carlotta walked *through* the dining room, collecting the bottles of ketchup left on the tables."

- *Thru* is an abbreviated spelling of *through* that should be avoided in formal writing.

### to, too, two

- *To* is a preposition: "Carol went *to* the store."
- *Too* means "also"; think of too many *o's.* "Casey went to the store, *too.*"
- *Two* is the number. "*Two* friends went to the store."

### vain, vane, vein

- *Vain* means "conceited" or "useless": "The Olympic gold medalist was sometimes accused of being *vain*" or "The student made a *vain* attempt to fix his broken hollandaise sauce."
- *Vane* refers to the object on the top of the barn, often a rooster or an arrow, that indicates the direction of the wind: "The weather *vane* was rusty but effective."
- *Vein* is the tube that carries blood back into the heart: "The students removed the *veins* from the prawns." Or we could talk about a *vein* of precious metal or minerals running through a rock.

### weather, whether

- *Weather* is the rain, sleet, or snow reported on the news: "The bad *weather* discouraged many people from attending the outdoor concert."
- *Whether* is a subordinating conjunction that means "if": "The producers don't know *whether* their concert will be successful."

### where, were, we're

- *Where* is a relative pronoun that may sometimes be used like a subordinating conjunction.

  *Where* are you going? [pronoun]

  David told his mother *where* he was going. [conjunction]
- *Were* is the past tense of the verb "to be": "David and Diana *were* going to the concert."
- *We're* is the contraction for "we are": "*We're* going to the concert."

## who, which, that

- *Who* refers to people only, *which* to animals and things only, and *that* to either people or things.

  These are the students *who* have scored above 90 on the quiz.

  These are the quizzes, *which* happen to be printed on yellow paper.

  These are the quizzes *that* the students took yesterday.

- *Who* may begin a restrictive or nonrestrictive clause (see also Chapter 28):

  These are the students *who* have scored above 90 on the quiz. [restrictive]

  These students, *who* scored above 90 on the quiz, are already studying for their final exam. [nonrestrictive]

- *Which* begins a nonrestrictive clause:

  These are the quizzes, which happen to be printed on yellow paper.

- *That* begins a restrictive clause. Do not use a comma in this case (see Chapter 28).

  These are the quizzes, *that* the students took yesterday. [incorrect use of comma with restrictive clause]

  These are the quizzes *that* the students took yesterday. [correct]

## who, whom

- Use *who* whenever the subjective case is required: "I know who you are."
- Use *whom* when the objective case is required: "I know whom you are with." [*Whom* is the object of the preposition *with.*]

## whose/who's

- *Whose* is the possessive form, as in the sentence "The hostess is speaking with the server *whose* customers have finished their appetizers."
- *Who's* is a contraction of *who* and *is.* For example, "The hostess is speaking with the server *who's* getting ready to go home."

### woman, women

One letter here changes both the meaning and the pronunciation of the word that refers to 51% of the population! Just like the words *man* and *men*, the word with *a* is singular (*woman*), and the word with *e* is plural (*women*). "The *woman* [one woman] at the front of the line bought tickets for the three *women* behind her."

### would have/would of

Do not use *of* in expressions such as *could of, would of,* or *should of*; the correct usage is *could have, would have,* or *should have.* "The customers w*ould have* skipped dessert had they not been tempted by the tiramisu."

### your/you're

- *Your* is the possessive form; it means "belonging to you." For example, "*Your* pastries are delicious."
- *You're* is a contraction; it means "you are." For example, "*You're* fortunate to live near this pastry shop."

appendix IV

# Basics of Business Writing

A business letter or a résumé tells a story, your story, whether you're applying for a job, soliciting investors for your restaurant, or looking for another source of lobsters. The language of a letter or résumé, like that of an essay, should be fresh, clear, and accurate. Each has its own format, however, which is briefly explained in this appendix.

## BUSINESS LETTERS

A business letter has five parts: the heading, address, salutation, body, and closing. The **heading** contains the name and address of your business and the date of the letter. If you are using stationery with letterhead, that is, with the name and address of your business already printed on it, then you can write just the date in the heading. Begin the date at the left-hand margin, and do not abbreviate the name of the month. If you are not using letterhead, tab over to the center of the page and write your street address on the first line. Write your city, state, and zip code on the second line and the date on the third line (Figure IV.7). Some writers skip a line between the address and the date (Figure IV.4). See the discussion of block and modified block format below.

The **address** gives the name of the person and/or business that you are writing to and should be flush with the left margin. It begins with the person's or company's name, followed by the street address on the next line, and the city, state, and zip code on the following line. If the person's title is long, write it on a separate line. Study the example below:

| | |
|---|---|
| Person's name and title: | James Mitchell, Manager |
| Company name: | Jim's Bar and Grill |
| Street address: | 555 Westchester Avenue |
| City, state, zip code: | Smithtown, New York 12345 |

You may choose to use the official post office abbreviation for the state name (Figure IV.1).

The **salutation** or greeting should be flush with the left margin and two lines below the address. If you know the name of a particular person, use it. In the example above, you might write "Dear Mr. Mitchell." Use "Ms." for women, unless you know they prefer another title. If you

**FIGURE IV-1** **Postal Codes for States, District of Columbia, and Puerto Rico.**

| | | |
|---|---|---|
| AL Alabama | AK Alaska | AZ Arizona |
| AR Arkansas | CA California | CO Colorado |
| CT Connecticut | DE Delaware | DC District of Columbia |
| FL Florida | GA Georgia | HI Hawaii |
| ID Idaho | IL Illinois | IA Iowa |
| KS Kansas | KY Kentucky | LA Louisiana |
| ME Maine | MD Maryland | MA Massachusetts |
| MI Michigan | MN Minnesota | MS Mississippi |
| MO Missouri | MT Montana | NE Nebraska |
| NV Nevada | NH New Hampshire | NJ New Jersey |
| NM New Mexico | NY New York | NC North Carolina |
| ND North Dakota | OH Ohio | OK Oklahoma |
| OR Oregon | PA Pennsylvania | PR Puerto Rico |
| RI Rhode Island | SC South Carolina | SD South Dakota |
| TN Tennessee | TX Texas | UT Utah |
| VT Vermont | VA Virginia | WA Washington |
| WV West Virginia | WI Wisconsin | WY Wyoming |

know the person's job title, you may use that as well. Or you may use both names: "Dear James Mitchell."

> Dear Professor Fromm:
>
> Dear President Washington:
>
> Dear Chef Waters:
>
> Dear Dr. Kildare:

If you don't know the name of a specific person, you may address the company generally ("Dear First National Bank") or a job title within the company ("Dear Manager," "Dear Sales Associate"). In any case, follow the salutation with a colon.

The **body** of the letter starts two lines below the salutation. In general, each paragraph is single spaced, with double spacing between paragraphs. As with essays, the first paragraph or introduction of a letter should both indicate the subject or purpose of the letter and catch the reader's attention. Particularly when the recipient doesn't know you personally, as in a job application, the first paragraph has to do something to ensure he or she reads the whole letter. Unlike essays, however, business letters typically address the readers directly (*you*) and strive to make their points very briefly. If an idea or proposal requires fuller development, it would more likely be included in an attached report. As in all your formal writing, avoid clichés, slang, and sexist language in a business letter.

Most business letters should be no longer than one page. If you must use a second page, however, be sure to choose plain paper of the same quality as the first page. If you've used letterhead for the first page, use a plain sheet for the second. Type a heading on the second page that includes the name of the recipient, the date, and the page number. There should be at least two lines of text on the second page.

The **closing** begins two lines below the body. Close a formal letter with "Sincerely yours" or "Yours truly." If you know the recipient of the letter personally, you may use "Best wishes" or "Regards." Follow these words with a comma. Then, four lines below, write your name and your title, if appropriate. Again, if your title is long, write it on the line below.

> Julia Fernandez, Manager

or

> Julia Fernandez
>
> Director of Sales and Marketing

**FIGURE IV-2 The Closing, Signature, and Additional Notes.**

Best regards,

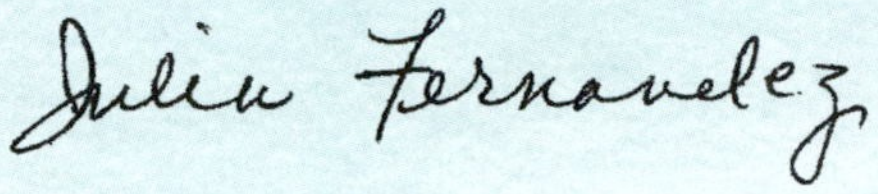

Julia Fernandez

Julia Fernandez
Director of Sales and Marketing
Enclosure
cc: Anna Vitelli, Vice President
Sales and Marketing

You will sign the letter above your name; that is the purpose of the four spaces between it and the closing phrase. Sign your full name, unless you are on a first-name basis with the recipient of the letter.

Business letters may also have additional notes following the closing and the signature that indicate that material has been enclosed with the letter or that copies of the letter have been sent to other people. In general, begin these notes four lines below the closing; if you need to shorten a long letter, use only two lines (see Figure IV-2).

Most business letters follow one of two formats on the page. With letterhead stationery, writers tend to use the **block format**, in which every line is flush with the left-hand margin (Figure IV.3). On plain paper, you may use a *modified* block format, in which the heading and closing begin at (or slightly to the right of) the center line, while all other lines are flush with the left-hand margin (Figure IV.5). Some writers also indent the first line of each paragraph when using a modified block format (Figure IV.7). In any case, the letter should be centered on the page and should make an attractive picture.

The address on the envelope of a business letter should be the same as the address inside the letter.

**FIGURE IV-3 Business Letter in Block Format.**

Bistro Urbano
67 Main street
urbanville, NY 19901

February 27, 2006

Andrea Palmer
The Coffee Company
123 East Market Street
Urbanville, New York 19901

Dear Ms. Palmer:

It was a pleasure to speak with you on the phone last week. We are very interested in trying out the two new varieties you spoke of, the Mocha Madness and Vanilla Vice.

Please send three pounds of each, and bill our account.

We look forward to expanding our business with you.

Sincerely yours,

*Gerald Abernathy*

Gerald Abernathy, Executive Chef

## EXERCISE IV.1 Business Letter Format

Write a brief but formal business letter to your bank or utility company regarding a problem (real or imaginary). Be sure to include the action you wish the company to take.

**FIGURE IV-4 Modified Block Format.**

Bistro Urbano
67 Main Street
Urbanville, NY 19901

February 27, 2006

Andrea Palmer
The Coffee Company
123 East Market Street
Urbanville, New York 19901

Dear Ms. Palmer:

It was a pleasure to speak with you on the phone last week. We are very interested in trying out the two new varieties you spoke of, the Mocha Madness and Vanilla Vice.

Please send three pounds of each, and bill our account.

We look forward to expanding our business with you.

Sincerely yours,

Gerald Abernathy

Gerald Abernathy, Executive Chef

## WRITING E-MAIL MESSAGES

E-mail is a quick and convenient way to keep in touch with friends and family, and in this context you can feel pretty free to write as you would like. However, when you're using e-mail in a business setting, you should take care to present yourself and your company professionally. Just as in a business letter, your tone in a work-related e-mail should be professional, and your writing should be clear and concise.

In order to maintain a professional tone, be sure to include both a salutation and a closing, even within your own organization. In a formal e-mail to outside recipients, both the salutation and closing should be similar to those in a business letter: *Dear Mr. Mitchell* or *Dear Ms. Fernandez* and *Yours truly* or *Best regards*. Within your own company, however, the salutation may be less formal, for example, *Good morning, Ms. Grey* or *Hello, Anne.* Similarly, the closing need not be as formal as *Yours truly;* rather, it might be as simple as *Best, Jane Doe* or *Thanks, Adam.* Unless the recipient is a close personal friend, always include both a salutation and a closing.

Some companies ask their employees to use a **signature block**, that is, to sign e-mails with full name and title, department, company, street address, telephone and fax numbers, and Web site, if applicable. Signature blocks may be programmed to conclude every e-mail but are especially appropriate when the recipients are outside the company. Look at the example below:

Gerald Abernathy, Executive Chef
Bistro Urbano
67 Main Street
Urbanville, NY 19901
(999) 555-7721

**www.bistrourbano.com**

The subject line of an e-mail should—obviously—indicate what the communication is about. Within the body of the e-mail message, take care that your organization is clear, with an introduction that includes the subject and purpose of the communication, a concise explanation of the specifics, and a conclusion that indicates what action is desired. If a fuller account of the details is necessary, attach it in a separate file. Workers may receive hundreds of e-mail messages per day. Respect their time; keep your e-mails brief and to the point.

Another aspect of professionalism here is following the rules of standard written English for grammar and spelling. Use complete, correct sentences. Check the spelling of every e-mail document. Avoid informal abbreviations such as *IMHO* (in my humble opinion) or *PMFJI* (pardon me for jumping in). In addition to the informality, the expressions themselves are too stale and worn for any professional correspondence. It is also wise to avoid *emoticons,* the "faces" created with punctuation marks such as :) for a smile or :( for a frown, unless you are writing to someone you know well.

Additional and compelling reasons to be professional in your e-mail correspondence concern privacy. E-mail messages are often stored indefinitely by organizations, may be read by supervisors, and can be subpoenaed by a court of law. Therefore, resist the temptation to bash the boss electronically or to shoot an angry e-mail to a colleague. Be smart, and take a moment to think. If you wouldn't put it in a business letter, don't put it in a business e-mail.

A final word about business e-mail: your readers will notice unclear, unprofessional, or misspelled messages. They don't have to be English teachers to identify a sloppy e-mail, just as they don't need to be gourmet chefs to recognize a cracked plate!

### EXERCISE IV.2 E-mail Messages

Rewrite the following e-mail in a more professional manner: "Jay whats shakin got the specs yet Bob." Add a signature block.

## RÉSUMÉS

A résumé is a summary of your qualifications for a job, both in terms of training and of experience. Like business letters, résumés generally contain certain types of information in a typical order. The heading should clearly state your name and address, as well as ways to contact you, such as your telephone number and e-mail address. The rest of the résumé should be tailored specifically to the job for which you are applying. The next sections typically describe your education and work experience, followed by any other relevant skills or activities. At the end of the résumé you often list the names and addresses of references or note that you have such references available.

However, a résumé is not simply a list of schools you've attended and jobs you've had. Once you identify a job you'd like to apply for, you review your training and experience and decide which aspects are most closely aligned with the job requirements. Sometimes your education will be the most important factor; at other times it will be your work history. The sample résumé in Figure IV.5, for example, emphasizes the applicant's education, while the sample résumé in Figure IV.6 emphasizes the applicant's experience.

Unpaid work can also be important. For example, if you were applying for a job as a restaurant manager, your ten years of "volunteering" in your family's establishment would be significant. On the other hand, don't include information that is not relevant to the particular job you're applying for. The fact that you were a rocket scientist will not get you hired as a pastry chef.

**FIGURE IV-5 Sample Résumé Featuring Education.**

Anjelica Garcia-Jones
123 Orchard Lane, Urbanville, NY 19901
cell phone: (999) 555-1222
e-mail: agarciajones@hotmail.com

Career Objective: To obtain a supervisory position at a restaurant where I can utilize my professional degrees, front of house management experience, and wines expertise.

EDUCATION

***Smithfield Culinary College***, Marketville, NY

Bachelor of Professional Studies in Hospitality Management — June 2005

- Future Leader in the Industry Award, 2005
- Dean's List, junior and senior years
- Member, Sommelier Society

***Urban Business School***, Urbanville, NY

Bachelor of Arts in Business Administration — June 2001

- Graduated with honors
- Member, Student Council, senior year
- Junior year abroad, Lima Business School, Peru
- Most Promising Freshman Scholarship, 1997-1998

WORK EXPERIENCE

Bistro Urbano, Urbanville, NY — 6/05-present

***Dining Room Manager & Sommelier***

- Supervised front of house operations.
- Developed new systems of scheduling and evaluating wait staff.
- Expanded wine list to include South American vintages.

Bistro Urbano, Urbanville, NY — 7/01-8/03

***Bookkeeper***

RELEVANT SKILLS AND INTERESTS

Fluent in Spanish

Member, Women Chefs & Restaurateurs

**FIGURE IV-6** **Sample Résumé Featuring Experience.**

David Jones
123 Orchard Lane, Urbanville, NY 19901
cell phone: (999) 555-1221
e-mail: limajones@hotmail.com

WORK EXPERIENCE

Bistro Urbano, Urbanville, NY 5/01-present

*Sous Chef*

- Supervised staff of 8 at small urban café.
- Managed purchasing and inventory.
- Developed daily lunch special in South American cuisine.

**Downtown Diner**, Urbanville, NY 6/98-5/01

*Sous Chef*

- Worked sauté and grill stations at busy city diner.
- 150 lunch covers daily.
- Promoted to sous chef in 2001.

**The Four Seasons**, Chicago, IL 4/97-8/97

*Tournant*

- Rotated through each station as an intern.
- Trained my replacement at this four-star/five-diamond restaurant.

**Jones Bar & Grill**, Urbanville, NY 6/92-8/96

*Line Cook*

- Line cook in family-owned restaurant serving American and South American cuisines.

EDUCATION

***Smithfield Culinary College***, Marketville, NY
Associate of Occupational Studies in Culinary Arts May 1998

PROFESSIONAL SKILLS AND MEMBERSHIPS

Fluent in Spanish
Member of American Culinary Federation and National Restaurant Association

Like a menu, a résumé will probably be studied for only two or three minutes, especially when there are many applicants for a position. Therefore, ensure that the important information is visible and concise. In addition to your contact information, employers need to know the types of establishments you've worked in and the specific tasks and responsibilities you've performed. Begin these descriptions with active verbs like *supervised* or *designed*. Avoid full sentences that begin with *I* or *My responsibilities were.*

Your résumé should be printed on high quality paper and should be free of typographical errors. This document represents you in a competition. If it appears sloppy or is difficult to follow, your potential employers may not even take the time to read it! On the other hand, try to make your résumé stand out in a positive way with a line, a color, or an unusual (but legible) font. Show intelligence and creativity in designing this "menu" for potential employers.

An important point about your résumé is *accuracy*. Do not misrepresent or inflate your information. If you haven't received the degree yet, don't mislead your prospective employer into thinking you have. If you were a line cook at your previous job, don't say you were the sous chef. Even if you were to be hired, this false information could get you fired immediately.

## EXERCISE IV.3 Writing A Résumé

Describe the "dream job" you'd like to have in two to five years. Then write the résumé that would prove you're the best candidate.

Finally, résumés are typically accompanied with a cover letter in which you introduce yourself to the prospective employer and specify the job for which you are applying (Figure IV.7). Never send a résumé without a cover letter! The cover letter should be clear and specific. Are you applying for a particular job? Are you interested in future opportunities?

The role of the cover letter is to make the reader look at your résumé. Do some research on the employer, and use that information in your cover letter. For example, you might say something like "I became particularly interested in your establishment when I learned of the opening of your Asian-themed snack bar." Use the cover letter to highlight the most important and interesting information about you that is *relevant* to this specific job, particularly the information that

**FIGURE IV-7** Sample Cover Letter.

123 Orchard Lane
Urbanville, NY 19999
March 11, 2006

Paola Allende, Executive Chef
*Café Conquistador*
85 Wisteria Lane
Suburbia, Washington 99999

Dear Chef Allende:

I am relocating to the Seattle area and would like to apply for the position of sous chef at *Café Conquistador* advertised in *The Seattle Times*.

My wife's brother lives in Seattle and has told me of your innovative establishment. I am particularly intrigued by the many South American dishes on your menu since my mother is from Peru and our family restaurant includes several items from her native country. Further, this cuisine is a specialty of the restaurant where I am currently working as sous chef.

I will be in Seattle next week and hope to visit your restaurant at that time. I will telephone you on Monday morning to inquire about scheduling an interview. A copy of my résumé is enclosed.

You can reach me on my cell phone at 555–1221 or via e-mail at limajones@hotmail.com. I look forward to meeting you soon.

Sincerely yours,

David Jones

David Jones

Enc.

reflects the requirements of the position. If an advertised job requires fluency in Spanish, and you are fluent in Spanish, be sure to put that fact in the cover letter.

The résumé and cover letter are important pieces of *persuasive* writing. Think about who your readers are and what they need to know about you to make a decision.

## RECIPE FOR REVIEW Business Writing

### Business Letters

1. Business letters have five parts: the heading, the address, the salutation, the body, and the closing.
2. Business letters may use a block or modified block format.
3. Use a formal vocabulary and correct grammar in your business letters. Proofread carefully for typos.

### E-mail Messages

1. Like business letters, e-mail messages outside your company should include a formal salutation and closing, as well as a signature block.
2. Use a formal vocabulary and correct grammar in your e-mails. Proofread carefully for typos.

### Résumés

1. Résumés are typically one-page summaries of the information a potential employer needs to know about you. They include a heading with your contact information, your education and training, work experience, related interests and activities, and references. Some résumés state a career objective.
2. Design your résumé for a specific job.
3. Send a cover letter with your résumé.

### Ideas for Writing

1. Describe three situations in which you would need to write a business letter. Choose one, and write the letter, paying particular attention to format and style.
2. Write a cover letter in which you explain why you are the best candidate for the dream job you described in Exercise IV.3.

3. Find a job listing online or in the newspaper, and design your résumé for that particular position.

4. If you have an existing résumé, rewrite it to fit your present education and experience. Can you make it more creative or attractive? Are the outlines of your duties crisp and vivid? What can you do to make your résumé stand out in a crowd?

appendix VI

# Creating A Persuasive Menu

Menus, and the research that goes into developing them, are an important part of a restaurant's success. The design and quality of the menu should reinforce the "brand identity" of the restaurant, the unique nature of the dining experience at this particular establishment, and this in turn can influence the customers' likelihood to return. Routine analysis of each menu item's performance allows management to identify items that are not turning a profit and should be deleted. Such analysis also identifies more promising items that should be moved to a more prominent location on the page. With its multi-purpose ends to inform, entertain, and persuade, the menu is a complex document requiring a number of different skills.

Menu pricing, for example, demands an understanding of recipe costing, analysis of competitors, seasonality, and psychology. Costing out every written recipe is the best way to ensure that you don't lose money needlessly—and with the average lifespan of a restaurant now well under one year, every dollar you can save will be important. Then look around at your competitors. What are they charging for an appetizer of quesadillas? Be sure to factor in the relative portion size and quality of the dish as well. Pay attention to the seasonal changes in the quality and cost of your ingredients. February in the American northeast is not the best time to highlight fresh peaches, for example, while in August you can have customers standing in line for dishes that feature the locally grown fruit. On the other hand, although apples are also out of season in February, they are tough enough to travel, so that a New Zealand variety can be as fresh and delicious in midwinter as a local variety is in October.

The psychology of menu pricing is especially intriguing, as again you attempt to understand your audience and use the most persuasive prices. For example, be sure to take into account that customers have certain expectations regarding pricing. They would be quite surprised if a humble dish of roast turkey with mashed potatoes and gravy cost more than a ten ounce filet mignon! In addition, retailers of all kinds know that $8.95 looks less expensive than $9.00. Finally, although in many menus the prices are set off in their own column or in a special typeface, consider whether prices that simply follow the menu description might keep your customers focused on the food rather than on their finances. Although you may not want the customers to focus on the cost, you yourself should regularly analyze the profit history of each menu item and consider deleting those that perform poorly.

The menu's presentation is also important as it in many ways embodies the particular dining experience you're offering. Like the experience itself, the menu should be entertaining. It should be attractive and reflect the same theme as the restaurant. It would be silly, for example, to use an inexpensive, informal menu design at a trendy, upscale bistro. Yet that inexpensive, informal menu might be just right for a small diner. Photographs of food items can be very persuasive. You do want to be certain, however, that your staff can routinely produce items as perfect in appearance as the glossy images on the menu; if not, customers may send the dish back to the kitchen! Another popular visual tool is a little icon beside particular menu items, like a heart that indicates "heart-friendly" dishes low in fat and sodium.

Some menu designers recommend using three colors: one for general categories such as Pasta or Dessert, a second for the majority of the menu descriptions, and a third to highlight one or two items in each category. Type font styles and sizes can used in the same way. In addition, pay attention to where individual items are placed. Those at the top of any category and those on the right-hand page tend to attract the most attention. Menu inserts and table tents are also effective in directing customers' notice to particular items. Be sure to use all of the available space, without crowding, of course. The "sundries" you list on the back page might add a nice chunk to the check!

In terms of the menu copy itself, use everything you know about writing to create vivid, tempting descriptions of each item. Be clear about the nature of your restaurant and the kinds of customers you hope to attract;

then write content that will appeal to your particular audience. Avoid using culinary jargon or foreign words, unless you translate them. Customers would rather not order an item at all than feel foolish asking you what it is! You should make each dish sound appealing, although you may want to spend more time on the especially profitable ones. Be concise, and especially try to avoid such wordy clichés as “accompanied by,” “served with,” and “atop a bed of.” Focus on specific nouns, sensory adjectives, and active verbs.

**EXERCISE V.1** **Writing Menu Descriptions**

Go back to the two restaurants from Chapter 1, Bistro Urbano and Downtown Diner. Write menu descriptions for each restaurant for the following items: a cup of soup, a ground beef dish, and a slice of chocolate cake. While using similar ingredients (two vegetable soups, for example), your descriptions should reflect the differences in audience and purpose between the two establishments.

Organize the menu into logical categories. Customers typically spend only three minutes reading it; be sure they can find what they are looking for—as well as the items you particularly want them to find! Finally, proofread your menu very, very carefully for spelling and grammar. The care with which you create the menu should reflect the same care with which you create the meal.

**EXERCISE V.2** **Restaurant Visit**

Visit a restaurant, and analyze the effectiveness of the menu in terms of the factors discussed above.

## RECIPE FOR REVIEW

1. Menu pricing requires an understanding of recipe costing, analysis of competitors, seasonality, and psychology.
2. The menu should be entertaining and attractive and should reflect the same theme as the restaurant.

3. Use colors, fonts, and layout to direct customers' notice to certain items.
4. Write vivid, appealing descriptions of each food item. Avoid culinary jargon and clichés.
5. Proofread your menu carefully! See Chapter 30.

*appendix* VI

# Annotated Research Paper

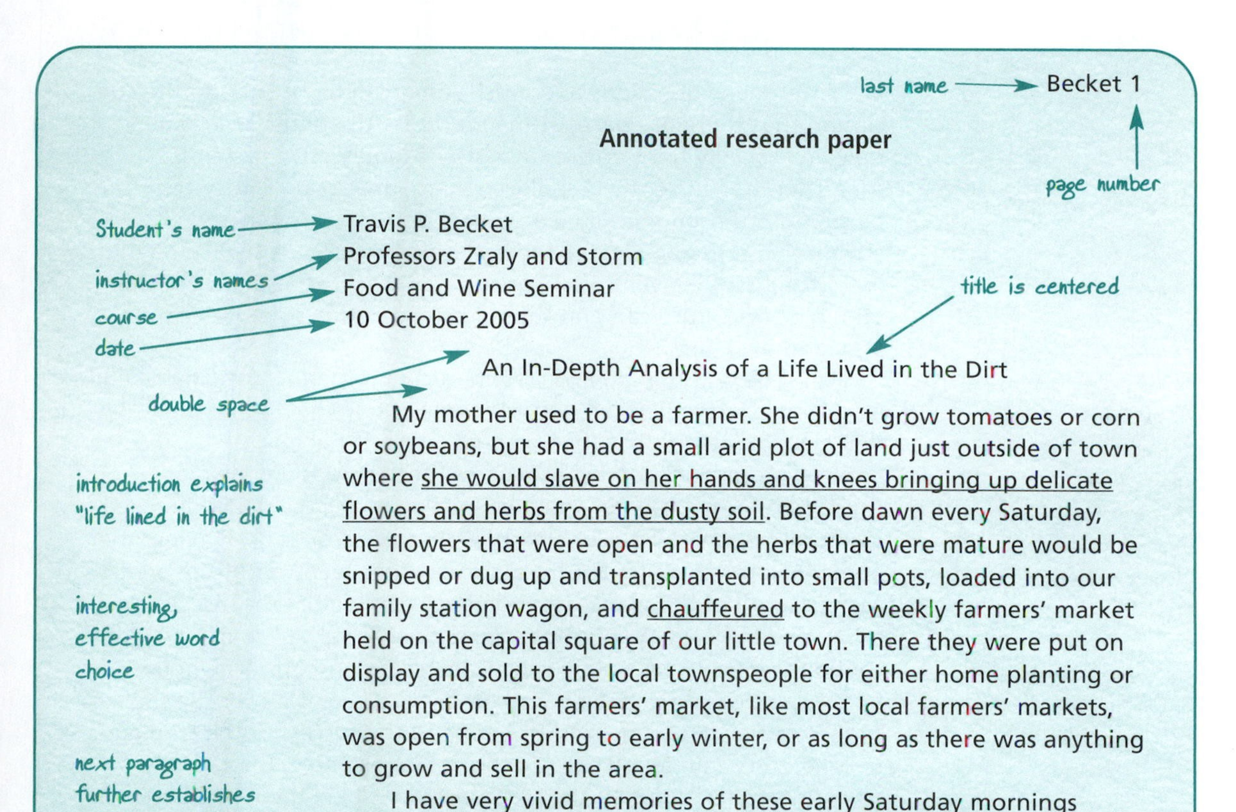

Becket 1

**Annotated research paper**

Travis P. Becket
Professors Zraly and Storm
Food and Wine Seminar
10 October 2005

An In-Depth Analysis of a Life Lived in the Dirt

My mother used to be a farmer. She didn't grow tomatoes or corn or soybeans, but she had a small arid plot of land just outside of town where she would slave on her hands and knees bringing up delicate flowers and herbs from the dusty soil. Before dawn every Saturday, the flowers that were open and the herbs that were mature would be snipped or dug up and transplanted into small pots, loaded into our family station wagon, and chauffeured to the weekly farmers' market held on the capital square of our little town. There they were put on display and sold to the local townspeople for either home planting or consumption. This farmers' market, like most local farmers' markets, was open from spring to early winter, or as long as there was anything to grow and sell in the area.

I have very vivid memories of these early Saturday mornings because I would participate in them with my mother. I remember anxious car trips in the pre-dawn light, hoping that the plants wouldn't

*(continues)*

Becket 2

tip over during the ride to the square and that once we arrived, the plants would be sold. I remember eating the herbs that weren't sold, along with vegetables that my mother traded for the plants that weren't sold. I remember eating items that had been submerged in the very unappetizing dirt of my mom's field just a few hours prior. I remember wondering if everybody else in America ate as well as I did.

*excellent transition*

I wondered this same idea last month at the University of California at Santa Cruz while eating a lunch that, once again, consisted of foods that had been submerged in dirt just a few hours prior. The apprentices who worked the fields agreed with me about the superior quality of the organic food they had grown. As I saw the way that the apprentices relied on organic foods and how they supported organic farming and eating, a new question arose. "If this is the way that food should be produced in America, is it possible to produce enough for everybody?" For the rest of my time at UCSC, this was the thought in my mind and the question on my lips. If organics is best for the soil and the body, can it be utilized at a level that will feed everybody?

*the research question*

After many questions and much research I have come to the conclusion that there are parts of the world that benefit environmentally and socially from environmentally-friendly (EFA) systems. America is not currently in a place where large-scale EFA systems can be supported economically, but with the proper growth and advancement in knowledge and technology, combined with the ultimate futility of conventional agricultural practices, organic and sustainable agriculture can and should be the nation's predominant agricultural system.

*answer to the question*

Let it be said that an EFA system, such as organics, sustainable and biodynamic farming, is defined generally as one that "relies on ecosystem management rather than external agricultural inputs" for the growth, protection, and overall success of the crops (Organic Agriculture). An example would be instead of using a pesticide to kill a natural insect predator of a crop, a different crop is grown near the primary crop, attracting a different insect that feeds off of the predator insect. All of the plants are indigenous, as well as the insects they bring, and the ecosystem is not affected negatively. In contrast, a conventional agricultural system is one that relies on "off-farm inputs" such as pesticides, herbicides, and other agro-chemicals for the success of the crop, with the main focus being on the success of the crop and not the ecosystem as a whole. In the earlier example, the farmers would spray pesticides on the crops in order to kill the insect pest (Organic Agriculture).

*definition of key term*

*parenthetical citation directs readers to works cited page*

*even short phrases are sometimes quoted*

*no page numbers available for this web site*

*period follows parentheses*

Becket 3

In general, the areas of the world that are currently best suited for EFA systems are underdeveloped or poor regions, or small self-sustained communities. This is because EFA systems tend to rely on the natural resources that are currently available, as opposed to expensive inputs such as chemicals and machinery.

*where EFA systems are currently successful*

Burkina Faso is in an area of sub-Sahara Africa where desertification and soil erosion is a major problem, stealing millions of hectares of arable land every year. Desertification is the spread of desert-like conditions to areas that were once arable, caused by overgrazing of livestock, improper irrigation, and planting too many crops in one area (World Factbook). A traditional method of sustainable farming was rediscovered in the area, and now the traditional method of "Zaï" is being applied to these damaged lands. According to Nicholas Parrott and Terry Marsden of Cardiff University, since this process was started in 1990, the food supply has become less vulnerable as the process of desertification is in reverse and the land is producing more (38). Zaï "involves making seed holes 20–30 cm wide and deep and using the earth to make a raised 'demi-lune' barrier on the downslope side. Compost and/or natural phosphate is placed in each hole and sorghum or millet seeds planted when it rains" (Parrott and Marsden 39). The process allows the seeds to grow in a reservoir of fertilized water in what would otherwise be barren land. As a result of using the Zaï method, more water seeps into and is retained in the earth because of its enhanced composition and proliferating insect population. To prevent the earth from wearing away, rocks are saved to build retaining walls around the fields (Parrott and Marsden 39).

*first example*

*tag line*

*a direct quote is an exact copy of the original's words and punctuation*

*a paraphrase restates information in your own words*

*both authors appear in parenthetical citation*

This form of EFA has had innumerable positive effects on the environment of Burkina Faso. Since the early 1990s, the return of over 100,000 hectares to arable condition has improved crop production by 35%. Further, production has stabilized between dry and wet years (Parrott and Marsden 39).

There have also been many social benefits in Burkina Faso because of Zaï. Like other organic farming methods, it requires manpower and thus has increased employment for men who might otherwise have moved from these rural areas to find jobs in the larger cities. This combined with the greater yield on crops and the increased return on sales has strengthened the food security for the people; they now have more reliable food sources year-round. Neighboring nations such as Niger, which was once known for abuse of agrochemicals, have adopted this traditional farming method and met with the same success, restoring 5,800 hectares of degraded land in recent years (Parrott and Marsden 39).

*effective transition*

Zaï has strong cultural importance as well. As the soil continues to improve through this ancient, environmentally-friendly method, both crop production and jobs have expanded while the movement of the population away from the countryside and toward the cities has

*each paragraph is documented, even when the source is identical*

*(continues)*

Becket 4

decelerated (Parrott and Madsen 39). The use of Zaï has helped to validate the importance of local knowledge and culture. Knowing that they relied on traditional practices and "indigenous knowledge" to succeed where technology was failing, the citizens of Burkina Faso now have the self-confidence to address other environmental and social issues; as a nation they can be more self-reliant. Because of these events Burkino Faso was the first African country to host the International Federation of Organic Agricultural Movements (IFOAM) International Scientific Conference, where it was stated that "organic agriculture in developing countries is not a luxury but a precondition for attaining food security" (qtd. in Parrott and Marsden 38).

*documenting an indirect source*

*transition to second example*

Another account of a beneficial agricultural transfer from conventional farming to EFA comes from The Maikaal Bio-Cotton Project, in Madhya Pradesh, India. Cotton is a crop that is especially prone to insect infestation, and the typical method of defense is the application of bountiful amounts of pesticides and insecticides. This was how farmers in Madhya Pradesh dealt with the whitefly pest, despite the evident health damage it was causing the ecosystem and the workers. Despite repeated applications of pesticides, the whitefly problem did not relent, as the fly had developed pesticide resistance to the specific pesticide used on the plants. The solution for most farmers was either to switch to a new pesticide or to a new cash crop. Many chose the latter due to "declining returns and toxicity problems" (Parrott and Marsden 24).

*explanation of EFA project*

Rather than completely shut down, in 1992 several cotton farmers joined together with their local spinning mill to start India's first certified organic agriculture project. One technique used was to determine a natural insect predator of the whitefly, and then to grow plants that attracted this natural predator to close proximity to the cotton. The natural predators consume the whitefly, reducing the insect problem. Other techniques such as crop-rotation with wheat, soybeans and chili, mating pattern disruption of the whitefly, and organic compost-spreading have been implemented. In seven years, the project has grown from several farms worth of land to almost 1000 farmers and over 15,000 acres of land. The project now uses many other biodynamic techniques that have been developed by organic chemists that work for the project (Parrott and Marsden 24-5).

*results of project*

As a whole, the farms involved in the project report a 20% higher yield of organic cotton, soybeans, wheat, and chili than other farms in the area that are not participating. The higher yield means higher income as well due to the crops' certified organic status. Composting has resulted in soil that retains moisture more efficiently and has cut down the amount of weeds that grow around the crops. In addition, "labour requirements are substantially reduced and production costs for organic cotton are 30–40% of those for conventional production" (Parrott and Marsden 25).

Becket 5

One final benefit of the Maikaal Bio-Cotton Project was that it opened India up to the relatively new organic cotton export market, where the majority of the demand came from industrialized nations like the USA. Since the cotton was certified organic, it carried a much higher price tag than conventional cotton and the farmers received a larger payment (Parrott and Marsden 25).

*effective transition*

By these two accounts I have strived to show that an organic or sustainable agriculture system can have manifold benefits on ecosystems that are unbalanced or altogether destroyed, and on the economy and culture of the region in which they are grown. Obviously different growing regions present different environmental and social challenges, but the essence of EFA is that it strengthens the relationship between the ecosystem and the economy. The economy of a region that practices EFA inherently relies on the overall stability and health of the ecosystem, and as the ecosystem detoxifies itself and returns to a more balanced state, it naturally produces a higher yield of more resilient crops, which then catch a higher market price (Posey).

*discussion of first two examples*

What about America and other industrialized nations that already have a strong conventional farming system built on efficiency and maximum output? In the short run, there is very little chance that organic agriculture can sustain the demand needed to feed everybody. The problem does not lie in whether or not an EFA system can produce the needed volume of food on the same amount of land that is currently being farmed conventionally, but in that the logistics of distribution make it economically unfeasible (Organic Agriculture).

*application to other nations*

The reason the price of organic food is often so much higher in stores is because EFA farms operate on much smaller scales of output, but they still pay the same price for storage and transportation that conventional farmers do. In some cases distribution costs are higher for organics, since by law they must still be segregated from conventional foods during transportation and holding. This increases the price of the food substantially, but since organic foods are still relatively limited in terms of supply, the demand raises the price even higher. Large farming organizations in America offset the incredibly high price of storing and shipping crops with the large volumes of crops that they handle. This is called economies of scale (Organic Agriculture).

*explanation of barriers to EFA in the United States*

Currently the most economically successful organic systems in America are small self-sustained communities, similar to the one I grew up in. This is because labor and transportation costs are virtually nonexistent due to the incredible small volume of production. The costs for labor, handling, and distribution stay within the economy, and they achieve efficiency because often times the transportation costs are no more than paying for gas in the family station wagon (Posey).

*a lecture is also cited*

*(continues)*

Becket 6

this is a term paper published on the Internet

this paragraph stresses the value of EFA and the rishs of conventional practices

tag line

Shapiro was quoted in another source

Also, entities that sell less than $5,000 a year or less in organic food do not have to be certified "Organic" by the USDA. This "flying beneath the radar" allows these local farmers to dramatically reduce their initial overheads (Massiello).

Although it is not economically feasible currently, a sustainable system of agriculture very well may be necessary for the continual production of food in this country. As the population of the nation and world grow, more stress is put on conventional forms for higher yields of crops. This stress often comes in the form of chemical fertilizers and other agrochemicals. According to biotechnology giant Monsanto CEO Robert Shapiro, the end result of these stressors along with conventional farming methods is "loss of topsoil, of salinity of soil as a result of irrigation, and ultimate reliance on petrochemicals" (qtd. in Vasilikiotis). Shapiro has also stated that "the commercial industrial technologies that are used in agriculture today to feed the world . . . are not inherently sustainable. They have not worked well to promote either self-sufficiency or food security in developing countries." Feeding the world with a continual reliable food source "is out of the question with current agricultural practice" (qtd. in Vasilikiotis). In the long run, conventional agricultural systems that are reliant on agrochemicals produce weak soil, weak plants, and the need for more agrochemicals to sustain yields.

transition to third example this one in the U.S.

abbreviation saves space

One last research project worth mentioning comes from the University of California-Davis, called the Sustainable Agriculture Farming Systems Project (SAFS). This eight-year project was conducted during the 1990's to compare "conventional farming systems with alternative production systems that promote sustainable agriculture" (Vasilikiotis). It measured the yields of two-year and four-year conventional farming systems, organic systems, and low-input systems. The crops grown were tomatoes, corn, safflower, and beans (SAFS).

results of the U.S. study

At the end of eight years, the conventional, organic, and low-input systems all had generally comparable yields. The yields of the organic system were lower for the first three years, but once the soil was detoxified of the agrochemicals used in previous years, the yields increased to the same level as those of the conventional system, and then ultimately surpassed the conventional system by the last years of the reports (the study is still active). All throughout this test period the nitrogen, organic carbon, and nutrient content levels of the organic and low-input systems increased, while the nutrient content levels in the conventional system decreased (Vasilikiotis).

results here suggest the following course of action

What this study shows is that although the costs to raise the organic crops were higher, they returned a higher yield than the conventional two- and four-year systems. There was also a visible increase in the soil quality as compared to the conventional system, and when the organic crops were sold at a premium price, there was a significantly higher gross return over the conventional system (SAFS). For the future of our nation's food source, we should be looking towards organics, sustainable agriculture, and other EFA systems. Despite the initial cost of production, they offer what appears to be the only viable option for reliably feeding all the people in the nation. In the event that society realizes its need for a self-sustainable agriculture system, organics could fill the total production level that would be left by conventional farming. If the nation were dedicated enough to invest in large-scale EFA production systems, economies of scale could be reached in terms of the volume of crops that would be transported at one time. In volume, shipping and distribution could be cost-effective. This would dramatically reduce the market price for organic foods, making them a practical option.

results here suggest the following course of action

overcoming barriers to EFA in the U.S.

The current place in our world for organic, sustainable, and biodynamic agriculture is very uncertain. It has been proven that they are successful in areas of the world that suffer from environmental pollution or degradation, and where conventional farming practices are no longer an option due to pests developing immunities to agrochemicals. EFA systems restore life to the land, and can increase the economical stability of poor regions by creating new jobs, increasing income levels, and helping to assure food security. Unfortunately, due to the prices of production and distribution compared to the volume being distributed, EFA systems are too expensive for American farmers and consumers to support on a large scale. Thus, the maximum benefit of many organic systems goes to the local community. Organics and other EFA systems must be taken seriously when it comes to long-term food production. They have proven to produce high yields and long-term stability.

beginning to wind down

summary of EFA benefits

summary of EFA barriers

It is often said that "United we stand, divided we fall." I believe that this old adage holds true for humanity's relationship with the environment. It was Kate Posey, the wistful tour guide at Santa Cruz, who said, "We see things so neatly arranged in grocery stores, sometimes we forget what the plant looks like." She was speaking of the disconnect between human beings and their food, how we as a society often forget that food comes not from a grocery store, but from being submerged in very unappetizing dirt. It is the connection with the dirt that makes us appreciate the taste of the food, and understand our place in the ecosystem. Organics brings society, economy, and ecology back into a state of interdependence and strength. But just as with soil quality, it will take a few years to detoxify the society and economy and to reach maximum yield.

Conclusion emphasizes and expands topic's significance

tag line

quotes add interest but don't replace author's words

looks bact to introduction and author's childhood

final statement of main idea

*(continues)*

Becket 8

page numbers continues

begins on new page

centered

Works Cited

double space

Massiello, Geneva. “Organic Food and Global Sustainability.” Term paper. University of Florida, 2002. 07 Oct. 2005 <http://plaza.ufl.edu/gmassie/termpaper.htm>.

Organic Agriculture at FAO. Food and Agriculture Organization of the United Nations. 07 Oct. 2005 <http://www.fao.org/organicag/ fram11-e.htm>.

list is alphabetized

Parrott, Nicholas, and Terry Madsen. The Real Green Revolution: Organic and Agroecological Farming in the South. London: Greenpeace Environmental Trust, 2002. 24-25, 37-39, 41. Greenpeace. 05 Oct. 2005 <http:// www.greenpeace.org.uk/ MultimediaFiles/Live/ FullReport/4526.pdf>.

cite online as well as print source

Posey, Kate. “Organic Nature.” Talk given at University of California Santa Cruz, Organic Garden, 14 Sept. 2005.

Sustainable Agriculture Farming Systems Project (SAFS). “Economic Viability of Organic and Low-Input Farming Systems.” Sustainable Agriculture Farming Systems Project Sep. 1997. University of California Davis. O7 Oct. 2005 <http://safs.ucdavis.edu/newsletter.econ.pdf>.

"hanging" paragraph format

Vasilikiotis, Christos, Ph.D. “Can Organic Farming Feed the World?” November 2000. University of California, Berkeley. 07 Oct. 2005 <http://www.cnr.berkeley.edu/ ~christos/articles/ cv_organic_farming.html>.

punctuation follows MLA style

World Factbook. 2005. Central Intelligence Agency. 07 Oct. 2005 <http://www.cia.gov/cia/ publications/factbook/docs/ notesanddefs.html>.

break URLs at slash only

include date of access for electronic sources

appendix VII

# A Taste for Reading

## KITCHEN CONFIDENTIAL: ADVENTURES IN THE CULINARY UNDERBELLY* *by Anthony Bourdain*

*Born in 1956 in New York City, Anthony Bourdain graduated from the Culinary Institute of America and worked as a professional chef for twenty years before writing the bestselling exposé of the restaurant industry* Kitchen Confidential. *His travels around the world were detailed in* A Cook's Tour, *also a Food Network series, and he is a frequent contributor to* eGullet.com. *Bourdain lives in New York City, where he is executive chef at Brasserie Les Halles.*

My first indication that food was something other than a substance one stuffed in one's face when hungry—like filling up at a gas station—came after fourth grade in elementary school. It was on a family vacation to Europe, on the *Queen Mary*, in the cabin-class dining room. There's a picture somewhere: my mother in her Jackie O sunglasses, my younger brother and I in our painfully cute cruisewear, boarding the big Cunard ocean liner, all of us excited about our first transatlantic voyage, our first trip to my father's ancestral homeland, France.

It was the soup.

It was *cold*.

*Reprinted by permission of Bloomsbury USA.

This was something of a discovery for a curious fourth-grader whose entire experience of soup to this point had consisted of Campbell's cream of tomato and chicken noodle. I'd eaten in restaurants before, sure, but this was the first food I really noticed. It was the first food I enjoyed and, more important, remembered enjoying. I asked our patient British waiter what this delightfully cool, tasty liquid was.

"Vichyssoise," came the reply, a word that to this day—even though it's now a tired old warhorse of a menu selection and one I've prepared thousands of times—still has a magical ring to it. I remember everything about the experience: the way our waiter ladled it from a silver tureen into my bowl; the crunch of tiny chopped chives he spooned on as garnish; the rich, creamy taste of leek and potato; the pleasurable shock, the surprise that it was cold.

I don't remember much else about the passage across the Atlantic. I saw *Boeing Boeing* with Jerry Lewis and Tony Curtis in the *Queen*'s movie theater, and a Bardot flick. The old liner shuddered and groaned and vibrated terribly the whole way—barnacles on the hull was the official explanation—and from New York to Cherbourg, it was like riding atop a giant lawnmower. My brother and I quickly became bored and spent much of our time in the "Teen Lounge," listening to "House of the Rising Sun" on the jukebox, or watching the water slosh around like a contained tidal wave in the below-deck saltwater pool.

But that cold soup stayed with me. It resonated, waking me up, making me aware of my tongue and, in some way, preparing me for future events.

## **TORTILLAS*** *by José Antonio Burciaga*

*José Antonio Burciaga was born in El Paso, Texas, in 1940. After serving in the United States Air Force, Burciaga worked as a graphic illustrator and in 1974 began writing newspaper articles, short stories, and poems. "Tortillas" was published in 1988. Burciaga was also a painter, particularly of mural art, including the "Last Supper of Chicano Heroes" at Casa Zapata, a dormitory at Stanford University in which half the students were Chicano. As both an artist and an activist, Burciaga exposed discrimination and worked for social justice until his death in 1996.*

My earliest memory of *tortillas* is my *Mamá* telling me not to play with them. I had bitten eyeholes in one and was wearing it as a mask at the dinner table.

As a child, I also used *tortillas* as hand warmers on cold days, and my family claims that I owe my career as an artist to my early experiments with *tortillas*. According to them, my clowning around helped me develop a strong artistic foundation. I'm not so sure, though. Sometimes I wore a *tortilla* on my head, like a *yarmulke*, and yet I never had any great urge to convert from Catholicism to Judaism. But who knows? They may be right.

For Mexicans over the centuries, the *tortilla* has served as the spoon and the fork, the plate and the napkin. *Tortillas* originated before the Mayan civilizations, perhaps predating Europe's wheat bread. According to Mayan mythology, the great god Quetzalcoatl, realizing that the red ants knew the secret of using maize as food, transformed himself into a black ant, infiltrated the colony of red ants, and absconded with a grain of corn. (Is it any wonder that to this day, black ants and red ants do not get along?) Quetzalcoatl then put maize on the lips of the first man and woman, Oxomoco and Cipactonal, so that they would become strong. Maize festivals are still celebrated by many Indian cultures of the Americas.

"In the *mercado* where my mother shopped . . ."

*Courtesy of University of Texas–Pan American Press.

When I was growing up in El Paso, *tortillas* were part of my daily life. I used to visit a *tortilla* factory in an ancient adobe building near the open *mercado* in Ciudad Juárez. As I approached, I could hear the rhythmic slapping of the *masa* as the skilled vendors outside the factory formed it into balls and patted them into perfectly round corn cakes between the palms of their hands. The wonderful aroma and the speed with which the women counted so many dozens of *tortillas* out of warm wicker baskets still linger in my mind. Watching them at work convinced me that the most handsome and *deliciosas tortillas* are handmade. Although machines are faster, they can never adequately replace generation-to-generation experience. There's no place in the factory assembly line for the tender slaps that give each *tortilla* character. The best thing that can be said about mass-producing *tortillas* is that it makes it possible for many people to enjoy them.

In the *mercado* where my mother shopped, we frequently bought *taquitos de nopalitos,* small tacos filled with diced cactus, onions, tomatoes, and *jalapeños.* Our friend Don Toribio showed us how to make delicious, crunchy *taquitos* with dried, salted pumpkin seeds. When you had no money for the filling, a poor man's *taco* could be made by placing a warm *tortilla* on the left palm, applying a sprinkle of salt, then rolling the *tortilla* up quickly with the fingertips of the right hand. My own kids put peanut and jelly on *tortillas,* which I think is truly bicultural. And speaking of fast foods for kids, nothing beats a *quesadilla,* a *tortilla* grilled-cheese sandwich.

Depending on what you intend to use them for, *tortillas* may be made in various ways. Even a run-of-the-mill *tortilla* is more than a flat corn cake. A skillfully cooked homemade *tortilla* has a bottom and a top; the top skin forms a pocket in which you put the filling that folds your *tortilla* into a taco. Paper-thin *tortillas* are used specifically for *flautas,* a type of taco that is filled, rolled, and then fried until crisp. The name *flauta* means *flute,* which probably refers to the Mayan bamboo flute; however, the only sound that comes from an edible *flauta* is a delicious crunch that is music to the palate. In México *flautas* are sometimes made as long as two feet and then cut into manageable segments. The opposite of *flautas* is *gorditas,* meaning *little fat ones.* These are very thick small *tortillas.*

The versatility of *tortillas* and corn does not end here. Besides being tasty and nourishing, they have spiritual and artistic qualities as well. The Tarahumara Indians of Chihuahua, for example, concocted a corn-based beer called *tesgüino,* which their descendants still make today. And everyone has read about the woman in New Mexico who was cooking her husband a *tortilla* one morning when the image of Jesus Christ miraculously appeared on it. Before they knew what was happening, the man's breakfast had become a local shrine.

Then there is *tortilla* art. Various Chicano artists throughout the Southwest have, when short of materials or just in a whimsical mood, used a dry *tortilla* as a small, round canvas. And a few years back, at the height of the Chicano movement, a priest in Arizona got into trouble with the Church after he was discovered celebrating mass using a *tortilla* as the host. All of which only goes to show that while the *tortilla* may be a lowly corn cake, when the necessity arises, it can reach unexpected distinction.

## CUTTING GREENS* *by Lucille Clifton*

*Born in 1936, Lucille Clifton was raised in Depew, New York, and was employed by state and federal agencies, including the Office of Education in Washington, D.C., until 1971. Since then she has held positions as a writer in residence at various universities, including Columbia University School of the Arts, George Washington University, and St. Mary's College of Maryland, and published a dozen books of poetry, two of which were finalists for the Pulitzer Prize in 1988. Her poems look at objects, events, people—even her own body—in an attempt to dig into deeper political issues, questions of identity. Lucille Clifton has also written over twenty children's books. "cutting greens" was published in 1987.*

curling them around

i hold their bodies in obscene embrace

thinking of everything but kinship.

collards and kale

strain against each strange other

away from my kissmaking hand and

the iron bedpot.

the pot is black,

the cutting board is black,

my hand,

and just for a minute

the greens roll black under the knife,

and the kitchen twists dark on its spine

and i taste in my natural appetite

the bond of live things everywhere.

---

*Lucille Clifton, "cutting greens" from *Good Woman Poems and a Memoir 1969–1980.* 

# A STUDENT RESPONDS TO "CUTTING GREENS" IN A JOURNAL ENTRY

**STUDENT WRITING** Alicia Lacey

The poem "cutting greens" by Lucille Clifton is about the feeling the author gets when she cuts greens. Typically collards and kale are mentioned in this poem. Honestly, I like the poem. It is just something I can relate to. Greens is always something I can look forward to eating during the holidays. One thing that I disliked about the poem was that it was too short. I needed more information. There was a lot of visual imagery in the poem for such little information. The one question I have about the poem is "What did the author mean when she was talking about her hand?" She described a black cutting board, a black knife, and her hand. My assumption is that she was making a statement that she is African-American.

Collard greens are a tradition in my family. The history of this type of vegetable goes back in time since slavery. When the slaves didn't have too much to eat, they ate whatever was edible. Collard greens is a very bitter food; it takes a lot of seasoning for it to become a delicious entrée or side dish. My mother wouldn't let me cook collards because she thought that I couldn't season them right. One day when I was cooking dinner, I snuck some frozen greens in a pot to cook. My mother came into the kitchen and saw the box on the counter. She asked me why I was cooking the greens, when she told me not to. I told her that I wanted to make them and I knew she would like them. Sure enough, she tasted my greens, they were good she said, but they weren't better than hers. I can agree with her on that.

## **BORDERLAND*** *by M. F. K. Fisher*

*Mary Frances Kennedy Fisher (1908–1992) published over twenty books that intertwine culinary topics with those of history and culture and with events of her own life. "Borderland," for example, which appeared in her first book,* Serve It Forth, *in 1937, explores the author's "secret eatings" within the context of her life in France with her first husband, Alfred Fisher. The soldiers who are mentioned briefly in this essay bring to mind the city's uneasy position on the border with Germany, just prior to its occupation by Hitler during World War II.*

Almost every person has something secret he likes to eat. He is downright furtive about it usually, or mentions it only in a kind of conscious self-amusement, as one who admits too quickly, "It is rather strange, yes—and I'll laugh with you."

Do you remember how Claudine used to crouch by the fire, turning a hatpin just fast enough to keep the toasting nubbin of chocolate from dripping off? Sometimes she did it on a hairpin over a candle. But candles have a fat taste that would taint the burnt chocolate, so clean and blunt and hot. It would be like drinking a Martini from silver.

Hard bitter chocolate is best, in a lump not bigger than a big raisin. It matters very little about the shape, for if you're nimble enough you'll keep it rolling hot on the pin, as shapely as an opium bead.

When it is round and bubbling and giving out a dark blue smell, it is done. Then, without some blowing all about, you'll burn your tongue. But it is delicious.

However, it is not my secret delight. Mine seems to me less decadent than Claudine's, somehow. Perhaps I am mistaken. I remember that Al looked at me very strangely when he first saw the little sections lying on the radiator.

That February in Strasbourg was too cold for us. Out on the Boulevard de l'Orangerie, in a cramped dirty apartment across from the sad zoo half full of animals and birds frozen too stiff even to make smells, we grew quite morbid.

---

*M. F. K. Fisher. "Borderland." *The Art of Eating.* 

Finally we counted all our money, decided we could not possibly afford to move, and next day went bag and baggage to the most expensive *pension* in the city.

It was wonderful—big room, windows, clean white billows of curtain, central heating. We basked like lizards. Finally Al went back to work, but I could not bear to walk into the bitter blowing streets from our warm room.

It was then that I discovered how to eat little dried sections of tangerine. My pleasure in them is subtle and voluptuous and quite inexplicable. I can only write how they are prepared.

In the morning, in the soft sultry chamber, sit in the window peeling tangerines, three or four. Peel them gently; do not bruise them, as you watch soldiers pour past and past the corner and over the canal towards the watched Rhine. Separate each plump little pregnant crescent. If you find the Kiss, the secret section, save it for Al.

Listen to the chambermaid thumping up the pillows, and murmur encouragement to her thick Alsatian tales of *l'intérieure.* That is Paris, the interior, Paris or anywhere west of Strasbourg or maybe the Vosges. While she mutters of seduction and French bicyclists who ride more than wheels, tear delicately from the soft pile of sections each velvet string. You know those white pulpy strings that hold tangerines into their skins? Tear them off. Be careful.

Take yesterday's paper (when we were in Strasbourg *L'Ami du Peuple* was best, because when it got hot the ink stayed on it) and spread it on top of the radiator. The maid has gone, of course—it might be hard to ignore her belligerent Alsatian glare of astonishment.

After you have put the pieces of tangerine on the paper on the hot radiator, it is best to forget about them. Al comes home, you go to a long noon dinner in the brown dining-room, afterwards maybe you have a little nip of *quetsch* from the bottle on the *armoire.* Finally he goes. Of course you are sorry, but—

On the radiator the sections of tangerines have grown even plumper, hot and full. You carry them to the window, pull it open, and leave them for a few minutes on the packed snow of the sill. They are ready.

All afternoon you can sit, then, looking down on the corner. Afternoon papers are delivered to the kiosk. Children come home from school just as three lovely whores mince smartly into the *pension's* chic tearoom. A basketful of Dutch tulips stations itself by the tram-stop, ready to tempt tired clerks at six o'clock. Finally the soldiers stump back from the Rhine. It is dark.

The sections of tangerine are gone, and I cannot tell you why they are so magical. Perhaps it is that little shell, thin as one layer of enamel on a Chinese bowl, that crackles so tinily, so ultimately under your teeth. Or the rush of cold pulp just after it. Or the perfume. I cannot tell.

There must be some one, though, who knows what I mean. Probably everyone does, because of his own secret eatings.

## BLACKBERRY EATING* *by Galway Kinnell*

*Galway Kinnell was born in Rhode Island in 1927. After his graduation from Princeton University and extensive travels in Europe and the Middle East, he returned to the United States and became involved with the Civil Rights Movement and antiwar protests. One of the most influential poets writing in America today, Kinnell is on a quest for spirituality, whether through an important event like 9/11 or through ordinary moments like hearing the footsteps of his child or eating blackberries.*

I love to go out in late September

among the fat, overripe, icy, black blackberries

to eat blackberries for breakfast,

the stalks very prickly, a penalty

they earn for knowing the black art

of blackberry-making: and as I stand among them

lifting the stalks to my mouth, the ripest berries

fall almost unbidden to my tongue,

as words sometimes do, certain peculiar words

like *strengths* or *squinched*,

many-lettered, one-syllabled lumps,

which I squeeze, squinch open, and splurge well

in the silent, startled, icy, black language

of blackberry-eating in late September. [1980]

*"Blackberry Eating" from *Mortal Acts, Mortal Words* by Galway Kinnell. 

## EPITAPH FOR A PEACH: FOUR SEASONS ON MY FAMILY FARM* *by David Mas Masumoto*

*David Mas Masumoto is a third generation farmer, growing organic peaches, nectarines, grapes, and raisins near Del Rey, California.* Epitaph for a Peach *chronicles a year in his life in which he must decide whether or not to continue growing the Sun Crest, a high-quality variety that has had little commercial success. In language as fragrant and delicate as one of his own peaches, Masumoto explores his relationship with his heritage and his family, as well as with the land.*

### The Furin

A small *furin* hangs on our farmhouse porch. Its miniature bell delicately jingles with the slightest breeze. A long strip of paper captures the air currents and translates the movement into sound. I can peer out over the fields, watching the advancing spring season with its green blankets of foliage, and hear the wind.

Nikoko likes the fragile sounds. The metal chime rings like a whisper, the voice tiny like a child's. Occasional spring winds in the valley blow strong enough to snap the outstretched vine canes. Most of the time soft breezes brush our cheeks with such subtleties that we ignore their presence. A *furin* reminds grown-ups what children already sense. Niki says she hears the wind singing.

I spend the spring battling nature, trying to farm differently, hoping somehow I am contributing to the quest to save my peach. The more I struggle, the more the burden seems to weigh. Each new approach generates more questions; the complexity of working with nature slips into a growing pattern of chaos.

I remember a Japanese saying about the power of bamboo. Its strength is not found in a rigid structure that blocks the wind; instead, the stalks bend with the wind. Their power resides in their very flexibility. I'm working on becoming like bamboo. I've abandoned my attempts to control and compete with nature, but letting go has been a challenge.

---

I'm trying to listen to my farm. Before, I had no reason to hear the sounds of nature. The sole strategy of conventional farming seems to be dominance. Now, with each passing week, I venture into fields full of life and change, clinging to a belief in my work and a hope that it's working.

As I recall the past spring from my porch, the ringing of the *furin* helps me understand as it flutters in a subtle breeze. For the first time in my life, I see the wind.

## MONSOON DIARY: A MEMOIR WITH RECIPES* *by Shoba Narayan*

*In* Monsoon Diary: A Memoir with Recipes, *Shoba Narayan tells stories of her family and her childhood in South India and shares recipes like this one for "Soft Idlis." Ms. Narayan graduated from the Columbia Graduate School of Journalism and later won the M. F. K. Fisher Award for Distinguished Writing. She lives in New York City.*

### Soft Idlis

My grandfather fell in love with my grandmother over *idlis*. As a child bride, Nalla-ma was put to work on the granite grinding stone (*aatukal* in Tamil). She was twelve and spent her morning turning the stone to make *idli* batter. Enter my grandfather, a strapping lad of twenty-two. Desperate to ease the burden of his beautiful bride yet fearful of being taunted as a henpecked husband if caught beside her, he came up with an ingenious solution. He donned a sari, covered his head like any dutiful daughter-in-law, sat down beside my grandmother, and turned the heavy granite stone himself. They gazed into each other's eyes, didn't say a word, and together made the fluffiest *idli* batter imaginable.

Good *idlis* are soothing and filling. The trick is in the batter's proportion and consistency. The *urad* dal makes its soft, while the rice flour gives it heft. The batter has to be thick enough to hold its shape, yet thin enough to ferment. After many questions and experiments, I came upon the perfect *idli* recipe. Here it is.

SERVES 4

*1 cup* urad *dal*

*1 teaspoon fenugreek seeds*

*2 cups cream of rice (available in Indian stores and also called* idli rawa*; this is not cream of rice cereal, which is cooked)*

*2 teaspoons salt*

*1 teaspoon plain yogurt, lowfat or nonfat*

---

*From *Monsoon Diary* by Shoba Narayan. 

1. Combine the dal with fenugreek in a bowl and cover with enough warm water to cover by one inch. Soak for 2 hours. In a separate bowl, mix the cream of rice with just enough water to form a paste.

2. Grind the dal and fenugreek seeds, using a grinder, adding a little water if necessary, so that it is the consistency of cake batter. Add the salt and yogurt. Put the cream of rice mixture into a grinder and blend it well with the dal batter until it's the consistency of honey—viscous, not too thin but not too thick either.

   Since the batter will ferment and rise to about twice its initial volume, pour it into a bowl large enough to accommodate this. Cover tightly and keep it in a warm place. Do not look at it for the next twenty-four hours.

3. When you uncover the bowl, the batter should have fermented and risen to about twice its height. There should be bubbles on top and a salty, fermented smell. Beat the dough, using a long spoon. Ladle it out into an *idli* stand, available in most Indian stores. Put the stand into a steamer, stock pot, or pressure cooker, and steam it (without pressure) for 10 minutes.

4. To test if done, stick a knife into the *idli*. When you lift it out, there should be no batter sticking to its sides. Eat with coconut chutney and onion *sambar*.

## FAST FOOD NATION: THE DARK SIDE OF THE ALL-AMERICAN MEAL* *by Eric Schlosser*

*Eric Schlosser was born in New York City in 1959 and attended Princeton and Oxford Universities. Originally intending to pursue a career as a playwright, he became a successful journalist and author, writing first for* Atlantic Monthly. *Schlosser lives in California with his wife, Shauna Redford, daughter of actor/director Robert Redford.* Fast Food Nation, *a biting exposé of the fast food industry, is his most well-known book.*

Over the last three decades, fast food has infiltrated every nook and cranny of American society. An industry that began with a handful of modest hot dog and hamburger stands in southern California has spread to every corner of the nation, selling a broad range of foods wherever paying customers may be found. Fast food is now served at restaurants and drive-throughs, at stadiums, airports, zoos, high schools, elementary schools, and universities, on cruise ships, trains, and airplanes, at K-Marts, Wal-Marts, gas stations, and even at hospital cafeterias. In 1970, Americans spent about $6 billion on fast food; in 2001, they spent more than $110 billion. Americans now spend more money on fast food than on higher education, personal computers, computer software, or new cars. They spend more on fast food than on movies, books, magazines, newspapers, videos, and recorded music—combined.

Pull open the glass door, feel the rush of cool air, walk in, get on line, study the backlit color photographs above the counter, place your order, hand over a few dollars, watch teenagers in uniforms pushing various buttons, and moments later take hold of a plastic tray full of food wrapped in colored paper and cardboard. The whole experience of buying fast food has become so routine, so thoroughly unexceptional and mundane, that it is now taken for granted, like brushing your teeth or stopping for a red light. It has become a social custom as American as a small, rectangular, hand-held, frozen, and reheated apple pie.

This is a book about fast food, the values it embodies, and the world it has made. Fast food has proven to be a revolutionary force in

---

*Excerpt from *Fast Food Nation: The Dark Side of the All-American Meal* by Eric Schlosser. 

American life; I am interested in it both as a commodity and as a metaphor. What people eat (or don't eat) has always been determined by a complex interplay of social, economic, and technological forces. The early Roman Republic was fed by its citizen-farmers; the Roman Empire, by its slaves. A nation's diet can be more revealing than its art or literature. On any given day in the United States about one-quarter of the adult population visits a fast food restaurant. During a relatively brief period of time, the fast food industry has helped to transform not only the American diet, but also our landscape, economy, workforce, and popular culture. Fast food and its consequences have become inescapable, regardless of whether you eat it twice a day, try to avoid it, or have never taken a single bite.

The extraordinary growth of the fast food industry has been driven by fundamental changes in American society. Adjusted for inflation, the hourly wage of the average U.S. worker peaked in 1973 and then steadily declined for the next twenty-five years. During that period, women entered the workforce in record numbers, often motivated less by a feminist perspective than by a need to pay the bills. In 1975, about one-third of American mothers with young children worked outside the home; today almost two-thirds of such mothers are employed. As the sociologists Cameron Lynne Macdonald and Carmen Sirianni have noted, the entry of so many women into the workforce has greatly increased demand for the types of services that housewives traditionally perform: cooking, cleaning, and child care. A generation ago, three-quarters of the money used to buy food in the United States was spent to prepare meals at home. Today about half of the money used to buy food is spent at restaurants—mainly at fast food restaurants.

The McDonald's Corporation has become a powerful symbol of America's service economy, which is now responsible for 90 percent of the country's new jobs. In 1968, McDonald's operated about one thousand restaurants. Today it has about thirty thousand restaurants worldwide and opens almost two thousand new ones each year. An estimated one out of every eight workers in the United States has at some point been employed by McDonald's. The company annually hires about one million people, more than any other American organization, public or private. McDonald's is the nation's largest purchaser of beef, pork, and potatoes—and the second largest purchaser of chicken. The McDonald's Corporation is the largest owner of retail property in the world. Indeed, the company earns the majority of its profits not from selling food but from collecting rent. McDonald's spends more money on advertising and marketing than any other brand. As a result it has replaced Coca-Cola as the world's most famous brand. McDonald's operates more playgrounds than any other private entity in the United States. It is responsible for the nation's bestselling line of children's clothing (McKids) and is one of the largest distributors of toys. A survey of American schoolchildren found that 96 percent could identify Ronald McDonald. The only fictional character with a higher degree of recognition was Santa Claus. The impact of McDonald's on the way we live today is hard to overstate. The Golden Arches are now more widely recognized than the Christian cross.

In the early 1970s, the farm activist Jim Hightower warned of "the McDonaldization of America." He viewed the emerging fast food industry as a threat to independent businesses, as a step toward a food economy dominated by giant corporations, and as a homogenizing influence on American life. In *Eat Your Heart Out* (1975), he argued that "bigger is *not* better." Much of what Hightower feared has come to pass. The centralized purchasing decisions of the large restaurant chains and their demand for standardized products have given a handful of corporations an unprecedented degree of power over the nation's food supply. Moreover, the tremendous success of the fast food industry has encouraged other industries to adopt similar business methods. The basic thinking behind fast food has become the operating system of today's retail economy, wiping out small businesses, obliterating regional differences, and spreading identical stores throughout the country like a self-replicating code.

America's main streets and malls now boast the same Pizza Huts and Taco Bells, Gaps and Banana Republics, Starbucks and Jiffy-Lubes, Foot Lockers, Snip N' Clips, Sunglass Huts, and Hobbytown USAs. Almost every facet of American life has now been franchised or chained. From the maternity ward at a Columbia/HCA hospital to an embalming room owned by Service Corporation International—"the world's largest provider of death care services," based in Houston, Texas, which since 1968 has grown to include 3,823 funeral homes, 523 cemeteries, and 198 crematoriums, and which today handles the final remains of one out of every nine Americans—a person can now go from the cradle to the grave without spending a nickel at an independently owned business.

## **HOT DOG*** *by Jeffrey Steingarten*

*Twenty years after graduating from Harvard Law School, Jeffrey Steingarten wrote his first article about food for* Vogue Magazine. The Man Who Ate Everything, *his first collection of essays published in 1997, was a* New York Times *best-seller and won the Julia Child Book Award. In 1994 he was created Chevalier in the Order of Merit for his writing on the culinary art of France.*

Everybody's in a panic these days about the dangers of eating raw shellfish. But I have a plan. I've decided to give up skiing this winter so that I can eat my fill. By my calculations, the chance of suffering a substantial injury in one day of skiing is ten times worse than the chance of getting sick from eating a plate of cold, plump, briny, succulent raw oysters or clams. It follows that if I give up ten days of skiing, I can feast on oysters twice a week for the entire year.

To be perfectly truthful, I've never skied a day in my life or eaten less than my fill of anything. My plan was born at supper with an unfortunate friend, fresh from the slopes and hospitals of Aspen, where he had broken his shoulder by crashing into a shrub on a downhill run. He wore a brace on his upper torso and needed help turning the pages of his menu. I immediately recovered from excessive feelings of sympathy as I watched my friend choose his food with a superstitious adherence to every modern nutritional fad and rumor he had ever heard. For the life of me, I cannot understand why some people are eager to take on all sorts of dangers and then go paranoid over a much less risky endeavor—especially when that endeavor is dinner.

We *do* have ample cause to be worried about seafood safety. An investigation in the February 1992 issue of *Consumer Reports* found that a full 44 percent of the seafood its staff purchased at supermarkets and fish stores contained unacceptable levels of fecal coliform bacteria, which can cause all sorts of gastrointestinal illnesses. The federal government has shirked its duty to ensure the safety of our seafood, and proposals are now before Congress to remedy the situation.

Most bacteria and viruses are destroyed by cooking, which is why the federal Food and Drug Administration recommends that fish be

---

cooked to 145 degrees Fahrenheit or until it flakes easily at the center near the bone; oysters and clams should be boiled for four to six minutes. These are reliable recipes for cataclysmically overcooked seafood.

Raw shellfish is where most of the danger lurks. In 1991 the FDA conducted a risk assessment of fish and shellfish in cooperation with the Centers for Disease Control and discovered that, when raw or partially cooked mollusks (mussels, clams, and oysters) are excluded, only one illness results from every two million servings of seafood. This is an extremely low number compared to the danger of eating chicken, with one illness in every twenty-five thousand servings.

But when raw or partially cooked shellfish is added in, the risk jumps eightfold. Raw clams, oysters, and mussels account for 85 percent of all seafood-borne illnesses. One in every two thousand servings of raw mollusks is likely to make somebody ill.

As high as this number seems, it means that if you eat a plate of raw oysters every week, you will get sick once in forty years or twice in a full and happy lifetime. And you can reduce the risk further by avoiding the main threat—raw mollusks taken from March to October in the Gulf of Mexico, when they are likely to be infected with *Vibrio vulnificus.* The warmer the water and the higher the temperature at which oysters are shipped and stored, the greater the danger. This is the principal rationale for the old rule of thumb that oysters should be eaten only in months whose name contains an *r,* because these are the cold-weather months from September through April. (A second reason is epicurean: oysters spawn in warmer weather, depleting their tasty glycogen and losing their succulence.) These days, Gulf oysters are safe, if at all, only from November through February.

For the very young, the very old, and people with weakened immune systems, including those who are HIV-positive, an infection by *Vibrio vulnificus* from a contaminated oyster can lead to death. But for most diners, the worst outcome is a day or two of unpleasant and unsightly gastrointestinal distress.

If raw shellfish makes you sick once in every two thousand servings, how does this compare to the hazards of going skiing? The statistics are elusive—the skiing industry does not encourage the collection and publication of data. But there seems to be general agreement that a substantial injury occurs once in every 250 days of skiing or, at the least, once in 400. These include leg fractures, spine fractures, contusions, lacerations, and knee injuries. A study in Munich found at least one minor injury in every 59 days of skiing and a really serious disaster in every 500; it defined "serious" as meaning that the skier would be off the slopes for at least 3 days. My last bad oyster kept me from table for only one. And most accident surveys leave out gondola crashes, skiers' smashing into each other in the subarctic cold; the danger of radiation (the yearly risk of cancer from cosmic rays is two-thirds greater at the altitude of Denver than at sea level, where oysters live); and injuries that blossom after the skier returns home, like the newly popular sprain of the ulnar collateral ligament of the metacarpophalangeal joint of the thumb. To say that a day of skiing is ten times more dangerous than a delicious plate of oysters is, I think, an act of generosity to the sport and its hapless participants.

Ski apologists point out that skiers suffer fewer *fatalities* than swimmers, cyclists, or equestrians, and that skiing is, on an hourly basis, no more dangerous than junior-high-school football. This may be a welcome consolation prize to the skiing industry, but it is even better news to me. It means that if I am willing to give up junior-high-school football this fall, I can happily devour all the sushi, sashimi, and ceviche that my heart desires.

October 1992

# Glossary of Terms

**access, date of**—in the list of works cited, the date of access is the day, month, and year on which you last viewed an electronic source

**acronymn**—a word constructed from the first letters of terms we are trying to learn; for example, *WOFT* in Chapter 18 takes its *W* from *with*, its *O* from *of*, and so on

**action verb**—a verb that tells what the subject of a sentence is doing; see also *verb*

**active voice**—see *voice*

**address**—in a letter, the name of the person and/or business that you are writing to, followed by the street address, city, state, and zip code

**adjective**—a word that describes or modifies a noun

**adverb**—a word that describes or modifies a verb, adjective, or another adverb; see also *conjunctive adverb*

**alternating format**—in a compare and contrast essay, a type of organization in which you focus on the first characteristic of the two items, then the second characteristic, and so on; also called *point-by-point format*

**antecedent**—a word that comes before another word, as in a pronoun must agree with its antecedent

**APA style**—the format for documentation developed by the American Psychological Association

**apostrophe**—the raised "comma" that is used to form possessives and contractions

**appositive**—a word or phrase that sits next to a noun and renames it

**article**—the parts of speech *a*, *an*, and *the; a* and *an* are *indefinite articles; the* is a *definite article*

**audience**—those who receive a communication, for example, through reading an essay, listening to a speech, or watching a movie

**auxiliary verb**—see *helping verb*

**base form**—with verbs, the word that follows *to* in an infinitive phrase; also called the *stem*

**block format**—(1) in a compare and contrast essay, a type of organization in which you write all about the first item, then all about the second; (2) the layout of a business letter in which every line is flush with the left margin

**body**—the middle section of an essay, containing specific information that develops the main idea of the piece; or, the content of a letter between the greeting and the closing

**"box out"**—in this textbook, to put brackets around a word or phrase, especially prepositional phrases

**brackets**—a mark of punctuation used in pairs [ ] within quotations to set off words, phrases, or explanations that were not in the original text or to set off explanatory words and phrases within parentheses

**brainstorm**—to generate ideas and information through free writing, making lists or charts, talking to others, or doing research or experiments

**capitalization**—using a capital or uppercase letter at the beginning of a sentence or proper noun

**case**—(1) the name given to different forms of nouns and pronouns; the *subjective case* is used for the subject of a sentence (*Ivan* or *he*), the *objective case* for a direct or indirect object and with prepositions (*Ivan* or *him*), and the *possessive case* for modifiers (*Ivan's* or *his*); (2) one of two types of letters, *uppercase* (capital letters) and *lowercase*

**categorical order**—organization according to category or type

**causal chain**—a sequence in which one consequence is the cause of another, which in turn causes another consequence, and so on

**cause and effect**—a way of developing ideas by analyzing the causes and/or effects of an event or condition. An *immediate cause* occurs a short time before the event; a *remote cause* is farther away in time. A *main cause* is the most important or most powerful cause; a *contributing cause* is a less important or less powerful cause.

**Chicago style**—a documentation format laid out in *The Chicago Manual of Style*

**chronological order**—according to the sequence of events

**citation**—the facts that describe a source used in your work, including author, title, and publishing information

**cite**—to use and document an outside source in your work

**clause**—a group of words that contains a subject and a verb; clauses may be *independent* or *main* and stand alone as a complete sentence; or they may be *dependent, subordinate,* or *relative* and unable to stand alone

**cliché**—a phrase that has been used so often that it has lost both precision and interest, for example, *hungry as a horse*

**closing**—the line that ends a letter, for example, *Sincerely yours,* plus your name and title

**"coat hanger"**—in this textbook, a subordinating conjunction

**collective noun**—a word that names a group with several members but is usually treated as singular, for example, *team*

**colon**—the punctuation mark that looks like two periods, one above the other; used to introduce a list or quotation, or to follow the greeting in a business letter, as well as in expressions of time and in certain parts of bibliographic citations

**comma**—a punctuation mark formed like a curved hand and used to separate meaningful groups of words in a sentence

**comma splice**—two independent clauses joined by a comma only

**common noun**—a noun that refers to a category rather than to a specific individual

**communication**—any of several ways of sharing information, feelings, and ideas; *verbal communication* uses words, as in speaking or writing; *nonverbal communication* does not use words, as in gestures or facial expressions

**compare and contrast**—to develop an idea by looking for similarities and differences

**comparative form**—the comparative of an adjective or adverb is formed by adding the suffix *–er* or the word *more* (*tastier, more delicious*) and used to judge two items against each other

**complete predicate**—the part of the sentence that is not the *complete subject*; see *predicate*

**complete subject**—see *subject*

**complete thought**—one of the three elements of a sentence, in addition to subject and verb, and not to be confused with context or information; a complete thought suggests an independent structure

**compound adjectives**—adjectives that are modified by an adverb and that directly precede a noun

**compound number**—a number spelled with two words connected by a hyphen (*twenty-one*)

**compound subject**—two or more subjects that share the same verb

**compound verb**—two or more verbs that share the same subject

**conclusion**—the end of an essay, which may summarize its main points, state or restate the main idea, and provide the reader with a sense of closure

**conjunction**—a word that joins two or more words, phrases, or clauses; *coordinating conjunctions* (*for, and, nor, but, or, yet, so*) join equal elements; *subordinating conjunctions* (*because, although, while*) join a dependent to an independent clause

**conjunctive adverb**—a word or phrase that modifies an entire clause and suggests how its idea is related to that of another word or clause

**connotation**—the feelings or associations that move beyond a simple definition and make a word seem positive or negative, or simply neutral; see *denotation*

**contraction**—a combination of two words in which missing letters are indicated by an *apostrophe*

**coordinating conjunction**—see *conjunction*

**count nouns**—generally refer to concrete things and can be counted; see *noncount nouns*

**cut and paste**—(1) editing functions on the word processor; (2) to physically separate, reorganize, and tape together sentences and paragraphs from a paper copy of an essay

**dangling modifier**—an error in which a *modifier*, often an initial participial phrase, does not actually refer to a specific word in the sentence

**dash**—a punctuation mark that is similar to but longer than a hyphen and used variously like a comma, semicolon, or colon

**definite article**—*the*; see also *article*

**demonstrative pronoun**—one of a group of words (*this, these, that, those*) that directs the reader's attention to particular nouns, for example, *this page*

**denotation**—the basic definition of a word; see *connotation*

**dependent clause**—see *clause*

**description**—writing that "paints a picture," often using sensory details

**direct address**—speaking to the audience directly, as in "Joey, do your homework"

**direct object**—the *direct object* of a verb receives the action of the verb; compare *indirect object*

**direct quote**—an exact copy of the words and punctuation of the original text

**direct speech**—words quoted exactly as they were spoken

**directions**—instructions for performing a task, using an appliance, and so on

**draft**—(1) to write; (2) a version of an essay, for example, *rough draft* or *final draft*

**edit**—to correct the grammar, word usage, and punctuation of a text

**edition**—a specific version or printing of a book; second and later editions should be specified when documenting a source

**editor**—one who collects various essays or poems into a book and/or prepares a manuscript

**emphatic order**—organization according to the importance of the ideas or the emphasis you want to give them

**end marks**—punctuation marks that indicate the end of a complete sentence, including the *period, question mark*, and *exclamation point*

**essay**—a series of paragraphs that develop a single main idea

**euphemism**—a watered-down term used supposedly to spare the reader's feelings when handling such uncomfortable topics as death or sex

**exclamation point**—a punctuation mark that adds emphasis to a sentence or indicates a command

**exemplification**—the process of developing an idea through examples

**"FANBOYS"**—a common acronym for the seven coordinating conjunctions, *for, and, nor, but, or, yet,* and *so*

**figurative language**—a group of methods (for example, *simile* or *metaphor*) by which one item is compared with another to create images or to explore ideas; a *figure of speech* refers to a single method

**first person**—*I* (singular), *we* (plural)

**format**—the style or rules governing the layout of a paper or letter and the documentation of outside sources

**future tense**—formed with *will* + the base verb, it describes an action or condition that is expected but has yet to occur

**future perfect tense**—see *perfect tenses*

**future progressive** tense—see *progressive tenses*

**generic noun**—a noun that names the typical member of a group, such as *the average student*

**gerund**—the base form of the verb + *-ing*; used as a noun

**heading**—in a letter, your name and address and the date; on an essay, your name, the instructor's name, the course title, and the date

**helping verb**—in a verb phrase, one or more words that indicate the mood, tense, or voice of the main verb; also called *auxiliary verb*; see also *modal*

**hyphen**—the short line used within a word to connect compound numbers or adjectives

**imperative mood**—a sentence in this mood is a command, such as *Please sit down*; see also *mood*

**indefinite article**—*a* and *an*; see also *article*

**indefinite pronoun**—a pronoun that refers to a general rather than a specific person, place, or thing (*everyone*); often, it suggests an amount (*some*)

**independent clause**—see *clause*

**indicative mood**—a sentence in this mood is a simple statement or question; see also *mood*

**indirect object**—the person or thing *to* which or *for* which the action is performed; compare *direct object*

**indirect question**—a statement that contains the content but not the exact words of a question, for example, *He asked me what the homework was.* The exact words would be enclosed in quotation marks, as in *"What is the homework?" he asked.*

**indirect speech**—a statement that contains the content but not the exact words of a quotation, for example, *She mentioned that she has a sister.* The exact words would be enclosed in quotation marks, as in *"I have a sister," she said.*

**infinitive**—*to* + base form of verb, such as *to cook, to write,* or *to be*

**informative writing**—writing whose purpose is largely to communicate information rather than to persuade or entertain

**intensive pronouns**—pronouns that are formed by adding *-self* to the personal pronouns (*yourself*) and used to add emphasis

**interjection**—an exclamatory word such as *Hey!* that can be added to or deleted from a sentence without changing the sentence's structure

**interrogative pronoun**—a pronoun such as *who* or *what* used to ask a question

**introduction**—the beginning of an essay; an introduction should catch the reader's attention, state the topic of the paper, and perhaps include the main point

**italics**—the slanted type used for the titles of books, periodicals, movies, television series, works of art, and ships; for words in a foreign language; and when referring to a letter or word

**jargon**—the technical vocabulary specific to particular jobs, professions, or specialties

**linking verb**—a word that "links" or joins the subject to the rest of the sentence; linking verbs include forms of *to be, to appear, to seem*

**main clause**—see *clause*

**main idea**—the essential point you want to communicate, often expressed in a single sentence

**main verb**—the last word in a verb phrase, which reflects the meaning of the verb rather than properties of tense or mood

**metaphor**—a figure of speech in which one thing is said to be another; for example, *The kitchen was a cardboard box inferno.*

**misplaced modifier**—a word or phrase that is out of position relative to the word it describes

**MLA style**—a documentation format prescribed by the Modern Language Association

**modal**—a subset of *helping verbs,* including *can, could, may, might, must, shall, should, will,* and *would,* used with a main verb to express tense or mood

**modifier**—a word, phrase, or clause that describes, explains, or limits another word, phrase, or clause; see also *dangling modifier* and *misplaced modifier*

**mood**—a property of verbs including the *indicative mood,* which simply tells or asks without suggesting any hidden meaning; the *imperative mood,* which is used to give orders or commands; and the *subjunctive mood,* which is used mostly in formal situations to talk about a wish to make a statement that is not factual

**narration**—the process of telling a story

**narrator**—the character who is telling the story; often the narrator is an authorial figure outside the story itself

**noncount nouns**—include things that cannot be counted and do not have a plural form; see *count nouns*

**nonrestrictive clause**—see *restrictive clause*

**nonverbal communication**—communication without words, for example, gestures or facial expressions; compare *verbal communication*

**noun**—a word that names something—a person, place, thing, or idea; a *common noun* may refer to categories of persons, places, and things; a *proper noun* refers to a specific person, place, or thing and is always capitalized; a *generic noun* names the typical member of a group; see also *collective noun*

**noun fragment**—a group of words that renames or describes a noun but does not form a complete sentence

**number**—a property of subjects and verbs, which are either *singular* (one) or *plural* (more than one)

**object**—the *object of a preposition* is the noun or pronoun that follows it to form a prepositional phrase; the *direct object* of a verb receives the action of the verb; the person or thing *to which* or *for which* the action is performed is the *indirect object*

**objective case**—a noun or pronoun form used for the objects of verbs or preposition; see *case*

**outline**—(1) to put the main (and supporting) ideas of an essay in order; (2) a list or chart of the main (and supporting) ideas of an essay

**paragraph**—a group of sentences that develops a single main idea; a mini-essay

**paraphrase**—a restatement or translation of the original text; note that while the words and sentence structure must be different from the original, you must be certain that the paraphrase contains the same information and point of view; no quotation marks are used with a paraphrase

**parallelism**—when comparable ideas are expressed in the same grammatical forms

**parentheses**—a punctuation mark ( ) used to set off words and phrases that explain or refer to something within the main sentence

**parenthetical citation**—information about an outside source that is enclosed in parentheses and that leads the reader to the full citation on the Works Cited page

**participle**—a *prinicipal part* of the verb; the *present participle* adds *-ing* to the verb base; the *past participle* of regular verbs adds *-d* or *-ed*

**passive voice**—see *voice*

**past tense**—formed by adding *-d* or *-ed* [*cooked* or *basted*] to the verb base, it describes an action that occurred in the past or a condition that existed in the past; see also *tense*

**past perfect tense**—see *perfect tenses*

**past progressive**—see *progressive tenses*

**perfect tenses**—the *past perfect* indicates that one action in the past was completed before another; the *present perfect* indicates an action that occurred or a condition that existed at some indefinite time in the past; the *future perfect* describes an action that will be completed before another time in the future; see also *tense*

**period**—a punctuation mark that indicates the end of a sentence

**periodical**—a magazine, journal, or newspaper published at regular intervals

**person**—a characteristic of pronouns; see *first person, second person, third person*

**personal pronoun**—a word that refers to a specific person, place, thing, or idea: *I, we, you, he, she, it, they*

**persuasive writing**—writing that seeks to compel a response from the audience, whether it is sympathy, agreement, or action

**phrase**—a group of words that functions like a single word

**plural**—more than one, as in the *plural form* of a noun

**point-by-point format**—see *alternating format*

**point of view**—the narrator's position with regard to the story, generally first person (*I*) or third person (*they*)

**positive form**—the positive of an adjective or adverb is its most simple form, which describes a noun without comparing it to another one (*tasty, delicious*)

**possessive case**—a noun form that indicates possession, usually by adding *'s*; see *case*

**possessive pronoun**—the possessive form of a pronoun, such as *my, our, your, his, her, its, their*

**predicate**—the part of the sentence that is not the subject; the *simple predicate* is the verb

**prefix**—a word or syllable attached to the beginning of a word to make a new word

**preposition**—a word that shows the "position" of one noun in relation to another

**prepositional phrase**—a group of words that begins with a preposition and ends with a noun or pronoun

**present tense**—describes an action that is happening now, in the present; see also *tense*

**present perfect tense**—see *perfect tenses*

**present progressive** tense—see *progressive tenses*

**principal parts**—see *verb*

**process analysis**—an explanation of how a process is performed; a type of organization according to the sequence of steps in a process

**process narrative**—the story of how a process was performed

**progressive tenses**—indicates that an action was, is, or will be continuing; the *past progressive* is formed with *was/were* + present participle and indicates a continuous action in the past; the *present progressive* is formed with *is/are* + present participle and indicates the action is continuing in the present; the *future progressive* is formed with *will be* + present participle and indicates an action that begins in the future and continues for an indefinite length of time

**pronoun**—a word used in place of a noun; see also *personal pronoun, possessive pronoun, relative pronoun, reflexive pronoun, demonstrative pronoun, interrogative pronoun*

**proofreading**—the final check of a piece of writing for spelling and punctuation errors

**propaganda**—written or spoken material that seeks to compel a response from the audience through manipulation rather than open debate

**proper noun**—see *noun*

**punctuation marks**—symbols such as *periods, commas,* or *parentheses* that guide the reader's understanding of the structure and meaning of a sentence; see Chapters 27 to 29

**purpose**—the reason for communication, for example, to inform, entertain, and/or persuade

**quarterly**—a periodical published four times per year

**question mark**—the symbol that ends an interrogative sentence

**quotation marks**—symbols used to set off the titles of songs, poems, short stories, individual episodes in a television series, articles, and chapters

**reflexive pronoun**—a type of pronoun formed by adding *-self* to a pronoun (*myself, yourself*) and used when the subject of the sentence is also the object

**refutation**—an outline of some of the arguments *on the opposite side* and an explanation of why these arguments are unreasonable, unethical, or otherwise less persuasive than your own arguments

**relative clause**—see *clause*

**relative clause fragment**—a sentence fragment containing a relative pronoun (*who, which, that*)

**relative pronoun**—a word that "relates" to another noun and connects it to a dependent or relative clause; relative pronouns include *who, which,* and *that*

**research**—gathering information or evidence in order to answer a question

**restrictive clause**—a clause that limits the meaning of a particular word or is otherwise essential to the meaning of a sentence; a *nonrestrictive* clause is *not* essential to the meaning of a word or sentence

**revision**—the process of reevaluating and rewriting a piece of writing; literally *re-seeing*

**rhetorical modes**—ways of developing an idea and/or organizing an essay; rhetorical modes include narration, description, exemplification, compare and contrast, process analysis, and cause and effect

**rough draft**—the first version of an essay, also called the first draft

**run-on sentence**—two independent clauses joined without an appropriate conjunction or punctuation

**salutation**—the greeting in a letter

**second person**—*you* (singular and plural)

**semicolon**—a punctuation mark that looks like a comma with a period on top; used to join related independent clauses

**sensory details**—information that comes from the five senses (sight, smell, taste, touch, hearing)

**sentence fragment**—an incomplete sentence, missing a subject, a verb, and/or an independent structure or "complete thought"

**setting**—where and when a story takes place

**sexist language**—expressions that inappropriately specify one gender when both should be included, for example, *he* used to mean "people in general"

**signature block**—at the end of a business e-mail, the author's full name and title, department, company, street address, telephone and fax numbers, and Web site, if applicable

**simile**—a figure of speech in which one thing is said to be *like* another

**simple predicate**—see *predicate*

**simple subject**—see *subject*

**singular**—referring to one rather than many, as in a singular subject

**slang**—words or phrases that have become popular within a certain group of people but may not be recognized by a general audience

**spatial order**—according to the physical layout

**speaking**—a form of *verbal communication*; compare *nonverbal communication*

**specific details**—examples, descriptions, or factual information that develops, explains, or illustrates an idea

**stem**—see *base form*

**subject**—the word or group of words that is performing the action of the sentence or that is being described by the rest of the sentence; the *simple subject* consists of one or more nouns or pronouns (or phrases acting like nouns); the *complete subject* includes all the words that modify or describe the subject

**subjective case**—see *case*

**subjunctive mood**—used mostly in formal situations to talk about a wish or to make a statement that is not factual; see also *mood*

**subordinate clause**—see *clause*

**subordinate clause fragment**—a type of incomplete sentence that consists of a subordinating conjunction plus a subject and verb

**suffix**—a word or syllable attached to the end of a word to make a new word

**summary**—a condensed statement of a text's main idea(s) and key supporting points

**superlative form**—the superlative form of an adjective or adverb is formed by adding the suffix *-est* or the word *most* and is used to compare *more than two* items (*tastiest, most delicious*)

**supporting points**—information that explains, illustrates, or develops the main idea

**tag line**—a phrase that introduces outside material, often by naming the author and/or title of the source

**tense**—a characteristic of verbs that indicates the time that an action was performed or that a condition existed; see also *present tense, past tense, perfect tenses, progressive tenses*

**thesis statement**—the main idea of an essay; the topic of the paper plus the point the author hopes to make about that topic

**third person**—*he, she, it* (singular); *they* (plural)

**topic sentence**—the main idea of a paragraph

**transition/transitional expression**—a word or phrase that shows the connection between ideas; transitions can move the reader from one part of a sentence to the next, from one sentence to the next, or from one paragraph to the next

**URL**—an acronym for *Uniform Resource Locator* or Internet address

**verb**—an essential ingredient of a sentence, it is a word or phrase that tells what the subject of the sentence is doing or connects the subject with some information later in the sentence; *regular verbs* follow the same general rules as they change form, while *irregular verbs* have different forms that must be memorized; the *principal parts* of a verb include the infinitive or base form, the past tense, and the past (and sometimes the present) participle; see also *action verb, helping verb, linking verb*

**verb phrase**—a group of words that contains one or more helping verbs and a main verb

**verbal**—a word that looks like a verb but is used in a different way, for example, an infinitive phrase or a gerund

**verbal communication**—communication through words, for example, speaking or writing; compare *nonverbal communication*

**voice**—a characteristic of verbs; in the *active voice,* the subject of the sentence is performing the action of the verb, while in the *passive voice,* the subject of the sentence is *receiving* that action

**"WOFT?"**—in this textbook, an acronym used to remember the four common prepositions *with, of, for,* and *to*

**Works Cited page**—a list in MLA style of all the sources cited in your paper

**Works Consulted page**—a list of sources *read,* not all of which are necessarily *cited*

**writing**—a form of *verbal communication*

# Notes

[1]"Teens," *Studs Terkel: Conversations with America,* 2002, Chicago Historical Society, 11 Mar. 2006 <http://www.studsterkel.org/radio.php?gallery=sub—Teens>.

[2]For example, Peter Elbow's *Writing with Power: Techniques for Mastering the Writing Process* and *Writing without Teachers,* both reprinted by Oxford University Press in 1998.

[3]*Finding Forrester,* written by Mike Rich, dir. Gus Van Sant, perf. Sean Connery, Rob Brown, F. Murray Abraham, Columbia Pictures, 2000. Courtesy of Sony Pictures Entertainment.

[4]Quotations are from *Finding Forrester.* Courtesy of Sony Pictures Entertainment.

[5]Donald M. Murray, "The Maker's Eye: Revising Your Own Manuscripts," *The Writer* Oct. 1973, rptd. in *Language Awareness: Readings for College Writers,* ed. Paul Eschholz, Alfred Rosa, and Virginia Clark, 8th ed. (Boston: Bedford/St. Martin's, 2000) 161–65.

[6]For a fuller discussion of the dual nature of an essay's "shape," see Peter Elbow's "The Music of Form: Rethinking Organization in Writing," *College Composition and Communication* 57 (2006): 620–66.

[7]Screenplay by Robert Nelson Jacobs, based on the novel by E. Annie Proulx, dir. Lasse Hallström, perf. Kevin Spacey, Judi Dench, Julianne Moore, Miramax, 2001. Text from *The Shipping News* courtesy of Miramax Film Corp.

[8]*Parents for Megan's Law,* Megan's Law and Childhood Sexual Abuse Prevention Clearinghouse, 13 Jan. 2006 <http://www.parentsformeganslaw.com>.

[9]See Lynne Truss, *Eats, Shoots & Leaves* (New York: Gotham Books, 2003) back cover.

[10]Alan Dundes, "Seeing is Believing," *Interpreting Folklore* (Bloomington: Indiana UP, 1980).

[11]Monica Bhide, "A Question of Taste," *Best Food Writing 2005*, ed. Holly Hughes (New York: Marlowe, 2005) 103–104. Courtesy of Monica Bhide.

[12]Adapted from a classroom activity conducted by David Bourns at Oakwood School, Poughkeepsie, New York, 1971.

[13]For example, see Ann C. Noble, *The Wine Aroma Wheel*, 1990, 13 Aug. 2006 <http://www.winearomawheel.com>.

[14]For a fuller discussion of an essay's organization in both time and space, see Peter Elbow, "The Music of Form."

[15]Text from *The Shipping News* courtesy of Miramax Film Corp.

[16]Sharon Zraly, conversations at the Culinary Institute of America, Hyde Park, New York, October 2005.

[17]Garry Marshall, quoted in *Music from the Movies*, 22 Jan. 2006 <http://www.musicfromthemovies.com/feature.asp?ID=29>.

[18]"American Diner Slang," ed. Bernadette Lynn, *h2g2* 19 Mar. 2003, BBC, 24 Jan. 2006 <http://www.bbc.co.uk/dna/h2g2/alabaster/A890589>.

[19]For example, see "Jargon Buster: Winter Sports," *Virgin.net*, 19 Feb. 2006 <http://www.virgin.net/wintersports/jargonbuster.html> for the terms in the first sentence; "Wine Terms," *Tasting Wine*, 19 Feb. 2006 <http://www.tasting-wine.com/html/wine-terms.html> for the fourth; and "Glossary of Internet & Web Jargon," *UC Berkeley – Teaching Library Internet Workshops*, University of California at Berkeley, 10 Mar. 2006 <http://www.lib.berkeley.edu/TeachingLib/Guides/Internet/Glossary.html> for the fifth.

[20]This text follows *The New Food Lover's Companion: Comprehensive Definitions of Nearly 6,000 Food, Drink, and Culinary Terms* by Sharon Tyler Herbst (Hauppauge, NY: Barron's Cooking Guide, 2001) in spelling *hollandaise* with a small *h*. However, many writers and chefs, like Robert Hannon in this excerpt, spell *Hollandaise* with a capital *H*.

[21]Anthony Pratkanis and Elliot Aronson, *Age of Propaganda: The Everyday Use and Abuse of Persuasion* (New York: W. H. Freeman, 1991), qtd. in Aaron Delwiche, "Introduction: Why Think about Propaganda?" *Propaganda Critic*, 29 Sep. 2002, 16 Feb. 2006 <http://www.propagandacritic.com/articles/intro.why.html>.

[22]Delwiche, Aaron. "Introduction: Why Think about Propaganda?" *Propaganda Critic*, 29 Sep. 2002, 16 Feb. 2006 <http://www.propagandacritic.com/articles/intro.why.html>.

[23]Center for Science in the Public Interest (CSPI), "Fact Sheet: Lowering the Drinking Age Is a Bad Idea," *Alcohol Policies Project: Advocacy for the Prevention of Alcohol Problems,* Center for Science in the Public Interest, February 1998, 20 Sep. 2006 <http://www.cspinet.org/booze/mlpafact.htm>.

[24]"Talking Points/Arguments: Answering the Critics of Age-21," *Alcohol Policies Project: Advocacy for the Prevention of Alcohol Problems,* Center for Science in the Public Interest, February 1998, 20 Sep. 2006 <http://www.cspinet.org/booze/mlpatalk.htm>.

[25]Bonnie Liebman, "Designed to Sell," *Nutrition Action Healthletter,* October 2006: 8–9.

[26]Text from *The Shipping News* courtesy of Miramax Film Corp.

[27]Nicholas Parrott and Terry Marsden, *The Real Green Revolution: Organic and Agroecological Farming in the South* (London: Greenpeace Environmental Trust, 2002) 39, 05 Oct. 2005 <http://www.greenpeace.org.uk/MultimediaFiles/Live/FullReport/4526. pdf>. Courtesy of Greenpeace.

[28]Parrott and Marsden 39.

[29]Parrott and Marsden 39.

[30]Christos Vasilikiotis, "Can Organic Farming 'Feed the World'?" November 2000, University of California, Berkeley, 07 Oct. 2005 <http://www.cnr.berkeley.edu/~christos/articles/cv_organic_farming.html>. Courtesy of Christos Vasilikiotis.

[31]Quotes courtesy of Monica Bhide.

[32]Thanks to Sharon Zraly for introducing me to Comma Man.

[33]Elizabeth Dye, "Straight Out of Brooklyn," *Willamette Week Online* 16 Jun. 2004, 14 Dec. 2005 <http://www.wweek.com/editorial/3033/5198>.

[34]While *judgement* is an alternative spelling, *judgment* is generally preferred.

[35]This text follows *The New Food Lover's Companion: Comprehensive Definitions of Nearly 6,000 Food, Drink, and Culinary Terms* by Sharon Tyler Herbst (Hauppauge, NY: Barron's Cooking Guide, 2001) in spelling *hollandaise* with a small *h.* However, many writers and chefs (like Robert Hannon in Chapter 12) spell *Hollandaise* with a capital *H.*

# Works Consulted

## GENERAL*


Bauman, M. Garrett. *Ideas and Details: A Guide to College Writing.* 5th ed. Boston: Heinle, 2004.

Hacker, Diana. *The Bedford Handbook.* 6th ed. Instructor's Annotated Edition. Boston: Bedford/St. Martin's, 2002.

Kelly, William J., and Deborah L. Lawton. *Odyssey: A Guide to Better Writing.* 2nd ed. Boston: Allyn & Bacon, 2000.

Kirszner, Laurie G., and Stephen R. Mandell. *Patterns in College Writing: A Rhetorical Reader and Guide.* 7th ed. New York: St. Martin's, 1998.

Murray, Donald M. "The Maker's Eye: Revising Your Own Manuscripts." First published in *The Writer* Oct. 1973; rptd. in *Language Awareness: Readings for College Writers.* Ed. Paul Escholz, Alfred Rosa, and Virginia Clark. 8th ed. Boston: Bedford/St. Martin's, 2000, 161–65.

Warriner, John E., and Francis Griffith. *English Grammar and Composition.* Rev. ed. New York: Harcourt, Brace & World, 1965.

Yarber, Mary Laine, and Robert E. Yarber. *Reviewing Basic Grammar: A Guide to Writing Sentences and Paragraphs.* 5th ed. New York: Longman, 2001.


*In acknowledging works consulted for this project, special note must be made of the important influence of this first group of publications.

## CHAPTERS 1–17

"American Diner Slang." Ed. Bernadette Lynn. *h2g2* 19 Mar. 2003. British Broadcasting Corp. 24 Jan. 2006 <http://www.bbc.co.uk/dna/h2g2/alabaster/A890589>.

"American Slang." *h2g2* 28 Jul.1999. BBC. 24 Jan. 2006 <http://www.bbc.co.uk/dna/h2g2/alabaster/A128143>.

Beckson, Karl, and Arthur Ganz. *A Reader's Guide to Literary Terms: A Dictionary.* New York: Farrar, Straus and Giroux, 1960.

Bhide, Monica. "A Question of Taste." *Best Food Writing 2005.* Ed. Holly Hughes. New York: Marlowe, 2005. 102–104. First published in *The Washington Post* on February 21, 2005.

Bourns, David. Lemon exercise adapted from a classroom activity conducted at Oakwood School, Poughkeepsie, New York, 1971.

Clines, Raymond H., and Elizabeth R. Cobb. *Research Writing Simplified: A Documentation Guide.* 3rd ed. Boston: Addison Wesley Longman, 2000.

"A Cost of Corporate Jargon?" The Public Relations Society of America. *Strategist,* Spring 2003: 3. Qtd. in Douglas J. Swanson. Home Page. University of Wisconsin–LaCrosse. 18 Feb. 2006 <http://www.uwlax.edu/faculty/swanson/poorcommunication.htm>.

"A Cup of Joe." *h2g2* 10 Oct. 2003. BBC. 24 Jan. 2006 <http://www. bbc.co.uk/dna/h2g2/alabaster/A1300410>.

Delwiche, Aaron. *Propaganda Critic.* 29 Sep. 2002. 16 Feb 2006 <http://www.propagandacritic.com>.

Devine, Rachel. "Talk the Talk." *iVillage: The Website for Women.* iVillage Limited. 18 Feb. 2006 <http://www.ivillage.co.uk/workcareer/survive/opolitics/articles/0,,156475_164240,00.html>.

Elbow, Peter. "The Music of Form: Rethinking Organization in Writing." *College Composition and Communication* 57.4 (2006): 620–66.

Elbow, Peter. *Writing with Power: Techniques for Mastering the Writing Process.* New York: Oxford UP, 1981.

Eldred, Tony. "Let's Make It Hard for People to Buy Stuff." *Hospitality Management Knowledge Library.* 2006. Eldred Hospitality Pty Ltd. 24 Jan. 2006 <http://www.eldtrain.com.au/members/library/ sales19. htm>.

Elmore, Sam. "A Blank Page." *Get The News.net.* 1994. 16 Nov. 2005 <http://getthenews.net/Sam/ablankpage.htm>.

Foote, Jon. "Jargon." *E-mail Humor.* 2002. Planet Footey. 24 Jan. 2006 <http:bellsouthpwp.net/j/o/jonfoote/emailhumor/larry36.html>.

Friedman, Steven Morgan. "Incomprehensible Business Jargon." *WestEgg.com.* 10 Mar. 2006 <http://www.westegg.com/jargon>.

Gibaldi, Joseph. *MLA Handbook for Writers of Research Papers*. 6th ed. New York: MLA, 2003.

"Glossary of Internet & Web Jargon." *UC Berkeley – Teaching Library Internet Workshops*. University of California at Berkeley. 10 Mar. 2006 <http://www.lib.berkeley.edu/TeachingLib/Guides/Internet/Glossary.html>.

Hult, Christine A. *Researching and Writing across the Curriculum*. Needham Heights, MA; Boston: Allyn & Bacon, 1996.

Jacob, Dianne. *Will Write for Food: The Complete Guide to Writing Cookbooks, Restaurant Reviews, Articles, Memoir, Fiction and More* . . . . New York: Marlowe, 2005.

"Jargon Buster: Winter Sports." *Virgin.net*. 19 Feb. 2006 <http://www.virgin.net/wintersports/jargonbuster.html>.

Kemmer, Suzanne. "Modern Usage of English: Medical Jargon." Home Page. 2003. Rice University. 10 Mar. 2006 <http://www.ruf.rice. edu/~kemmer/Words04/usage/jargon_medical.html>.

List, Carla. *An Introduction to Information Research*. Dubuque, Iowa: Kendall/Hunt, 1998.

Maddox, Garry. "How Experts Get Movie Titles Wrong." *The Sydney Morning Herald Blogs: Entertainment*. 12 Oct. 2005. *The Sydney Morning Herald*. 22 Jan. 2006. <http://blogs.smh.com.au/entertainment/archives/box_office/002552.html>.

Noble, Ann C. *The Wine Aroma Wheel*.1990. 13 Aug. 2006 <http://www.winearomawheel.com>.

Peha, Steve. *Teaching That Makes Sense*. 2003. Teaching That Makes Sense, Inc. 10 May 2006 <http://www.ttms.org>.

Pinker, Steven. "Chasing the Jargon Jitters." *TIME* 13 Nov. 1995. 24 Jan. 2006 <http://pinker.wjh.harvard.edu/articles/media/1995_ 11_13_time.html>.

Rader, Walter. *Online Slang Dictionary*. Home Page. 2003. University of California at Berkeley. 24 Jan. 2006. <http://ocf.berkeley.edu/~wrader/slang/b.html>.

Rubino, Robert. "Let's Stop the Sports/War Jargon Swap." *Northern California Aces*. 3 Nov. 2003. Northern California ACES: A Chapter of the American Copy Editors Society. 19 Feb. 2006 <http://www.norcalaces.org/archives/2005/11/lets_stop_the_s.php>.

Schrobsdorff, Susanna. "Attack of the Weasel Words." *Newsweek*. 13 Jul. 2005. MSNBC.com. 18 Feb. 2006 <http://www.msnbc.msn.com/id/8514826>.

*Stone Writing Center @ Del Mar College*. Del Mar College. 21 Feb. 2006 <http://www.delmar.edu/engl/wrtctr/handouts/exemp.pdf>.

Walston, John. *BuzzWhack.com.* 18 Feb. 2006 <http://www.buzzwhack.com/index.html>.

Watson, Don. *WeaselWords.com.* 18 Feb. 2006 <http://www.weaselwords.com.au>.

"Wine Terms." *Tasting Wine.* 19 Feb. 2006 <http://www.tasting-wine.com/html/wine-terms.html>.

## FILMS

*About a Boy.* Dir. Chris Weitz and Paul Weitz. Perf. Hugh Grant, Nicholas Hoult, Toni Collette, Rachel Weisz. Based on the novel by Nick Hornby. Universal, 2002.

*Apollo 13.* Dir. Ron Howard. Perf. Tom Hanks, Bill Paxton, Kevin Bacon, Gary Sinise, Ed Harris, Kathleen Quinlan. Universal, 1995.

*Babette's Feast.* Dir. Gabriel Axel. Perf. Stéphane Audran. Story by Isak Dinesen. From Denmark. Orion Classics, 1987.

*Big Night.* Dir. Campbell Scott and Stanley Tucci. Perf. Stanley Tucci, Tony Shalhoub, Isabella Rossellini. Samuel Goldwyn, 1996.

*Chocolat.* Dir. Lasse Hallström. Perf. Juliette Binoche, Johnny Depp, Alfred Molina, Lena Olin, Judi Dench. Miramax, 2000.

*Crash.* Written and directed by Paul Haggis. Perf. Don Cheadle, Matt Dillon, Ryan Phillippe, Sandra Bullock, Terrence Howard, Thandie Newton. Lion's Gate, 2005.

*Dinner Rush.* Dir. Bob Giraldi. Perf. Danny Aiello, Edoardo Ballerini, Vivian Wu, John Corbett, Sandra Bernhard. Access Motion Picture Group, 2001.

*Eat Drink Man Woman.* Dir. Ang Lee. From Taiwan. Samuel Goldwyn, 1994.

*Finding Forrester.* Dir. Gus Van Sant. Perf. Sean Connery, Rob Brown, F. Murray Abraham. Written by Mike Rich. Columbia Pictures, 2000.

*Galaxy Quest.* Perf. Tim Allen, Alan Rickman, Sigourney Weaver, Tony Shalhoub. Dir. Dean Parisot. DreamWorks SKG, 1999.

*Memento.* Dir. Christopher Nolan. Perf. Guy Pearce, Carrie-Anne Moss, Joe Pantoliano. Newmarket Entertainment, 2001.

*Miss Congeniality.* Perf. Sandra Bullock, Michael Caine, Candice Bergen, Benjamin Bratt, William Shatner. Dir. Donald Petrie. Warner Brothers, 2000.

*Mystic River.* Dir. Clint Eastwood. Perf. Sean Penn, Tim Robbins, Kevin Bacon, Marcia Gay Harden, Laura Linney. Screenplay by Brian Helgeland. Based on the novel by Dennis Lehane. Warner Brothers, 2003.

*Pulp Fiction*. Written and directed by Quentin Tarantino. Perf. John Travolta, Samuel L. Jackson, Uma Thurman, Harvey Keitel, Tim Roth. Miramax, 1994.

*The Shipping News*. Dir. Lasse Hallström. Based on the novel by E. Annie Proulx. Perf. Kevin Spacey, Judi Dench, Julianne Moore, Cate Blanchett, Pete Postlethwaite. Miramax, 2001.

*Sudden Impact*. Dir. Clint Eastwood. Perf. Clint Eastwood, Sondra Locke, Pat Hingle, Bradford Dillman. Warner Brothers, 1983.

*SuperSize Me*. Dir. Morgan Spurlock. Showtime Independent Films, 2004.

*Terminator 2: Judgment Day*. Dir. James Cameron. Perf. Arnold Schwarzenegger, Linda Hamilton, Robert Patrick. TriStar Pictures, 1991.

*Tortilla Soup*. Dir. Maria Ripoll. Perf. Hector Elizondo, Elizabeth Peña, Raquel Welch. Based on *Eat Drink Man Woman*. Samuel Goldwyn, 2001.

*Tampopo*. Dir. Juzo Itami. Perf. Ken Watanabe. From Japan. New York Films, 1986.

*What's Cooking?* Dir. Gurinder Chadha. Perf. Alfre Woodard, Joan Chen, Mercedes Ruehl, Kyra Sedgwick, A. Martinez. Trimark Pictures, 2000.

## CHAPTERS 18–31 AND APPENDICES I–III

Allen, Richard. "Camera Movement in *Vertigo*." *The Alfred Hitchcock Scholars/'The MacGuffin' Web Page*. 12 Jan. 2006. *The MacGuffin*. 8 Mar. 2006 <http://www.labyrinth.net.au/~muffin/camera_movement.html>.

*The American Heritage Book of English Usage*. Houghton Mifflin, 1996. *Bartleby.com: Great Books Online*. 14 Nov. 2005 <http://www. bartleby.com/64/>.

Bantick, Christopher. "Spelling Sux, OK?" *The Age*. 8 Dec. 2003. 13 Dec. 2005 <http://www.theage.com.au/articles/2003/12/05/1070351785633.html>.

Bays, Jeff. "Hitchcock: Basic Film Techniques." *Borgus.com*. January 2006. Borgus Productions. 8 Mar. 2006 <http://borgus.com/think/hitch.htm>.

Brians, Paul. *Common Errors in English Usage*. Home Page. Washington State University. 13 Dec. 2005 <http://www.wsu.edu/~brians/errors>.

Capital Community College. *Guide to Grammar and Writing*. Capital Community College Foundation. 3 Aug. 2005 <http://grammar.ccc.commnet.edu/grammar>.

*The Chicago Manual of Style.* 15th ed. Chicago: U of Chicago P, 2003.

Christ, Henry I., and J. C. Tressler. *Heath Handbook of English.* Boston: D. C. Heath, 1961.

"Culinary Terms." *Lowfat Lifestyle.* 2002. 13 Dec. 2005 <http://www.lowfatlifestyle.com/culinaryterms.htm>.

Germano, William. "Passive Is Spoken Here." *The Chronicle of Higher Education* 22 Apr. 2005: B20.

"Grammar Guide: Subject-Verb Agreement III." *GrammarStation.com.* 2002. 18 Nov. 2005 <http://www.grammarstation.com/servlet/GGuide?type=SVUI>.

Greenbaum, Sidney, and Randolph Quirk. *A Student's Grammar of the English Language.* Longman Group UK, 1990.

Kimball, Cornell. "A Study of Some of the Most Commonly Misspelled Words." *David Barnsdale.* 13 Dec. 2005 <http://www.barnsdle.demon.co.uk/spell/error.html>.

"List of Commonly Misspelled English Words—Advanced Vocabulary." ESLDesk. 2005. 13 Dec. 2005 <http://www.esldesk.com/esl-quizzes/misspelled-words/misspelled-words.htm>.

McAdams, Mindy. "A Spelling Test." Home Page. 21 Nov. 1995. Sentex Communication Corps. 13 Dec. 2005 <http://www.sentex.net/~mmcadams/spelling.html>.

*OWL Online Writing Lab.* 2004. Online Writing Lab at Purdue University. 6 May 2005 <http://owl.english.purdue.edu/sitemap.html>.

Simmons, Robin L. "The Relative Clause." *Grammar Bytes! Grammar Instruction with Attitude.* 3 Aug. 2005 <http://www.chompchomp.com/terms/relativeclause.htm>.

Stanley, Karen, ed. "Perspectives on Plagiarism in the ESL/EFL Classroom." *TESL-EJ Teaching English as a Second or Foreign Language* 6:3 (2002). 6 Sep. 2005 <http://www-writing.berkeley. edu/TESL-EJ/ej23/fl.html>.

Straus, Jane. *The Blue Book of Grammar and Punctuation.* 8th ed. 2004. Pub. Jane Straus. 18 Nov. 2005 <http://www.grammarbook.com>.

Strunk, William, Jr. "Words Often Misspelled." *The Elements of Style.* 1918. *Bartleby.com: Great Books Online.* 13 Dec. 2005 <http://www.bartleby.com/141/strunk4.html>.

*The University of West Florida Writing Lab.* 2003. The University of West Florida. 18 Nov. 2005 <http://uwf.edu/writelab>.

Warriner, John E. *English Composition and Grammar.* Benchmark ed. Orlando: Harcourt Brace Jovanovich, 1988.

"100 Most Often Misspelled Words in English." *YourDictionary.com: The Last Word in Words.* 2005. yourDictionary.com Inc. 13 Dec. 2005 <http://www.yourdictionary.com/library/misspelled.html>.

## APPENDIX IV

Alred, Gerald J., Charles T. Brusaw, and Walter E. Oliu. *The Business Writer's Handbook*. 6th ed. Boston: Bedford/St. Martin's, 2000.

Bureau of Labor Statistics. "Chefs, Cooks, and Food Preparation Workers." *Occupational Outlook Handbook, 2006-2007 Edition*. 20 Dec. 2005. The U.S. Department of Labor Bureau of Labor Statistics. 3 Mar. 2006 <http://stats.bls.gov/oco/content/ocos161.stm>.

"Job Search: Resumes and Vitae." *Career Services @ Virginia Tech*. 26 Jan. 2005. Virginia Tech. 21 Feb. 2006 <http://www.career. vt.edu/JOBSEARC/Resumes/Resume1.htm>.

"Job Search: Cover Letters." *Career Services @ Virginia Tech*. 14 Feb. 2006. Virginia Tech. 21 Feb. 2006 <http://www.career.vt. edu/JOBSEARC/coversamples.htm>.

Lockley, JoLynne. "Resume Tips for Culinary Professionals." *CookingSchools.com*. ALLSchools.com. 21 Feb. 2006 <http://www.cookingschools.com/articles/resume-tips/>.

"Resumes." *StarChefs Job Finder*. 2005. StarChefs: The Magazine for Culinary Insiders. 21 Feb. 2006 <http://www.starchefsjobfinder.com/career_center/resume.php>.

"Resume-Writing 101: Get Your Resume in Shape for Jobs and Internships." *CollegeBoard*. 2006. The College Board. 21 Feb. 2006 <http://www.collegeboard.com/student/plan/high-school/36957.html>.

"Writing Your Resume." *Okanaga College*. Cooperative Education, Graduate and Student Employment Center, Okanaga College. 21 Feb. 2006 <http://www.okanagan.bc.ca/Page10694.aspx>.

## APPENDIX V

Arnett, Alison. "The Secret Life of a Restaurant Critic." *The Boston Globe* 09 Oct. 2005. 16 Feb. 2006 <http://boston.com/news/globe/magazine/articles/2005/10/09/the_secret_life_of_a_restaurant_critic/>.

Hedden, Jenny. "Maximize Menu Merchandising Power." *Restaurants USA* May 1997. National Restaurant Association. 16 Feb. 2006 <http://www.restaurant.org/business/magarticle.cfm?ArticleID=477>.

Lipsky, Linda. "Designing Profitable Menus." *Restaurant Report* 2005. 16 Feb. 2006 <http://www.restaurantreport.com/features/ft_menudesign.html>.

Rolfe, Simon. “Tasty Profits—Maximizing Your Bottom-line via a Professionally Designed Restaurant Menu.” *Hotel News Resource* 08 Aug. 2005. 16 Feb. 2006 <http://www.hotelnewsresource.com/article 17898.html>.

## APPENDIX VII

Bourdain, Anthony. *Kitchen Confidential: Adventures in the Culinary Underbelly.* New York: Bloomsbury USA, 2000. 9–10.

Burciaga, José Antonio. “Tortillas.” *Weedee Peepo.* Edinburg, Texas: University of Texas–Pan American Press, 1988.

Clifton, Lucille. “cutting greens.” *Good Woman: Poems and a Memoir 1969-1980.* Rochester, NY: BOA Editions, 1987.

Fisher, M. F. K. “Borderland.” 1937. *The Art of Eating.* New York: John Wiley & Sons, 2004. 26–28.

Kinnell, Galway. “Blackberry Eating.” *Mortal Acts, Mortal Words.* Boston: Houghton Mifflin, 1980.

Masumoto, David Mas. “The Furin.” *epitaph for a peach: four seasons on my family farm.* New York: HarperCollins, 1995. 65–66.

Narayan, Shoba. *Monsoon Diary.* New York: Villard Books, 2004. 57–58.

Schlosser, Eric. *Fast Food Nation: The Dark Side of the All-American Meal.* Boston: Houghton Mifflin, 2001. 3–5.

Steingarten, Jeffrey. “Hot Dog.” *The Man Who Ate Everything.* New York: Alfred A. Knopf. 1997. 89–91.

# Index

## D

## T